QUASIBRITTLE FRACTURE MECHANICS AND SIZE EFFECT: A FIRST COURSE

QUASIBRITTLE FRACTURE MECHANICS AND SIZE EFFECT: A FIRST COURSE

ZDENĚK P. BAŽANT
Northwestern University

JIA-LIANG LE
University of Minnesota

MARCO SALVIATO
University of Washington

OXFORD
UNIVERSITY PRESS

Great Clarendon Street, Oxford, OX2 6DP,
United Kingdom

Oxford University Press is a department of the University of Oxford.
It furthers the University's objective of excellence in research, scholarship,
and education by publishing worldwide. Oxford is a registered trade mark of
Oxford University Press in the UK and in certain other countries

Published in the United States of America by Oxford University Press
198 Madison Avenue, New York, NY 10016, United States of America

British Library Cataloguing in Publication Data

Data available

Library of Congress Control Number: 2021937989

ISBN 978–0–19–284624–2

DOI: 10.1093/oso/9780192846242.001.0001

Printed and bound in the UK by
TJ Books Limited

Links to third party websites are provided by Oxford in good faith and
for information only. Oxford disclaims any responsibility for the materials
contained in any third party website referenced in this work.

**Dedicated to the memory of G. I. Barenblatt (1927–2018)
and to our loving and supportive wives
Iva, Miao, and Rossella**

G. I. Barenblatt, known to his friends as Grisha, was a giant of mechanics and applied mathematics. Within his life span, he was a rare scientist who made major contributions to both solid and fluid mechanics. In fracture mechanics, he made a transformative advance by inventing the cohesive crack model while a young scientist in Moscow. After the collapse of the USSR, he continued a productive career in the West, first as G. I. Taylor professor at the University of Cambridge and then for almost two decades as a professor of Mathematics in Residence at the University of California, Berkeley. He was a foreign member of the U.S. National Academy of Sciences, American Academy of Arts and Sciences, and the Royal Society of London and received many honors.[1]

[1] Born in 1927 in Moscow, USSR, Grigory Isaakovich Barenblatt (in Russian Григо́рий Исаа́кович Баренбла́тт) received his Ph.D. from the Lomonosov Moscow State University under the advisorship of A. N. Kolmogorov. Over his remarkable academic career, he contributed to almost all branches of continuum mechanics. He also contributed to the solutions of a number of problems in the theory of polymers, biology, chemistry and geophysics. His results in the fields of fracture mechanics, hydraulic fracturing, filtration of fluid and gas in a porous medium, mechanics of non-classical deformable solids, turbulence, self-similarities, nonlinear waves, and intermediate asymptotics are well known. His research led to a number of well-known monographs and about 250 papers. For his outstanding achievements, he received numerous awards, among them the G. I. Taylor award from the Society for Engineering Science, the Maxwell Prize of the International Committee on Applied and Industrial Mathematics, the Lagrange Prize of the Accademia Nazionale dei Lincei, Modesto Panetti Prize and Medal, and the Timoshenko Medal of the ASME. He worked scientifically until his death in Moscow in 2018.

Foreword

This book on fracture mechanics will serve as a monograph for professionals in the field of engineering materials, solids, and structures, and as a text for graduate students and aspiring professionals. The book approaches the subject of fracture mechanics from the collected and vast perspectives, experiences, and contributions of the authors with the fracture of quasibrittle materials such as concrete and fiber-reinforced composites, and natural materials such as ice and rock, which traditionally receive second shrift in most texts on fracture mechanics. The book is a notable contribution to fracture mechanics in many respects, including a chapter that considers probabilistic interpretations of fracture data and a chapter that analyzes and discusses an excellent set of practical applications that clearly illustrate the importance of size effects in fracture. Basic aspects of fracture mechanics are introduced and covered in the book, with extensive and well-chosen problems at the end of each chapter for students using the book as a text. In many of the topics, particularly those relevant to quasibrittle materials, the book brings the reader in contact with the most recent published research. One of the analytical approaches that receives prominent attention in the book owing to the manner in which many quasibrittle materials fail is cohesive zone modeling. The book will be an essential addition to the library of practitioners of fracture mechanics for coverage of these aspects alone. While many books lose steam towards the end, not this book. The penultimate chapter titled "Quasibrittle Size Effect Analysis in Practical Problems" deals with nine excellent current examples, including my two favorites: structural failure of the Malpasset Dam and tensile fracture of sea ice.

John W. Hutchinson
School of Engineering and Applied Sciences, Harvard University

Preface

The present book provides a comprehensive treatment of quasibrittle fracture mechanics and includes a concise but rigorous and complete treatment of linear elastic fracture mechanics, the foundation of all fracture mechanics. The book is designed for graduate and upper-level undergraduate university courses. It presents the fundamental principles of both linear and nonlinear fracture mechanics, the transition from distributed cracking damage to sharp cracks with the associated energetic and statistical size effects, the effects of the triaxial stress state in the fracture process zone, the probabilistic fracture mechanics, and nonlocal continuum modeling of distributed damage. It also points out various practical implications for engineering structures and includes a number of exercise problems. The presentation strives for conciseness without distracting the reader by excessive verbiage.

Quasibrittle materials, a.k.a. heterogeneous brittle materials, are the most widely used materials in engineering today, by both volume and cost. They include concrete as the archetypical case, fiber-polymer composites, tough ceramics, many rocks and stiff soils, fiber-reinforced concrete, cold asphalt concrete, masonry, wood, dental ceramics, rigid foams, particulate nanocomposites, bones, most architectured and printed materials, nacre, and various biomimetic materials. On the nano- and micrometer scales, virtually all the materials become quasibrittle, including silicon or thin metallic films. Fracture mechanics of these materials is important not only for structural safety and reliability, but also for durability, which has environmental implications. For example, the current CO_2 emissions from cement production are about to exceed those from all the cars and trucks in the world, and if, hypothetically, the lifetimes of concrete structures and pavements could be doubled, this would halve the emissions due to cement in the long run.

Fracture mechanics courses have been taught for decades in the departments of mechanical and aeronautical engineering. But they have mostly dealt with the classical fracture mechanics intended for metals and other brittle homogeneous materials. Such courses have had little relevance to the civil engineering students studying concrete and geotechnical engineering, as well as to the engineering students of mechanical and aerospace engineering studying fiber composites. Currently, the former gains in importance because the American Concrete Institute has just introduced the fracture mechanics-based size effects into its 2019 design code (ACI-318), and thus implicitly accepted the applicability of quasibrittle fracture mechanics to concrete structures.

This book is designed to serve as a text, and as a source of exercise problems, for a semester-long course (of about 35 class-hour duration). It can also be used in a quarter-long course (of 30 class-hour duration), in which case Chapters 6 and 7 are taught only in a review fashion. Chapter 2 on the foundations of linear elastic fracture mechanics is self-standing and can form, as a whole or as an initial part, a two to four-week segment in a broader undergraduate course on structural or continuum mechanics. Chapters 2, 3, and 7 can also be used as a segment of various specialty courses on concrete structures, composite structures, rock mechanics, pavement engineering, design of airframes, automobiles and ships, nuclear structures, and microelectronic mechanical systems.

The authors wish to express their appreciation of the academic environments at their institutions, Northwestern University, University of Minnesota, and University of Washington, conducive to scholarly pursuits and innovative teaching. The initial partial draft of this book was the result of the first author's teaching, since 1986, of a quarter-length graduate course on Fracture of Concrete, later renamed and broadened as Quasibrittle Fracture and Scaling. The second and third authors have for several years taught similar courses in the Civil Engineering department of the University of Minnesota and in the Aerospace Engineering department of the University of Washington, and contributed their valuable experience to this book. The new results covered in the book arose from researches funded by the National Science Foundation, Army Research Office, Office of Naval Research, Air Force Office of Scientific Research, Department of Energy, Department of Transportation, Los Alamos Scientific Laboratory, Argonne National Laboratory, and Boeing, Ford, and Chrysler companies. Grateful acknowledgment is due to all these sponsors.

Last but foremost, we wish to thank our wives Iva, Miao, and Rossella for their loving support while working on this book.

Zdeněk P. Bažant, Jia-Liang Le, and Marco Salviato
Evanston, Minneapolis, and Seattle
December 2020

Contents

1
Introduction

Fracture is viewed as something bad but, once understood, becomes beautiful.

Material failures that cause structural collapse are basically of two types—ductile and brittle. The latter is the subject of fracture mechanics and is far more dangerous since the structural collapse is usually sudden and without warning. A transition from the latter to the former exists, and the way to describe and control it is of keen interest.

Fracture mechanics is the field of mechanics that deals with crack initiation and propagation. The analysis of fracture does not fit the classical stereotype of boundary value problems of partial differential equations, since the geometry of body boundaries varies with the progress of fracture instead of being specified as fixed. This might look to some mathematicians as a formidable difficulty, but effective simple methods to tackle the important practical problems of fracture initiation and propagation have been developed over the span of the last hundred years. Brief pedagogical exposition of the main ideas is the goal of the present book.

The origin of fracture mechanics dates back to Griffith's work in 1921 (Griffith, 1921). The first study of its applicability to a macro-heterogeneous material, concrete, appeared in 1952, in the report by Bresler and Wollack (Bresler and Wollack, 1952) at the University of California, Berkeley. In 1961, Kaplan (Kaplan, 1961) published the first journal article on this subject. In 1962, Clough, in his report (Clough, 1962)[1] at the University of California, Berkeley, used finite element analysis with a large interelement crack to judge the safety of a large dam. This was an epoch-making contribution which may be seen today as the foundation of both the finite element method and of the computer simulation of fracture in a quasibrittle material. In 1968, again in Berkeley, Rashid (Rashid, 1968) noted that interelement cracks were cumbersome if crack propagation should be followed, and analyzed fracture propagation in a pre-stressed concrete reactor vessel by deleting the elements in which the strength limit was reached. Rashid's analysis was the harbinger of the crack band approach although the crucial problem of spurious mesh sensitivity was not even discussed for another

[1] website: https://nisee.berkeley.edu/elibrary/semm/1962.

Quasibrittle Fracture Mechanics and Size Effect: A First Course. Zdeněk P. Bažant, Jia-Liang Le and Marco Salviato, Oxford University Press. © Zdeněk P. Bažant, Jia-Liang Le, Marco Salviato 2022. DOI: 10.1093/oso/9780192846242.003.0001

decade. The stream of fracture studies of quasibrittle materials became a torrent in the late 1980s and progress continues until today (for literature reviews with extensive reference lists, see e.g. Bažant (1982); Bažant and Planas (1998); Bažant (2005)).

1.1 Why Fracture Mechanics?

Up to the early 20th century, it was universally believed that a structure fails when the stress reaches its limiting value—the material strength. In some fields, for example, the design of concrete structures or fiber-polymer composites, this belief is widespread even today.

In reality, though, aside from the material strength, there are other mechanical properties that govern structural failure. This calls for a deeper theory, called fracture mechanics. What, exactly, are the reasons to turn to fracture mechanics? They are six:

1. Formation of a crack requires energy supply to separate interatomic bonds and create new surfaces. If sufficient energy is not supplied, no crack forms even if the strength limit is reached. Hence, energy analysis must be part of the criterion of crack growth.

2. In structures made of concrete, composites, and various other materials, a strong nonstatistical size effect is observed and has been amply demonstrated. Specifically, the nominal strength of structure, σ_N, which is defined as the maximum load divided by the characteristic cross-section area, is not constant but decreases with increasing structure size D, much more so than could be explained by material randomness.

3. The results of computer analysis of structures must be objective. In particular, they must not change significantly upon changing the finite element (FE) size. It has been amply demonstrated that unless the material is elastoplastic, objectivity of finite element computations is not achieved with the material strength criterion alone.

4. Unlike plastic limit analysis, the equilibrium load in many types of failure decreases as the deflection increases after the peak load. In absence of buckling, this phenomenon can be explained only by crack growth.

5. For impact, earthquake and other dynamic loads, the energy absorption capability of structure is important. When fracture forms, it cannot be correctly estimated by using plastic limit analysis or strength criterion alone. Only fracture mechanics can predict the correct energy absorption of the structure.

6. Lastly, one cannot ignore dimensional analysis. The stress has the dimension of N/m^2, while the surface energy of a crack has a different dimension, namely J/m^2 or N/m. So, if both matter for failure, the notion of material strength, with the dimension of N/m^2, is insufficient. The ratio of these two dimensions has a dimension of length. Dimensional analysis thus shows that a material characteristic length must be considered, too. This is the hallmark of fracture mechanics.

1.2 Three Kinds of Fracture Mechanics

Today there are three kinds of fracture mechanics, each suitable for different situations.

First kind: Since Griffith's discovery (Griffith, 1921) until the 1950s, there was only one kind of fracture mechanics, the *linear elastic fracture mechanics* (LEFM), which is also called the mechanics of (perfectly) brittle fracture.

The theory of elasticity predicts an infinite stress at the tip of a sharp crack. Therefore, the material at the crack tip region must exhibit inelastic deformation. Depending on the constitutive behavior of the material, the crack tip region could experience plastic yielding or damage. The LEFM considers that the nonlinear zone in which fracture is formed, called the fracture process zone (FPZ), is so small compared to the structural dimensions that it can be considered as point-wise (Fig. 1.1a). The transverse stress profile along the crack line has a discontinuity such that the stress drops suddenly to zero right behind the crack tip. This is a good approximation for glass, fine-grained ceramics, glassy polymers, and metals embrittled by fatigue or hydrogen, because the FPZ size is in the order of micrometers. Exceptions, for these materials, are the micrometer-scale devices such a MEMS, various electronic components and thin metallic films (which are quasibrittle, as explained next).

In LEFM, the entire volume of the body is linearly elastic, and so analytical solutions can be obtained by methods of elasticity. But various LEFM approximations serve also as the basis of the other two kinds of fracture mechanics. Thus, the LEFM is the foundation of the entire discipline of fracture mechanics.

As for the size effect on the structural strength, in LEFM it is almost trivial, and normally not even discussed. When the specimens and the cracks are geometrically similar, the size effect of LEFM is a simple power law: structural strength $\propto$ (specimen size)$^{-1/2}$. But real brittle structures normally fail while the critical crack is still very small and has a length that is independent of structure size. In that case there is no size effect, which means that the failure load divided by a characteristic cross section area does not depend on the structure size.

The first linear elastic FE fracture analysis was apparently conducted by R.W. Clough at University of California, Berkeley, in 1962 (Clough, 1962). Simulating Norfolk Dam, Arizona, and its orthotropic foundation, he obtained the stress redistribution due to a large interelement crack and compared the stresses to material strength (at that time, of course, mesh size sensitivity was not yet even discussed). Today, there are powerful FE softwares for LEFM, such as XFEM (Dolbow and Belytschko, 1999). The recently popular phase-field model with a scalar damage law is also approximately equivalent to LEFM (Borden *et al.*, 2012, 2014).

Second kind: The second kind of fracture mechanics was developed from the 1950s to the 1980s. It is the ductile, or elastoplastic, fracture mechanics, also called the small-scale yielding fracture mechanics or (ductile) *cohesive fracture mechanics*. This kind applies to most non-fatigued metals and other plastic materials, except at micrometer scale and nano-scale.

In such materials, there is a long and wide plastic hardening nonlinear zone, or yielding zone, in front of the crack tip. Its size is not negligible compared to normal structural dimensions, while the FPZ is still negligible compared to these dimensions, virtually point-wise (except for micro- and nanoscale devices); (Fig. 1.1b). The stress profile along the crack line has a flat or mildly rising plateau ending with a sharp stress drop at the crack tip.

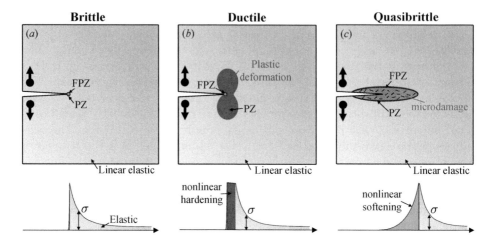

Fig. 1.1 Types of structural fracturing behavior: (a) brittle, (b) ductile and (c) quasi-brittle. Trends of the stress distributions along the crack line are shown at the bottom of each figure.

In metals, polymers, glasses, and other materials homogeneous on the macroscale, the width of the yielding zone (transverse to crack plane) is usually not important for the overall structural response, and the zone is then replaced by a cohesive zone in front of the crack tip, occupying a line segment of a certain characteristic length. Within that segment, the diagram of transverse normal stress versus the separation of the opposite crack faces (i.e., the crack opening) is considered to have a horizontal yield plateau, terminating with a sudden stress drop (Fig. 1.1b). For metals (in contrast to concrete and other materials), the unloading of the plastically yielded material on the sides of the cohesive line segment follows an unreduced elastic modulus. Thus the entire volume of the structure can still be considered as elastic, and the nonlinearity is introduced only through the boundary (or interface) condition for the cohesive segment at crack front. Obviously, this feature is advantageous for approximate analytical solutions.

In contrast to LEFM, the ductile fracture mechanics is endowed with a finite material characteristic length, l_0 (introduced by Irwin (Irwin, 1958)). The reason is that the stress, of dimension N/m^2, is needed to characterize the plasticity at the crack tip, while the formation for crack surfaces must be described by energy per unit area, which has a different dimension, N/m (or J/m^2). Obviously, the ratio of l_0 to the structure size D must cause a size effect (as shown by Palmer and Rice (Palmer and Rice, 1973) for cohesive shear fracture propagation in clay slopes).

Third kind: In the 1980s, a third kind emerged—the quasibrittle fracture mechanics, also called the cohesive softening fracture mechanics, which is the focus of this book. Here again there is a long and wide nonlinear zone, FPZ, in front of the crack tip, but this zone is not plastic. Rather, it undergoes progressive softening damage in the form of randomly distributed microcracking, frictional micro-slips, and grain interlock. The stress profile along the crack line has no plateau but descends towards the crack tip (Fig. 1.1c). There is no sudden stress drop at the tip of the open crack in the tail of this zone, except when the fracture starts from a sharp notch or pre-existing fatigued crack.

The quasibrittle materials are defined as heterogenous materials with brittle constituents. But for structures, the concept of quasibrittleness is relative. When the structure size is far larger than the maximum inhomogeneity size, for example, the grain size, or the yarn spacing in a woven composite, every quasibrittle structure, consisting of a quasibrittle material, becomes perfectly brittle. Vice versa, when a perfectly brittle homogeneous structure becomes sufficiently small, as in micro- or nano-scale devices, it transits to a quasibrittle fracture.

The archetypical quasibrittle material is concrete, and that is where the development of quasibrittle fracture mechanics began in the 1970s (Bažant, 1976; Bažant and Cedolin, 1979, 1980; Bažant, 1982; Bažant and Oh, 1983; Pietruszczak and Mróz, 1981; Bažant *et al.*, 1984). One reason is that, in normal concrete, the length of the FPZ is about 0.5 m, so that the quasibrittless is conspicuous on a convenient scale, the scale of large laboratory tests. Another is that the early finite element simulations of failure of concrete dams and nuclear containment shells (e.g. (Rashid, 1968; Zienkiewicz *et al.*, 1972)) in which the elements where deleted upon reaching the material strength limit, gave ambiguous failure loads depending on the chosen mesh size. This spurious mesh-size sensitivity was explained by unstable localization of cracking damage modeled by strain-softening constitutive laws Bažant (1976).

Aside from concretes, mortars, and fiber-reinforced concretes, quasibrittle materials comprise fiber-polymer composites, tough or toughened ceramics, dental ceramics, bone, cartilage, stiff clays and silts, cemented sands, cold asphalt pavement concrete, grouted soils, sea ice, refractories, rigid foams, consolidated snow, wood, particle board, paper, carton, cast iron, modern tough alloys, biological shells (e.g. nacre), filled resins or polymers, porous printed materials, and virtually all materials on approach to micro- or nano-scale, such as polysilicon, nanotubes, and graphene or graphene oxide sheets. Many of them are materials increasingly demanded for high-tech applications.

Depending on the type of quasibrittle material, the FPZ length and width varies enormously. In embrittled metals it is of micrometer dimensions, in textile polymer composites about 1 cm, in shale about 1 cm, in high-strength or normal concretes about 5 cm or 50 cm, in sea ice about 5 m, and in large-scale thermal bending fracture of the ice cover of the Arctic Ocean (consisting of thick mile-size floes connected by thin ice leads), about 10 km.

For structures not much larger than the typical FPZ size of the material, which represents a small multiple of the maximum aggregate size in concrete, the damage is such that microcracks that microslips cannot localize, because of heterogeneity. The consequence is that the failure behavior of small quasibrittle structures is quasi-plastic, with almost no size effect and a long plateau at the top of the load-displacement. For quasibrittle structures much larger than the FPZ size, the failure behavior is almost brittle, with a steep load drop after reaching the maximum load. As the structure size increases, the transition from ductile to brittle response is gradual, typically spread over several orders of magnitude of structure size. It is this transitional behavior between plasticity and brittle fracture that is of interest for quasibrittle materials and structures, and that is the source of the transitional size effect—the quasibrittle size effect.

1.3 Crack-Parallel Stresses and Tensorial Damage as Quasibrittle Fracture Basis

The cohesive fracture mechanics with line cracks has also been widely used for quasibritle materials such as concrete, fiber composites, coarse ceramics, or geomaterials in which FPZ is not only long but also wide, because of large grain or inhomogeneity size. However, a new type of experiment, called the gap test, revealed, and computer simulations corroborated by Nguyen *et al.* (Nguyen *et al.*, 2020*a,b*), a strong effect of crack-parallel normal stress on fracture propagation, causing the energy required for crack propagation in concrete to almost double or drop near zero.

This effect cannot be captured by the line crack models. They can be used, as an approximation, only when the crack-parallel stresses are *a priori* known to be small. For monotonic proportional loading, the fracture criterion can be adjusted by an approximate formula. But the effect of nonproportional loading histories, which is pronounced, is then missed.

The crack-parallel stress effect severely limits the applicability of all the classical line crack models to situations with very small crack-parallel normal stresses, in-plane or out-of-plane. This recently discovered effect calls for reorienting quasibrittle fracture mechanics from scalar to tensorial modeling, in which the fracture is represented as a cracking band with progressive distributed damage in a zone of *finite length and width* at fracture front. The constitutive law for this damage is the foundation of quasibrittle fracture mechanics.

Nevertheless, the LEFM and the cohesive crack model (CCM) still remain the pillars of the edifice of fracture mechanics. They are essential for understanding and teaching fracture mechanics, and for providing accurate benchmark solutions of special cases which tensorial damage mechanics must match. They must be taught before extensions such as the crack-parallel stress effect are discussed. Therefore, emphasis on LEFM and CCM must be retained in this introductory book.

1.4 Size Effect Type and Role of Material Randomness

Since the 1980s, the size effect on the nominal structural strength has been intensely researched. So far, two basic types of size effect have been identified for quasibrittle structures:

1. When the structure fails only after long, stable growth of one dominant crack at increasing load, the energy release from the structure into the crack front dominates. The fact that strain energy release from the undamaged part of the structure into the fracture front grows quadratically with the structure size while the energy dissipated at fracture front grows linearly causes an energetic size effect on the mean structural strength. This fact was exploited to derive a simple approximate, but general, size effect law for the mean structural strength in type 2 failures (Bažant, 1984*b*; Bažant and Kazemi, 1990). This law is not limited to reinforced concrete structures and has been shown to apply to all quasibrittle structures.

2. When the structure fails right at the initiation of a macro crack at the weakest location among many, the energy release is not dominant. Then not only the FPZ size but also the randomness of the local material strength is important for the size effect on the mean structural strength, which is called the type 1 size effect. This size effect has both energetic and statistical components. The latter grows with the structure size while the former does not. Eventually, for large enough sizes, the statistical size effect component becomes insignificant, making the entire size effect fully statistical.

The reason why material randomness causes a size effect on the mean nominal strength of structure is that the minimum value of the random local material strength likely encountered in the structure increases with the structure volume. The concept of statistical size effect was first perceived, in a qualitative sense, in the 17th century by Mariotte (Mariotte, 1686). But two and half centuries elapsed until the first mathematical description of the statistical size effect was developed by Weibull (Weibull, 1939, 1951). It is evident that the material randomness would not play a role in the mean size effect of type 2 failures since the location of damage is predetermined by the crack formed prior to the peak load, while it has a strong influence on the mean size effect of type 1 failure. However, studies have shown that the Weibull theory alone is insufficient to explain the type 1 size effect of quasibrittle structures (Bažant and Xi, 1991).

The interaction of statistical and energetic components in type 1 failures is reflected in the finite weakest-link chain (Bažant and Pang, 2006, 2007; Bažant *et al.*, 2009; Le *et al.*, 2011; Bažant and Le, 2017). The resulting size effect for not-too-large structures is rather different from the classical Weibull size effect, since the Weibull size effect is modelled by an *infinite* weakest-link chain (Bažant and Pang, 2006, 2007). In type 1 failure, in which there is no stable macro crack growth (under load-controlled tests), the structure fails as soon as a macro crack initiates at one site in the structure, which obviously happens in the weakest RVE (relative to the stress). The same is obviously true for a chain of links. It, too, fails as soon as its weakest link fails. So both cases are statistically equivalent, which leads to the same kind of statistical distribution.

The finiteness of the weakest-link model causes a significant deviation from the classical Weibull distribution. The resulting structural strength distribution depends strongly on the structure size. For the smallest possible material element (i.e., the RVE), the distribution is Gaussian (or normal), with a remote Weibull tail grafted on the left at the probability of about 0.001. As the structure size increases, the Weibullian portion of the distribution function penetrates into higher probabilities until, at infinite structure size, the entire distribution becomes Weibullian. The consequence for the statistical size effect is that the Weibull power-law size effect gets modified by a big upward deviation for small sizes.

When the crack at failure is neither negligible nor long enough, there is a smooth transition between types 1 and 2. Its approximate description by a universal size effect law was recently obtained by multi-sided asymptotic matching. Not surprisingly, its formula is more complicated (Hoover *et al.*, 2013; Hoover and Bažant, 2013, 2014b).

1.5   Applications of Size Effect in Structural Analysis and Design

The importance of size effect for engineering practice differs significantly among different fields of engineering. Aircraft engineering needs only performance specifications but no design codes, because only a few large aircraft are designed in a decade. Likewise automotive engineering. So, if for instance the crack band model is used in the finite element analysis, the size effect is automatically accounted for.

But structural engineering is different. Tens of thousands of different structures are designed annually, and most of them are different and too big to be tested up to failure. Therefore, detailed design codes are indispensable and must be simple enough to allow thousands of designers to work quickly. The designs of some special structure are checked, for example, by the realistic finite element simulation which implies the size effect. But even then, the preliminary design of all concrete structures must be based on the code, and no design violating the code is legally defensible in court litigation in the case of collapse.

Unfortunately, the consequence is that new research results, fracture mechanics included, do not get used widely unless incorporated in the design code. And the process of adopting new code specification is inevitably slow since, in democratic practice, it requires reaching consensus in large committees whose majorities have practicing engineers and professors-consultants who have little expertise in the underlying theory.

The first significant evidence of a large non-statistical size effect in reinforced concrete was delivered by Kani's large beam tests in Toronto in the the late 1960s (Kani, 1967). The evidence was soon enhanced by the tests of Leonhardt and Walther (Leonhardt and Walther, 1962) in Stuttgart and of Bhal (Bhal, 1968) in Switzerland. Shioya and Akiyama (Shioya and Akiyama, 1994) in Tokyo tested record-size prestressed concrete beams, up to 3 m deep and 30 m long, which strengthened the evidence of size effect. Unfortunately, the mathematical interpretation for design codes was not clear because all these tests were not geometrically scaled, causing the simple size effect to be mixed with complicated shape effects. Introduction of a basic theoretical concept, such as the size effect along with quasibrittle fracture mechanics, into the concrete design code represents a fundamental change, which calls for the deepest scrutiny. In the American Concrete Institute (ACI), it necessitated detailed explanations, accumulation of experimental data, design demonstrations, statistical comparisons, discussions of various alternatives, and rounds of voting in several committees dominated by practicing engineers and practice-oriented professors. After the first proposal by Bažant and Kim, 1984, it took 35 years for the quasibrittle size effect factor to be introduced, in 2019, into the ACI design code, Standard ACI-318, to govern all the shear failure of reinforced concrete beams and slabs. ACI thus became the first concrete society to implicitly recognize quasibrittle fracture mechanics as the theoretical basis of design against brittle structural failures (interestingly, the previous major conceptual change in the ACI code was the introduction of plastic limit analysis for ductile flexural failures, which was proposed by Charles Witney in 1931 and adopted for ACI code in 1971).

In one particular case, though, fracture mechanics of the LEFM type was introduced into the ACI code earlier. It was the code provision for the pullout of anchors in concrete, which are small enough to obtain many test results cheaply. Based on the fracture analysis of Ballarini *et al.* (Ballarini *et al.*, 1986) at Northwestern University in the mid-1980s, the anchor design equation in the ACI and other codes includes the correct size effect.

The Japan Society of Civil Engineers (JSCE) was the first to introduce the size effect into its specification of shear failure of reinforced concrete beams (Japanese Society of Civil Engineers, 1991). It was the result of a visionary proposal of Okamura and Higai in 1980 (Okamura and Higai, 1980). However, it was based on the Weibull statistical size effect. This type of size effect is now known to be incorrect for shear failure of beams, but it must be acknowledged that in 1980 no other size effect theory was in existence.

In 1990, the CEB (Committée européan du béton) introduced a size effect formula for beam shear, but it was purely empirical and incorrect in extension to large sizes, for which the experimental evidence was too scant and too scattered. In 2010, after CEB was absorbed into the *fib* (Fédération internationale de béton), a different size effect formula was introduced for beam shear, based on a theory but, unfortunately, an incorrect theory (based on the intuitive truss mechanics, crack spacing, and the strain across the cracks). Its size effect factor has the form $(1 + Cd)^{-1}$ where $C =$ constant and $d =$ beam depth). This formula gives too little size effect for normal beam sizes but then rapidly changes toward an excessive and thermodynamically impossible asymptotic size effect proportional to d^{-1} (which coincides with the size effect proposed by Leonardo da Vinci for ropes); (cf. (Bažant and Sun, 1987; Yu *et al.*, 2016)). A similar incorrect size effect was earlier introduced into the Swiss Code provision for floor slab punching by columns.

The overall safety factors for concrete structures are, for various reasons, much larger; for example, 3.5 to 8 for beam shear (one reason, for example, is the fact that the design code is meant to apply to an enormous variety of concretes). With the safety factors being so large, a single mistake, such as ignoring the quasibrittle energetic size effect, will not alone suffice to cause failure of a large structure. Typically two or three simultaneous mistakes must combine to bring the structure down, and then it is easy for the investigating committee to attribute the failure to mistakes other than ignoring the size effect (a good example is the 1991 failure of the enormous Sleipner oil platform in a Norwegian fjord; a size effect must have reduced the shear strength of thick shell walls by about 40%, but the investigating committee identified two other problems and ignored the size effect in its verdict). By contrast, in aeronautical engineering, where the safety factor is only 1.5, a single mistake will cause a disaster, which makes disaster investigations far less ambiguous.

The fiber composites community has been as reluctant to accept the quasibrittle fracture concepts as was the concrete community. Observations of the size effect in woven composites pointed to a cohesive or nonlocal fracture behavior but, for about 50 years, no post-peak softening of compact tension fracture specimens has been observed until, in 2016, it was found that the problem was in the low stiffness and low mass of the standard testing grips (or fixtures) used for decades. Redesigning the grips (U.S. Patent 10,416,053) to increase their stiffness by two orders of magnitude, and their mass by one order of magnitude, has made it possible to observe gradual post-peak softening in a stable manner (Salviato *et al.*, 2016*b*). The area under the stabilized load-deflection curve with postpeak softening, as measured with these grips, agreed well with the fracture energy deduced from the tests of size effect on maximum loads.

In the ceramics community a full acceptance of fracture mechanics has also been hard to achieve. Like in concrete, the flexural strength test of beams and the biaxial flexure test of discs are embodied in the ASTM standard without recognizing that the measured tensile strength must be size dependent (Bažant and Novák, 2000a; Zi et al., 2014; Kirane et al., 2014). The two-parameter or three-parameter Weibull distributions are, in the ceramics community, widely accepted, in ignorance of the fact they lead to an incorrect statistical (Type 1) size effect. The correct strength distribution was shown to be a size-dependent Gauss–Weibull grafted distribution (Bažant and Pang, 2006, 2007; Bažant et al., 2009; Le et al., 2011). The ASTM standard for the flexural test of concrete also ignores the size effect, even though the type 1 size effect on the flexural strength test has been amply proven theoretically as well as experimentally (Bažant and Novák, 2000a).

On the hand other, it is interesting that the metals community has been far more receptive in accepting new theoretical results in fracture. For example, Rice's J-integral was promptly adopted as the basis of the ASTM test standard. The reason perhaps is that the collapses of metallic structures are easier to interpret, because of relatively smaller safety factors.

In summary, while the quasibrittle fracture mechanics has made great advances, applications in practice face roadblocks. A comprehensive, though compact, exposition of the theory, attempted in the forthcoming chapters, should help to overcome these roadblocks.

Exercises

E1.1. State six reasons for using fracture mechanics, particularly for quasibrittle materials such as concrete of fiber composite.

E1.2. Describe three basic kinds of fracture mechanics, with a sketch of the nonlinear behavior at the near-tip region.

E1.3. Indicate the typical lengths of the fracture process zone (FPZ) for concrete, shale, rock, and polymer composites.

E1.4. Describe the effect of crack-parallel stresses and explain why a finite width of the fracture process zone with tensorial damage description needs to be considered.

E1.5. Explain in words only the source of the energetic size effect on structural strength in terms of the dependence of energy release on the structure size.

E1.6. Explain in words only the source of the statistical size effect.

2

Fundamentals of Linear Elastic Fracture Mechanics

Nature is nonlinear, but linearity is a pillar of everything.

The theory of linear elastic fracture mechanics (LEFM) is the foundation of the entire discipline of fracture mechanics. Research on LEFM has a long and rich history dating back to Kirsch's and Inglis' work on elastic solution of stresses around a hole in an infinite plate (Kirsch, 1898; Inglis, 1913), which showed that an elliptical hole flattened into a line crack implies infinite stress at the tip (Fig 2.1). The LEFM originated with Griffith's idea of energy release rate as the driver of crack growth (Griffith, 1921). Over the past century, tremendous advances have been achieved in mathematical treatment of LEFM, and today LEFM is a well-established and complete subject.

As will be discussed later, the LEFM predicts an infinite stress at the crack tip. Since no materials can sustain infinite stress, there must exist a nonlinear zone at the crack tip, in which the material would experience plastic deformation or damage. The theory of LEFM is anchored by the assumption that this nonlinear zone is essentially point-wise, that is much smaller than the size of the structure or its cross-section, so that most of the structure volume remains elastic. This chapter presents the fundamentals of the LEFM theory. They include energy release rate, near-tip and remote fields, stress intensity factors, J-integral, calculation of elastic compliance, and extension to bimaterial interfacial cracks.

2.1 Energy Release Rate and Fracture Energy

2.1.1 Energy Balance Analysis

Consider a specimen of thickness b, which contains a straight crack of length a (Fig. 2.2), and assume that the applied load P is increased quasi-statically to cause crack advance δa in the crack direction, sweeping area $\delta A = b\delta a$. The energy required to create the new fractured area, δW^F, is:

Quasibrittle Fracture Mechanics and Size Effect: A First Course. Zdeněk P. Bažant, Jia-Liang Le and Marco Salviato, Oxford University Press. © Zdeněk P. Bažant, Jia-Liang Le, Marco Salviato 2022. DOI: 10.1093/oso/9780192846242.003.0002

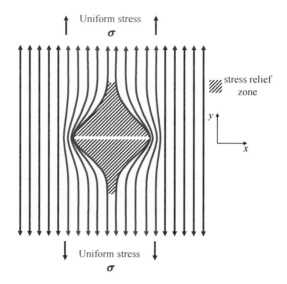

Fig. 2.1 Schematic of stress distribution in a cracked plate under tension (principal stress trajectories).

$$\delta W^F = R\delta A \quad \text{or} \quad \delta W^F = Rb\delta a \tag{2.1}$$

where R denotes the crack growth resistance, which can depend on the crack growth history (in heterogenous brittle materials, as recent research revealed, R depends also on the crack-parallel normal stress (Nguyen *et al.*, 2020*b*), which limits the applicability of LEFM). In the basic case where the crack growth resistance is a material constant, the symbol G_f, denoting the specific *fracture energy*, will be used instead of R.

Consider now the external work, δW, supplied to the structure during the fracturing process. A part of this work is stored in the structure in the form of elastic energy δU. The remainder is left to drive the fracture process and to generate kinetic energy δK_E. Since the process considered here is quasi-static, i.e. $\delta K_E \approx 0$, the remaining energy, δW^R, is what is available to drive the fracture process:

$$\delta W^R = \delta W - \delta U \tag{2.2}$$

Eq. 2.2 can be more conveniently written in terms of specific energies (energies per unit area of crack growth), as follows:

$$\mathcal{G}b\delta a = \delta W^R = \delta W - \delta U \tag{2.3}$$

where $\mathcal{G}$ is called the *energy release rate* which, as will be shown in the following sections, is a *state function*. This means that $\mathcal{G}$ depends only on the instantaneous crack geometry and boundary conditions but not on how they vary during the history of the fracture process.

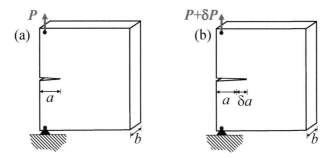

Fig. 2.2 Crack growth in a cracked specimen: (a) initial condition; (b) co-planar crack growth upon further loading.

Note that, in LEFM, the term "rate" generally refers to the derivative with respect to crack length a, and not with respect to time.

In a general incipiently dynamic situation (in which initial kinetic energy $\mathcal{K}_E = 0$ while kinetic energy increase $\delta\mathcal{K}_E \geq 0$):

$$\mathcal{G}\delta a = \mathcal{R}\delta a + \mathcal{K}_E/b \tag{2.4}$$

Since $\mathcal{K}_E \geq 0$, the foregoing energy analysis leads to the following three possible states of crack growth:

$$\begin{aligned}
&\text{If } \mathcal{G} < \mathcal{R} \quad \text{then} \quad \delta a = 0 \quad \text{and} \quad \delta\mathcal{K}_E = 0 \quad \text{No growth (stable equilibrium)} &&\text{(2.5a)}\\
&\text{If } \mathcal{G} = \mathcal{R} \quad \text{then} \quad \delta a \geq 0 \quad \text{and} \quad \delta\mathcal{K}_E = 0 \quad \text{Quasi-static growth possible} &&\text{(2.5b)}\\
&\text{If } \mathcal{G} > \mathcal{R} \quad \text{then} \quad \delta a > 0 \quad \text{and} \quad \delta\mathcal{K}_E > 0 \quad \text{Dynamic growth (unstable)} &&\text{(2.5c)}
\end{aligned}$$

In other words, if the energy available is less than required to break the material, then the crack cannot grow (i.e. the structure is stable). If the energy available equals the required energy then the crack can grow statically, i.e. with negligible inertia forces. If the energy available exceeds that required, then the structure is unstable and the crack will run dynamically (the excess of energy being transformed into kinetic energy).

Note that, based on (Nguyen *et al.*, 2020*b,a*), $\mathcal{R}$ in Eqs. 2.5a–2.5c must be considered to depend on crack-parallel normal stresses σ_{xx}, σ_{zz} unless these stresses are small enough compared to the strength limits in compression and tension.

2.1.2   Elastic Potential and Energy Release Rate

Consider the planar elastic specimen in Fig. 2.2 containing a crack of length a. Let P be the applied load and u the load-point displacement. For any incremental process, the work done by the external load P reads:

$$\delta W = P\delta u \tag{2.6}$$

Assuming that the specimen is in equilibrium, there exists a unique relationship between the force, $P = P(u, a)$, and the displacement, u, where $P(u, a)$ can be determined by a static elastic analysis. From this analysis, the stored elastic strain energy can be expressed as $U = U(u, a)$.

Now consider the case in which the body is subjected, in a static manner, to displacement variation δu causing the crack to extend by δa. Then, Eq. 2.3 can be written as follows:

$$\mathcal{G}\, b\, \delta a = P\left(u, a\right) \delta u - \left\{\left[\frac{\partial U(u, a)}{\partial u}\right]_a \delta u + \left[\frac{\partial U(u, a)}{\partial a}\right]_u \delta a\right\} \tag{2.7}$$

We recall the second Castigliano's theorem for an equilibrium displacement variation δu at $\delta a = 0$:

$$P\left(u, a\right) = \left[\frac{\partial U(u, a)}{\partial u}\right]_a \tag{2.8}$$

By substituting Eq. 2.8 into Eq. 2.7, we reach the equation:

$$\mathcal{G} - \mathcal{G}\left(u, a\right) - -\frac{1}{b}\left[\frac{\partial U(u, a)}{\partial a}\right]_u \tag{2.9}$$

This basic result indicates that the energy release rate $\mathcal{G}$ is a state function, i.e., depends only on the current boundary conditions and current specimen geometry.

Alternatively, we may consider the equilibrium load P as an independent variable instead of u. In that case, we must introduce a dual elastic potential, the complementary energy U^*, defined as:

$$U^* = Pu - U \tag{2.10}$$

(which is called the Legendre transformation). Substituting U from the foregoing expression into Eq. 2.2 and further considering Eq. 2.6, we obtain:

$$\delta W^R = \delta W - \delta U = P\delta u - \delta\left(Pu - U^*\right) = \delta U^* - u\delta P \tag{2.11}$$

The complementary elastic energy is a unique function of the applied load and crack length: $U^* = U^*(P, a)$. Similarly, $u = u(P, a)$. Now, for a general equilibrium process with varying P and a, we can rewrite Eq. 2.3 as

$$\mathcal{G}\, b\, \delta a = -u\left(P, a\right) \delta P + \left\{\left[\frac{\partial U^*(P, a)}{\partial P}\right]_a \delta P + \left[\frac{\partial U^*(P, a)}{\partial a}\right]_P \delta a\right\} \tag{2.12}$$

According to the first Castigliano's theorem, we have

$$u\left(P, a\right) = \left[\frac{\partial U^*(P, a)}{\partial P}\right]_a \tag{2.13}$$

Substitution of Eq. 2.13 into Eq. 2.12 yields the following alternative expression for the energy release rate $\mathcal{G}$:

$$\mathcal{G} = \mathcal{G}\left(P, a\right) = \frac{1}{b}\left[\frac{\partial U^*(P, a)}{\partial a}\right]_P \tag{2.14}$$

Note that, unlike Eq. 2.9, there is no minus in front of this expression.

The energy release rate can also be expressed in terms of the potential energy of the structure-load system, which is defined as $\Pi = U - \int_0^u P(u')\mathrm{d}u'$. Differentiation at constant u yields $[\partial U/\partial a]_u = [\partial \Pi/\partial a]_u$, and so the substitution into Eq. 2.9 yields

$$\mathcal{G} = \mathcal{G}\left(u, a\right) = -\frac{1}{b}\left[\frac{\partial \Pi(u, a)}{\partial a}\right]_u \tag{2.15}$$

Similarly, one may characterize crack equilibrium in terms of complementary potential energy, $\Pi^* = U^* - \int_0^P u(P')\mathrm{d}P'$, where the last term represents the complementary work of the applied load. Differentiation at constant P furnishes $[\partial U^*/\partial a]_P = [\partial \Pi^*/\partial a]_P$. Hence, substitution into Eq. 2.14 gives[1]

$$\mathcal{G} = \mathcal{G}\left(u, a\right) = \frac{1}{b}\left[\frac{\partial \Pi^*(P, a)}{\partial a}\right]_P \tag{2.20}$$

It is instructive to note that the aforementioned relation between the energy release rate and the elastic potential can be simply explained by a graphic representation of fracture process. Fig. 2.3a presents a quasi-static load-displacement response of the specimen (or structure). As the crack advances by an infinitesimal increment δa, the equilibrium state moves from point A to point B. The shaded area of triangle OAB represents the energy release available for this fracture process. Figs. 2.3b and c represent two virtual (non-equilibrium) fracture processes under constant displacement and under constant load, respectively.

---

[1]A shorter, though more formal, way to obtain $\mathcal{G}$ is to note that (for isothermal conditions) Π represents the Helmholtz free energy of the structure that needs to be supplied to create the crack (Bažant and Ohtsubo, 1977; Bažant and Cedolin, 1991); so,

$$\Pi = U(a, u) - \int_0^u P(u')\mathrm{d}u' + \int_0^a R(a')\mathrm{d}a' \tag{2.16}$$

The first variation of Π is

$$\delta\Pi = \left\{\left[\frac{\partial U}{\partial u}\right]_a - P\right\}\delta u + \left\{\left[\frac{\partial U}{\partial a}\right]_u + R\right\}\delta a \tag{2.17}$$

At equilibrium, at which the crack can propagate statically, $\delta\Pi$ must vanish for any combination of δu and non-negative δa. This requires that both expressions in Eq. 2.17 would vanish. The first yields the second Castigliano theorem, Eq. 2.8, and the second the energy release rate, Eq. 2.9 (the stability of crack equilibrium is indicated by the positiveness of the second variation, $\delta^2\Pi$; Bažant and Ohtsubo (1977)).

In an analogous way, note that (for isothermal conditions) the complementary work represents the Gibbs free energy of the structure. The Gibbs free energy of the structure-load system (Bažant and Cedolin, 1991), that is needed to create the crack, is expressed as:

$$\Pi^* = U^*(a, P) - \int_0^P u(P')\mathrm{d}P' - \int_0^a R(a')\mathrm{d}a' \tag{2.18}$$

The first variation is

$$\delta\Pi^* = \left\{\left[\frac{\partial U^*}{\partial a}\right]_P - u\right\}\delta P + \left\{\left[\frac{\partial U^*}{\partial P}\right]_a - R\right\}\delta a \tag{2.19}$$

At equilibrium, this variation must vanish for any combination of δP and non-negative δa. This requires that both expressions in Eq. 2.19 would vanish. The first yields the first Castigliano theorem, Eq. 2.13, and the second the energy release rate, Eq. 2.14 (but note that $\delta^2\Pi^*$ does not decide stability) (Bažant and Cedolin, 1991, ch. 10).

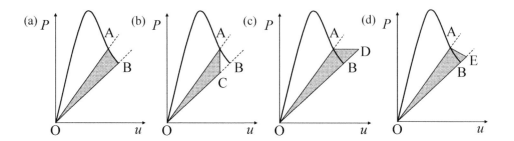

Fig. 2.3 (a) Actual (equilibrium) incremental fracture process. (b–d) Virtual (nonequilibrium) incremental fracture process: (b) at constant displacement; (c) at constant load; (d) at arbitrary $\Delta P/\Delta u$.

In Fig. 2.3b, the area of triangle OAC represents the decrease in strain energy, and in Fig. 2.3c the area of triangle OAD represents the increase in the complementary strain energy if the load is held constant. Since we consider the stress increments to be infinitesimal, the areas of triangles OAB, OAC, and OAD coincide except for second-order small term (which tends to zero for $\delta u \to 0$). Therefore, the energy release rate can be calculated as the limit of the ratio of the area of any of these triangles to the crack length increment. In fact, we may consider a virtual fracture process with an arbitrary $\delta P/\delta u$ as shown in Fig. 2.3d. The area of the triangle OAE also coincides with the area of any aforementioned triangle. This shows, again, that the energy release rate function is a state function independent of the loading path. It is evident that Fig. 2.3b corresponds to Eq. 2.9 and Fig. 2.3c corresponds to Eq. 2.14.

Example 2.1: *Double cantilever beam specimen*
Consider the double cantilever beam (DCB) shown in Fig. 2.4. If sufficiently slender, the cantilevers can be treated according to the engineering theory of bending, in which the cross-section remains plane and normal to the deflected beam axis. According to this theory, the cantilever ends at the crack tip behave as fixed. Based on the principle of virtual work, the load-point displacement u of each cantilever caused by load P at the end is

$$u = \int_0^a \frac{M\overline{M}}{EI}\,\mathrm{d}x = \frac{4Pa^3}{Ebh^3} \tag{2.21}$$

where $I = bh^3/12$, h = thickness of cantilevers, b = their width, $\overline{M} = x$, and $M = Px$. Based on Eq. 2.21, we obtain $P = Ebh^3u/4a^3$. The complementary energy can be calculated as $\Pi^* = 2 \cdot Pu/2 = 4P^2a^3/Ebh^3$. Based on Eq. 2.20, the energy release rate at the crack tip is

$$\mathcal{G} = \frac{1}{b}\left[\frac{\partial \Pi^*}{\partial a}\right]_P = \frac{12P^2a^2}{Eb^2h^3} \tag{2.22}$$

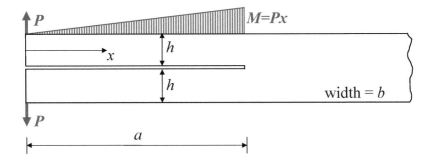

Fig. 2.4 Double cantilever beam specimen subjected to opening forces.

Following Eq. 2.5b, we may consider that the crack propagates when $\mathcal{G}$ reaches the fracture energy, G_f, of the material. By setting $\mathcal{G} = G_f$, Eq. 2.22 yields a relation between the applied load and the length of the propagating crack at equilibrium:

$$a = \frac{b}{P}\sqrt{\frac{EG_f h^3}{12}} \qquad (2.23)$$

Substituting Eq. 7.22 into Eq. 2.21 gives the complete equilibrium load-deflection curve during stable crack propagation

$$P = \left(\frac{8}{27}EG_f{}^3 h^3\right)^{1/4}\frac{b}{\sqrt{u}} \qquad (2.24)$$

Note that, to maintain equilibrium, the applied load will decrease with an increasing displacement. This is a general feature of specimens of the so-called positive geometry, which is defined as the geometry for which $\partial\mathcal{G}/\partial a > 0$.

Example 2.2: *Buckling of a surface layer demarcated by a crack*

Consider now a thick wall with a longitudinal crack of length $2a$ at small depth below the surface, subjected to uniform compressive stress $\sigma < 0$ (Fig. 2.5). Compression causes the layer between the surface and the crack to buckle as a sufficiently slender column which is considered to be of unit width and follows the engineering theory of bending, with fixed ends at the crack tips. The axial stress at buckling is

$$\sigma_{cr} = -E'I\pi^2/a^2 h \qquad (2.25)$$

where $I = h^3/12$. Before buckling, the layer is brought to some stress $|\sigma_0| > |\sigma_{cr}|$ and, due to buckling, the axial stress in the layer drops to σ_{cr}. This creates energy loss

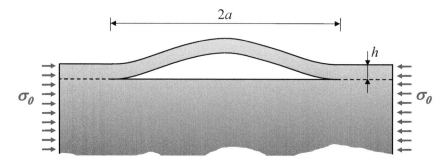

Fig. 2.5 Thick wall with a longitudinal crack at a shallow depth below the surface.

$U = -\Delta\Pi_0$ in the buckling layer, $-\Delta\Pi_0 = 2(\sigma_0 - \sigma_{cr})^2 ah/2E'$ (per unit width). The bending energy of the layer is zero because we deal with the onset of buckling. The crack will propagate when $-\partial(\Delta\Pi_0)/\partial a = 2G_f$ (in which we assume both two crack tips to propagate simultaneously due to symmetry). It follows that

$$\sigma_0^2 - 2\sigma_0(E'I\pi^2/a^2h) - 3(E'I)^2\pi^4/a^4h^2 = 2E'G_f/h \qquad (2.26)$$

from which one can solve the stress σ_0 at which the crack will propagate (note that if we assume only one crack tip to propagate, we would write $-\partial\Delta\Pi_0/\partial(2a) = G_f$, which would give the same σ_0). This problem illustrates some, but not all, aspects of delamination in layered composites (Sallam and Simitses, 1985; Yin *et al.*, 1986; Sallam and Simitses, 1987; Bažant and Grassl, 2007).

2.2 General Form of Near-Tip and Far Fields of a Notch

The previous analysis provides a means of determining the energy release rate at a crack tip, but not the near-tip stress field. To determine it, we evoke self-similarity and consider more generally a sharp V-notch (Fig. 2.6), whose limit case is a sharp crack.

Let r, θ be the polar coordinates centered at the tip of a crack or notch in a continuous homogeneous solid. Consider an annular region $r \in (r_1, r_2)$ surrounding the tip, in which $r_2 \gg r_1$ and $r_2 \ll$ the distance to the boundary or the body (other than the notch faces), and also much larger than any characteristic length of the material (e.g., the maximum inhomogeneity size). Since the boundary lies so far that it can have no effect, the transformation $r \to r'$, where $r' \in (r_1, r_2)$, creates a perfectly similar body. Further assume that the material constitutive law is also self-similar, which means that no material stress or strain limits are present, as is the case for elasticity (and also for the power-law plastic-hardening model of metals, not considered in this book). Then, within the annular domain, the field $f(r, \theta)$ of any physical variable must be self-similar with regard to such transformation. Therefore, we must have $f(r', \theta)/f(r, \theta) = \rho$, which is the scaling ratio depending on $\xi = r'/r$ only. This argument yields a functional equation:

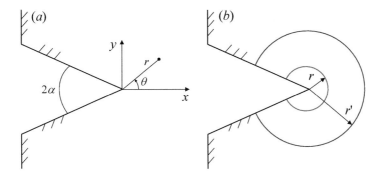

Fig. 2.6 (a) Sharp V-notch or crack ($\alpha = 0$) with a polar system of coordinates centered at the tip. (b) Circular domains of similar crack tip fields.

$$f(\xi r, \theta) = \rho(\xi) f(r, \theta) \qquad (2.27)$$

To solve it, we note it will be true for all r if and only if it is true for every infinitesimal coordinate interval $(r, r + \mathrm{d}r)$. So we first differentiate it with respect to ξ. This yields

$$r f_{,r}(\xi r, \theta) = \rho'(\xi) f(r, \theta) \quad \text{with} \quad f_{,r}(\xi r, \theta) = \mathrm{d}f(r,\theta)/\mathrm{d}r, \quad \rho'(\xi) = \mathrm{d}\rho(\xi)/\mathrm{d}\xi \quad (2.28)$$

Now consider $\xi \to 1$ and note that $\rho'(1) = \lambda = \text{constant}$. This furnishes an ordinary differential equation:

$$r \, \frac{\mathrm{d}f(r, \theta)}{\mathrm{d}r} = \lambda f(r, \theta) \qquad (2.29)$$

By applying separation of variables, Eq. 2.29 can be easily solved:

$$\ln f(r, \theta) = \lambda \ln r + \ln \psi(\theta) \qquad (2.30)$$

Here $\ln \psi(\theta)$ is the constant for integration with respect to r, which must depend on parameter θ. Eq. 2.30 can thus be rewritten as

$$f(r, \theta) = r^{\lambda} \, \psi(\theta) \qquad (2.31)$$

Based on this result, we conclude that any field variable near the tip of a crack or V-notch must have a separated form in polar coordinates, and that the radial dependence must be a *power function* (this is true not only in elasticity, but also for similar geometrical situations in hydraulics, electrostatics, electromagnetism, etc. On the other hand, when the field equations are nonlinear, the separated form in Eq. 2.31 may apply only asymptotically near the tip, even in infinite space, or not at all). Generalization to three dimensions with spherical coordinates is, of course, possible.

To apply the foregoing analysis to the separated near-tip stress and strain fields of an elastic body, we first write the near-tip displacement field, according to Eq. 2.31, as

$$u_i = r^\lambda F_i(\theta) \tag{2.32}$$

where the subscript i refers to a component in cartesian coordinates x_i ($i = 1, 2$). Then the strains (considered as the small, or linearized, strains) are

$$\epsilon_{ij} = \tfrac{1}{2}(u_{i,j} + u_{j,i}) \tag{2.33}$$

$$= \tfrac{1}{2}(u_{i,r}r_{,j} + u_{j,r}r_{,i}) + \tfrac{1}{2}(u_{i,\theta}\theta_{,j} + u_{j,\theta}\theta_{,i}) \tag{2.34}$$

$$= r^{\lambda-1}[g_{ij}(\theta) + r(F_{i,\theta}\theta_{,j} + F_{j,\theta}\theta_{,i})] \tag{2.35}$$

where $g_{ij}(\theta) = [r_{,j}F_i(\theta) + r_{,i}F_j(\theta)]\lambda/2 = $ functions of θ, $r_{,1} = dr/dx_1 = 1/\cos\theta$, and $r_{,2} = dr/dx_2 = 1/\sin\theta$ (and subscripts preceded by a comma denote partial derivatives). Note that the second term in the square bracket must be neglected because $r \to 0$. Hence, we have

$$\epsilon_{ij} = r^{\lambda-1}g_{ij}(\theta) \tag{2.36}$$

Furthermore, according to Hooke's law, the near-tip stress field can be expressed as

$$\sigma_{ij} = C_{ijkl}\epsilon_{kl} = r^{\lambda-1}f_{ij}(\theta) \tag{2.37}$$

where $f_{ij}(\theta) = C_{ijkl}g_{kl}(\theta)$ and $C_{ijkl} = $ tensor of elastic moduli.

2.3 Stress Singularities and Energy Flux at a Sharp Crack Tip

We will now determine the power-law exponent λ for a sharp crack in a homogenous elastic body. Consider that a crack extends from length a to length $a + h$. We can decompose this incremental fracture process into two steps:

1. Imagine, at first, that the crack extension h is cut in the material ahead of the crack while the crack faces are held fixed along this segment (Fig. 2.7a). We thus imagine that surface tractions equal to the stresses σ_{22} corresponding to the initial crack length a are externally applied along segment h to prevent it from opening.

2. Now imagine that the stresses, σ_{22}, externally applied along segment h, are gradually reduced to zero in proportion to $(1 - \tau)$ where τ is a parameter growing from 0 to 1. To maintain internal equilibrium in the solid surrounding the crack, the crack face displacements v corresponding in segment h to the new crack length $a + h$ must be simultaneously increased in proportion to the same parameter τ until, at $\tau = 1$, the displacement profile w corresponding to the new crack length $a + h$ is attained (Fig. 2.7b,c).

Based on Eqs. 2.32 and 2.37, we can write

$$\sigma_{22}^A = Ax^{\lambda-1}, \quad w^B = B(h - x)^\lambda \tag{2.38}$$

where A and B are some constants, x is the coordinate measured from the initial crack tip, and $h - x$ is the distance from the new crack tip. We can now calculate

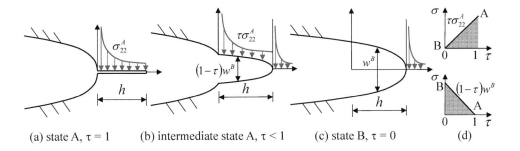

(a) state A, τ = 1   (b) intermediate state A, τ < 1   (c) state B, τ = 0   (d)

Fig. 2.7 Proportional release of closing tractions used to compute the energy release rate.

the work $\Delta \mathcal{W}$ done by the decreasing crack face stresses on the increasing crack face displacement in segment h;

$$\Delta \mathcal{W} = \int_{x=0}^{h} \int_{\tau=0}^{1} \left[A x^{\lambda-1} (1-\tau) \right] \left[B(h-x)^{\lambda} \tau \right] \mathrm{d}\tau \mathrm{d}x = \frac{ABI}{4\lambda} h^{2\lambda} \qquad (2.39)$$

where $I = \int_0^1 \xi^{\lambda-1} (1-\xi)^{\lambda} \mathrm{d}\xi$, and $\xi = x/h$. Integral I is a constant to be determined later. Also note that although the crack shapes pictured in Fig. 2.6 would not occur in reality, the calculation of $\Delta \mathcal{W}$ is exact because elasticity is path-independent and, therefore, only the initial and final states matter.

The incremental work $\Delta \mathcal{W}$ must be equal to the energy dissipation in the crack segment. Therefore, we can calculate the energy flux into the crack tip (i.e., energy dissipation per unit crack growth) as follows:

$$\mathcal{G} = \lim_{h \to 0} \frac{\Delta \mathcal{W}}{h} = \frac{ABI}{4\lambda} \lim_{h \to 0} h^{2\lambda-1} \qquad (2.40)$$

Since $\mathcal{G}$ can be neither zero nor infinite, we must have $2\lambda - 1 = 0$ or

$$\lambda = 1/2 \qquad (2.41)$$

Therefore $\mathcal{G} = ABI/2$. This indicates that the stress and strain fields at the tip of a sharp crack exhibit a singular behavior with a "$-1/2$" singularity, i.e. $\sigma_{ij}, \epsilon_{ij} \propto r^{-1/2}$ $(r \to 0)$.

Now we may ask whether we could extend the foregoing analysis of the energy flux to the near-tip stress field of a V-notch. The answer is no. The underlying reason is that the foregoing analysis assumes that, during the fracture process, the geometry of the crack remains the same. However, this is not the case for the V-notch. During the fracture process of a V-notch, a sharp crack will initiate from the notch tip (Fig. 2.8a). The change from a V-notch to a sharp crack indicates that in Eq. 2.38 the power-law exponents for σ_{22} and w must be different. The physical mechanism of this transition

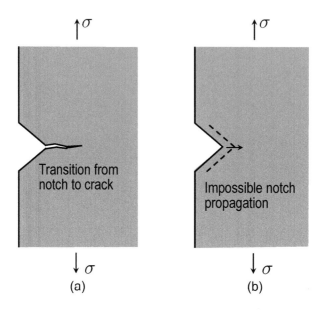

Fig. 2.8 Fracture of V-notch: (a) transition from V-notch to a propagating crack, and (b) unrealistic notch propagation.

involves a detailed analysis of crack initiation, which is not easy to treat in an explicit mathematical model.

Therefore, for a V-notch, it is simpler to directly use elastic analysis to calculate the near-tip fields. In fact, the elasticity solution shows that the stress singularity of the V-notch tip is weaker than the "−1/2" singularity, i.e. $\lambda > 1/2$. If we follow the foregoing analysis, Eq. 2.40 indicates that the energy flux into the notch tip is zero regardless of the applied load. This physically means that a V-notch will not propagate as a V-notch, which is independently made clear by Fig. 2.8b.

2.4   Westergaard's Solution for Crack in Infinite Body

As shown in Sec. 2.2 and Sec. 2.3, the self-similar solution and the energy flux analysis yield a general form of the near-tip fields. To obtain a detailed description of the stress and strain fields, we need to rely on the theory of elasticity. For plane elasticity problems, one effective way is to use complex variables. In this section, we present Westergaard's solution for a crack of length $2a$ in an infinite plane, subjected at infinity to tractions corresponding to a biaxial tensile stress σ (Fig. 2.9). The crack is assumed to lie on axis x and its center to be the origin of coordinates (x, y).

Based on the theory of elasticity (in two dimensions), the stress field may be obtained by solving a biharmonic equation

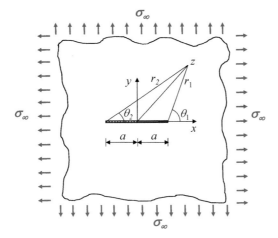

Fig. 2.9 Center-cracked infinite panel subjected to remote equiaxial tension.

$$\nabla^4 \Phi(x, y) = 0 \tag{2.42}$$

where $\Phi(x, y)$ is the Airy stress function, which is generally a complex function. The stresses are expressed as $\sigma_{xx} = \Phi_{,yy}, \sigma_{yy} = \Phi_{,xx}, \tau_{xy} = -\Phi_{,xy}$, which ensures that the differential equations of equilibrium are satisfied. Eq. 2.42 ensures that the compatibility conditions for strains and the isotropic Hooke's law are satisfied, too.

It is convenient to consider that $\Phi = \Phi(z) =$ real function of complex variable $z = x + iy$ representing the coordinate vector. According to Westergaard (Westergaard, 1939), for a class of boundary value problems including the present one,

$$\Phi = \text{Re } \hat{Z} + y \text{ Im } \bar{Z} \quad \text{where} \quad \mathrm{d}\hat{Z}/\mathrm{d}z = \bar{Z}, \quad \mathrm{d}\bar{Z}/\mathrm{d}z = Z \tag{2.43}$$

Since $Z(z)$ is a holomorphic function, ReZ and ImZ are harmonic functions, i.e., $\nabla^2(\text{Re } Z) = 0, \nabla^2(\text{Im } Z) = 0$, and it may be checked that Eq. 2.43 automatically satisfies Eq. 2.42. By calculating the derivatives, we obtain

$$\Phi_{,yy} = \sigma_{xx} = \text{Re } Z - y \text{ Im } Z', \tag{2.44}$$
$$\Phi_{,xx} = \sigma_{yy} = \text{Re } Z + y \text{ Im } Z' \tag{2.45}$$
$$-\Phi_{,xy} = \tau_{xy} = -y \text{ Re } Z' \tag{2.46}$$

Here $Z' = \mathrm{d}Z/\mathrm{d}z$, where Z can be any holomorphic function.

For the present problem (Fig. 2.9), Westergaard found that

$$Z(z) = \frac{\sigma z}{\sqrt{z^2 - a^2}} \tag{2.47}$$

where σ is the remote biaxial stress. Since this function is holomorphic, the field equation (Eq. 2.42) is automatically satisfied, except at the crack ($|x| \leq a, y = 0$). The

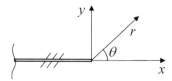

Fig. 2.10 Polar coordinate system centered at the crack tip.

boundary conditions, too, are satisfied because: (1) for $|z| \to \infty$, we have $\lim_{|z|\to\infty} Z = \lim \sigma z / \sqrt{z^2} = \sigma$; and (2) on the crack face, $y = 0$ and $|x| < a$, Eq. 2.47 gives

$$\sigma_{yy} = \text{Re}[Z]_{y=0} = \text{Re}\left[\frac{\sigma(x+iy)}{i\sqrt{a^2 - z^2}}\right]_{y=0} = 0, \quad \tau_{xy} = -[y\text{Re } Z']_{y=0} = 0 \quad (2.48)$$

For the crack extension line $(x > |a|,\ y = 0)$, we have

$$\sigma_{yy} = \text{Re}[Z]_{y=0} = \text{Re}\left[\frac{x+iy}{\sqrt{z^2 - a^2}}\right]_{y=0} - \frac{\sigma x}{\sqrt{x^2 - a^2}}, \quad \tau_{xy} = 0 \quad (2.49)$$

When the remote stress σ_{xx} is different from $\sigma = \sigma_{yy}$, the solution may be easily obtained by superposing a uniform field of stress equal to $\sigma_{xx} - \sigma_{yy}$ in the x direction, because this field has zero stress components on the plane $y = 0$. This indicates that Eq. 2.47 is independent of the normal stress in the x-direction.

Westergaard's solution describes the full elastic field for one particular body geometry. But we already know that the near-tip asymptotic field of Westergaard's solution must be valid for any body geometry. To find this field, we shift the origin of coordinates to the crack tip by replacing z by $z + a$ (Fig. 2.10). Then, considering $z \to 0$, Eq. 2.47 becomes

$$Z = \frac{\sigma(z+a)}{\sqrt{z(z+2a)}} \approx \sigma\sqrt{\frac{a}{2z}} \quad (2.50)$$

Since the dependence on a and z, must be valid for any structure geometry, we replace external load σ by a load-independent prefactor of the near-tip field, $K_I = \sigma\sqrt{\pi a}$, whose meaning will be discussed in the next section. K_I is commonly referred to as the *stress intensity factor*.

By introducing polar coordinates attached to the crack tip (Fig. 2.10), we have $z = re^{i\theta}$ and, from Eq. 2.50,

$$Z = \frac{K_I}{\sqrt{2\pi r}}\, e^{-i\theta/2}, \quad Z' = -\frac{K_I}{\sqrt{2\pi}}\frac{1}{2r\sqrt{r}}\, e^{-3i\theta/2} \quad (2.51)$$

Substituting Eq. 2.51 into Eqs. 2.44, 2.45 and 2.46, and noting Euler's relation $e^{ix} = \cos x + i\sin x$, we obtain the following expressions for the stresses near the crack tip (written with numerical subscripts for coordinates $x = x_1$ and $y = x_2$):

$$\sigma_{ij} = K_I\, r^{-1/2}\, f_{ij}(\theta) \quad (i = 1, 2,\ j = 1, 2) \quad (2.52)$$

where $f_{ij}(\theta)$ are dimensionless functions given by

$$\left.\begin{array}{c} f_{11}(\theta) \\ f_{22}(\theta) \end{array}\right\} = \frac{1}{\sqrt{2\pi}} \; \cos\frac{\theta}{2} \left(1 \mp \sin\frac{\theta}{2}\sin\frac{3\theta}{2}\right) \tag{2.53}$$

$$f_{12}(\theta) = \frac{1}{\sqrt{2\pi}} \; \sin\frac{\theta}{2}\cos\frac{\theta}{2}\cos\frac{3\theta}{2} \tag{2.54}$$

2.5   Stress Intensity Factor, Near-Tip Field, and Remote Field

From Eq. 2.52, we note that the stress field has indeed a separated form, as required by Eq. 2.37, and that the value $\lambda = 1/2$ (Eq. 2.41) is confirmed. By virtue of the self-similar behavior of near-tip fields (Eq. 2.37), the angular dependence functions f_{11} and f_{12} apply to any structure geometry and any type of loading, provided the shear stresses on the crack extension line vanish (which is condition of symmetry with respect to axis x). On the other hand, the expression for K_I varies with the structure geometry and loading configuration. Therefore, we can conclude that Eq. 2.52 with Eqs. 2.53 and 2.54 give a general expression of the near-tip stress field for the case where only the normal stress is present on the crack extension line.

It is evident that the stress intensity factor K_I must have the dimension of [stress] $\cdot$ [length]$^{1/2}$. Based on dimensional analysis, K_I can always be written as

$$K_I = \sum_{i=1}^{n} \frac{P_i}{bD} \sqrt{D} k_i(a_0/D) \tag{2.55}$$

where P_i ($i = 1, ..., n$) = applied loads, D = characteristic structure size, b = width of the structure in the transverse direction, a_0 = crack length, $k_i(\cdot)$ = dimensionless geometry-dependent function corresponding to load P_i. Note that the summation in Eq. 2.55 ensues from the linear superposition of the stress fields produced by each load P_i.

The corresponding near-tip strain field has the form of Eq. (2.35) with $\lambda = 1/2$, and it can be determined by substituting Eqs. 2.52, 2.53, and 2.54 into Hooke's law. By integrating the strains, we obtain the near-tip field of cartesian displacements in the form:

$$u_i = \frac{K_I}{E'} \; \sqrt{r} \; \phi_i(\theta) \qquad (i = 1, 2) \tag{2.56}$$

in which $E' = E$ = Young's modulus in the case of plane stress, and $E' = E/(1 - \nu^2)$ in the case of plane strain (ν = Poisson's ratio), and ϕ_i are dimensionless functions expressed as

$$\phi_1(\theta) = \sqrt{\frac{2}{\pi}} (1 + \nu) \cos\frac{\theta}{2} \left(1 - 2\nu + \sin^2\frac{\theta}{2}\right) \tag{2.57}$$

$$\phi_2(\theta) = \sqrt{\frac{2}{\pi}} (1 + \nu) \sin\frac{\theta}{2} \left(2 - 2\nu - \cos^2\frac{\theta}{2}\right) \tag{2.58}$$

In particular, the crack face opening profile is obtained from Eqs. 2.56 and 2.57 by setting $\theta = \pi$. In terms of the crack face displacement $v = u_2$, we have

$$v = \sqrt{\frac{8}{\pi}} \frac{K_I \sqrt{r}}{E'} (1 - \nu^2) = \sqrt{\frac{8}{\pi}} \frac{K_I \sqrt{r}}{E} \qquad (2.59)$$

for plane strain problems. The total crack opening width (also called the separation) is $w = 2v$.

An alternative way to obtain the near tip field is to substitute the strains calculated from Eq. 2.32 for $u_i(r, \theta)$ into Hooke's law. The resulting stresses are then substituted into the equilibrium equation, $\sigma_{ij,j} = 0$ as well as into the boundary conditions along the crack surface. This operation leads to two homogeneous linear differential equations for $F_1(\theta)$ and $F_2(\theta)$, with homogeneous boundary conditions for crack faces $\theta = \pm\pi$. This forms an eigenvalue problem, in which exponent λ plays the role of an eigenvalue. The analytical solution to this kind of problem was given (for isotropic materials) by Karp and Karal (Karp and Karal, 1962). Using finite difference approximations, one obtains a matrix eigenvalue problem, which allows later extension to cracks in anisotropic materials (as well as to dynamic crack propagation with inertia forces) (Achenbach and Bažant, 1975; Achenbach *et al.*, 1976a,b). The eigenvalue always is $\lambda = 1/2$.

Williams (Williams, 1952) pursued another way to derive the near-tip field. He considered a separated form of the Airy stress function, i.e.:

$$\Phi(r, \theta) = r^{\lambda+1} h(\theta) \qquad (2.60)$$

By substituting Eq. 2.60 into the biharmonic equation $\nabla^4 \Phi = 0$ with the boundary conditions on the crack faces or, more generally, the faces of a V-notch, he obtained a homogeneous linear differential equation for function $h(\theta)$, with homogeneous boundary conditions for $\theta = -\pi, \pi$. The power law exponent λ again represents the eigenvalue of the system, and Williams gave a widely used analytical solution, not only for cracks but also V-notches (Fig. 2.6) (Williams, 1952). for which the only difference is that the boundary conditions must be imposed at $\theta = -\alpha, \alpha$ where 2α = notch angle. For $\alpha > 0$, it is found that the singularity exponent $\lambda > 1/2$; function $\lambda(\theta)$, calculated by Williams, monotonically increases and reaches 1 for $2\alpha = \pi$, which is the case of a flat surface for which there is no singularity.

The foregoing approaches based on separation of variables yield the full asymptotic expansion of the displacement or stress field:

$$u_i = \sum_{\nu=1}^{\infty} a_\nu r^{\lambda_\nu} f_{i\nu}(\theta), \quad \sigma_{ij} = \sum_{\nu=1}^{\infty} b_\nu r^{\lambda_\nu - 1} f_{ij\nu}(\theta) \qquad (i = 1, 2) \qquad (2.61)$$

with coefficients a_ν, b_ν. For a crack, the eigenvalues are

$$\lambda_\nu = 1/2, \ 1, \ 3/2, \ 2, \ 5/2, \ 3, \dots \qquad (2.62)$$

Only the first, dominant, eigenvalue, $\lambda_\nu = \lambda = 1/2$, corresponds to the singular stress and strain fields, which prevails near the tip. Optimum matching of a suitable number of terms of the expansion to the boundary conditions can furnish an approximate solution to the complete boundary value problem.

It is interesting to mention that, using this approach in a little-known article (Knein, 1927), Knein was probably the first to obtain the near-tip field of a crack but his work was forgotten because his interpretation of the final result was incorrect (he incorrectly discarded the first term, corresponding to $\lambda_1 = 1/2$, thinking that infinite stresses at the crack tip cannot be allowed).

If body forces are present, one can show that their effect must asymptotically vanish for $r \to 0$, and so the near-tip fields are unaffected. The inertia forces, however, would change the angular dependence function (Achenbach and Bažant, 1975; Achenbach *et al.*, 1976a,b).[2]

2.6 Fracture Modes I, II, and III

So far, we have tacitly assumed that the stress field is symmetric with respect to the crack plane, which occurs when both the specimen geometry and the applied loads are symmetric. In this case there are no shear stresses on the crack extension line and no relative slip along the crack face. This mode of fracture is called mode I, or the opening mode (Fig. 2.11a).

In mode II (Fig. 2.11b), called also the in-plane shear mode, the planar stress field is antisymmetric with respect to the crack plane. In mode III (Fig. 2.11c), called also the anti-plane shear mode, the displacements are asymmetric in the z direction and shear stresses τ_{xz} and τ_{yz} exist on the crack plane. The elastic mode III analysis can be reduced to the Laplace differential equation and is usually the easiest to solve. For mode II, as well as III, the same kind of analysis as before can be carried out (Williams,

---

[2]In various problems such as the interaction of microcracks in micromechanical modeling, it is useful to know how the stress disturbance caused by formation of a crack in a field under stress σ decays with increasing distance from the crack and how it is angularly distributed. Following Bažant (1994), we may begin by writing Westergaard's solution (Eq. 2.47) in polar coordinates (r, θ) with the origin at crack center, i.e.:

$$Z = \sigma r e^{i\theta} \left(r^2 e^{2i\theta} - a^2 \right)^{-1/2} = \sigma \left[1 - (a/r)^2 e^{-2i\theta} \right]^{-1/2} \tag{2.63}$$

Note that $(1 - x)^{-1/2} \approx 1 + x/2$ when $x \ll 1$. For $r \gg a$, we have

$$Z \approx \sigma \left[1 + (a^2/2r^2) e^{-2i\theta} \right] = \sigma \left[1 + (a^2/2z^2) \right], \quad Z' \approx -\sigma a^2/z^3 \tag{2.64}$$

$$y \ \mathrm{Im} \ Z' = r \ \sin\theta \ \mathrm{Im}(-\sigma a^2 r^{-3} \ e^{-3i\theta}) = \sigma(a/r)^2 \sin\theta \ \sin 3\theta \tag{2.65}$$

Therefore, we can express the remote stresses as

$$\left. \begin{array}{c} \sigma_{xx} \\ \sigma_{yy} \end{array} \right\} = \mathrm{Re} \ Z \mp y \ \mathrm{Im} \ Z' = \sigma \ \frac{a^2}{r^2} \left(\frac{\cos 2\theta}{2} \mp \sin\theta \ \cos 3\theta \right) \tag{2.66}$$

$$\tau_{xy} = -y \ \mathrm{Re} Z' = r \sin\theta \ \frac{\sigma a^2}{r^3 \cos 3\theta} = \sigma \ \frac{a^2}{r^2} \ \sin\theta \cos 3\theta \tag{2.67}$$

By trigonometric rearrangements, we get $\sigma_{xx} = \sigma \ (a^2/2r^2) \ \cos 4\theta$, and $\sigma_{yy} = \sigma \ (a^2/r^2) \ (\cos 2\theta - \frac{1}{2}\cos 4\theta)$. Note that σ_{xx} is positive when $\cos 4\theta > 0$ or $\theta > 22.5°$, and negative when $\theta < 22.5°$. These regions correspond to the remote amplification and shielding sectors of σ_{xx}. For σ_y and τ_{xy} the sector angles are different, see Bažant (1994) for detailed analysis. It is worth noting from Eqs. 2.66 and 2.67 that, in an infinite solid, the stress changes caused by a crack decay as r^{-2}. In three dimensions, though, the stress changes caused by a penny-shaped crack decay as r^{-3}, and the σ_{xx} shielding cone has a different angle Bažant (1994).

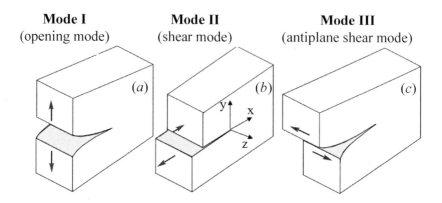

Fig. 2.11 Basic modes of fracture: (a) mode I or pure opening mode; (b) mode II or in-plane shear mode; and (c) mode III or antiplane shear mode.

1952; Sih, 1966; Neuber, 1985; Lazzarin *et al.*, 2007; Salviato and Zappalorto, 2016; Salviato *et al.*, 2018; Zappalorto *et al.*, 2019).

A general loading produces a combination of all three modes. By virtue of the principle of superposition, the near-tip fields can then written as

$$\sigma_{ij} = r^{\lambda-1}[K_I f_{ij}^I(\theta) + K_{II} f_{ij}^{II}(\theta) + K_{III} f_{ij}^{III}(\theta)] \tag{2.68}$$

$$u_i = r^\lambda \left[\frac{K_I}{E'}\phi_i^I(\theta) + \frac{K_{II}}{E'}\phi_i^{II}(\theta) + \frac{K_{III}}{2G}\phi_i^{III}(\theta)\right] \tag{2.69}$$

Note that exponent $\lambda-1$ is common to all the three modes, for a crack in a homogenous isotropic elastic solid. It should be noted that for cracks in general anisotropic solids, a separation of fracture modes is not possible.

A full description of the stress and displacement fields is provided by asymptotic expansions similar to Eq. 2.61, extended by superposition of all three modes. In view of Eqs. 2.52, 2.53, and 2.54, the stress intensity factors under general loading may be defined by the following equations for $\theta = 0$:

$$\begin{aligned}
K_I &= \lim_{x\to 0^+} \sigma_{yy}\sqrt{2\pi x}, \\
K_{II} &= \lim_{x\to 0^+} \tau_{xy}\sqrt{2\pi x}, \\
K_{III} &= \lim_{x\to 0^+} \tau_{xz}\sqrt{2\pi x}
\end{aligned} \tag{2.70}$$

Note that, in the literature prior to about 1960, the factor $\sqrt{2\pi}$ was absent from the foregoing equations. In that case, factor 2π would have to be inserted into Irwin's relation, to be presented next. The consensus prevailed to keep that fundamental relation simple.

It is evident that Eq. 2.55 can be used to express K_{II}, K_{III} for modes II and III loading cases. Meanwhile, we note that, due to the difference in the angular dependence functions of fracture modes I, II, and III (Eqs. 2.68 and 2.69), one cannot sum the stress intensity factors of different loading modes, for calculating the near-tip field.

2.7 Irwin's Relationship between Stress Intensity Factors and Energy Release Rate

The near-tip fields derived in Sec. 2.5 can now be inserted into the energy flux analysis presented in Sec. 2.3. Based on Eqs. 2.52, 2.53, and 2.59, we can calculate factors A and B in Eq. 2.38:

$$A = \frac{K_I}{\sqrt{2\pi}}, \quad B = \frac{4\sqrt{2}}{\sqrt{\pi}} \frac{K_I}{E'} \tag{2.71}$$

By substituting Eq. 2.71 into Eq. 2.40 and setting $\lambda = 1/2$, we have

$$\mathcal{G} = \frac{K_I^2}{E'} \tag{2.72}$$

This result represents a *fundamental equation of LEFM*, a major advance after Griffith in 1921, discovered in 1957 by George Irwin (Irwin, 1957, 1958). This equation can also be derived by the J-integral (defined in Sec. 2.8) along a circular path, but this leads to complex trigonometric transformations. Another derivation (Bažant and Planas, 1998, p. 93), based on a J-integral path along the crack surfaces in Fig. 2.7, is only slightly more involved than the present one.

As discussed in Sec. 2.1.1, we can set $\mathcal{G} = G_f$ as a fracture criterion, where G_f = fracture energy, which is treated in LEFM as a material constant (although in quasibrittle materials it depends strongly on the crack-parallel stress). Based on this fracture criterion, we obtain the corresponding critical value of the stress intensity factor, which is usually called the *fracture toughness*:

$$K_{Ic} = \sqrt{E'G_f} \tag{2.73}$$

The energetic fracture criterion presented in Eqs. 2.5a to 2.5c can now be re-written in terms of the stress intensity factor and the fracture toughness, i.e.:

If	$K_I < K_{Ic}$	then:	No crack growth (stable equilibrium)	(2.74a)
If	$K_I = K_{Ic}$	then:	Quasi-static growth possible	(2.74b)
If	$K_I > K_{Ic}$	then:	Dynamic growth (unstable)	(2.74c)

Note again that, based on Nguyen *et al.* (2020*b*,*a*), K_{Ic} for quasibrittle materials must here be considered to depend on crack-parallel normal stresses σ_{xx} and σ_{zz} unless these stresses are small enough compared to the strength limits in compression and tension.[3]

---

[3]Based on the fitting of test results for normal concrete at $\sigma_{zz} = 0$ and on their computational extrapolation, the following empirical formula for the critical value G_f has been obtained Nguyen *et al.* (2020*b*): $G_f/G_{f0} = 1 + a/(1 + b/\xi) - (1 + a + b)/(1 + b)\xi^c$ in which, for normal concrete, $a = 1.038, b = 0.245, c = 7.441, \xi = G_f/f_c, f_c =$ uniaxial compression strength of concrete, $G_{f0} = G_f$ value at zero crack-parallel normal stress.

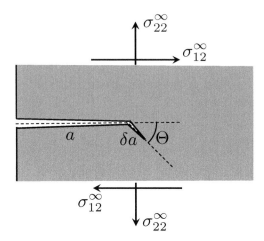

Fig. 2.12 Crack kinking under mixed-mode loading.

Irwin's relation leads to an important conclusion. K_I, just like $\mathcal{G}$, is again primarily a measure of the rate of energy release from the whole structure, and its role as the amplitude of the local singular near-tip stress field is only secondary. This field provides a precise means, required by the theory of elasticity, to deliver a given energy flux to the advancing crack tip.

Therefore, the value of stress intensity factor depends primarily on the change of stress field in the *whole* structure due to an infinitesimal fracture increment. By virtue of this fact, the stress field near the tip has only a minor effect on the K_I-value and need not be represented accurately in numerical computations.

One consequence, for example, is that the K_I values for cracks, asymptotically exact for slender enough beams, can be obtained by using the engineering theory of bending to calculate the energy release rate $\mathcal{G}$ (see Example 2.1), even though the hypothesis of planar cross-sections remaining plane cannot capture the near-tip field realistically. Another consequence is that while the strains near the crack tip are very large, the nonlinear finite strain theory is not needed.

When the structure is subjected to a combination of different loading modes, one can show that, in isotropic materials, the three fracture modes contribute through independent terms to the expression of $\mathcal{G}$. If the crack growth is coplanar, the foregoing analysis of energy flux in mode I can be applied to modes II and III, and the resulting total energy release rate can be expressed as

$$\mathcal{G} = \frac{K_I^2}{E'} + \frac{K_{II}^2}{E'} + \frac{K_{III}^2}{2G} \tag{2.75}$$

The formulation of a general propagation criterion for in plane mixed-mode cracks is more involved because, under mixed mode loading, the crack could kink out of its original plane (Fig. 2.12). The stress intensity factors K_I^t, K_{II}^t of this kinked crack segment can be related to K_I, K_{II}, and the so-called T-stress (understood as the non-singular normal stress component σ_{xx} acting parallel to the crack plane)(Hayashi and Nemat-Nasser, 1981; He and Hutchinson, 1989; He *et al.*, 1991); i.e.:

$$K_I^k = c_{11}K_I + c_{12}K_{II} + b_1 T\sqrt{\Delta a} \qquad (2.76)$$

$$K_{II}^k = c_{21}K_I + c_{22}K_{II} + b_2 T\sqrt{\Delta a} \qquad (2.77)$$

where Δa = kinking crack length (Fig. 2.12), and c_{ij}, b_i = constants, which depend on the kinking angle, Θ. The actual kinking angle, Θ_k, value can be determined by maximizing the energy release rate of a small kinked crack segment. At the initiation of kinking (i.e., for $\Delta a \to 0$), as it turns out, the Θ_k-value determined by using the maximum energy release rate condition coincides approximately with the kinking direction that corresponds to $K_{II}^k = 0$. In other words, the kinked crack will initiate in mode I. The fracture initiation condition for a mixed-mode crack can be written as

$$\mathcal{G}^k = \frac{1}{E'}(c_{11}K_I + c_{12}K_{II})^2 = G_f \qquad (2.78)$$

If the crack parallel T-stress is negligible or if its effect on fracture energy G_f is ignored (see Sec. 3.9), Eq. 2.78 may be written in terms of the energy release rate of the parent crack; that is:

$$\mathcal{G} = \frac{K_I^2}{E'} + \frac{K_{II}^2}{E'} = \eta G_f \qquad (2.79)$$

where parameter $\eta = \mathcal{G}/\mathcal{G}^k$. The dependence of η on the mode mixity can be found in (He *et al.*, 1991; Hutchinson and Suo, 1992)[4].

2.8   Rice's *J*-Integral

Aside from Griffith energy concept and Irwin's relation, the J-integral, proposed by J. R. Rice (Rice, 1968a,b), is the third fundamental concept of fracture mechanics. Rice showed that the J-integral equals the energy release rate $\mathcal{G}$ of LEFM, and so it can be used as an alternative way to assess crack propagation (Rice, 1968b). Moreover, the J-integral is also applicable to nonlinear elasticity, which can provide a good approximation to the yielding zone of plastic-hardening metals as long as there is no unloading. Here we present a simple and instructive derivation of the J-integral in Eulerian coordinates (x_1, x_2) moving with the crack tip.

Consider that a line integral is taken along a closed contour Γ surrounding the crack tip; see the solid curve in Fig. 2.13a. The crack faces are assumed to be traction free

---

[4]Note that, for several loads producing the same mode, one cannot sum the energy release rates $K_{I_i}^2/E'$. Rather, according to the principle of superposition, one must obtain the total $K_I = \sum_i K_{I_i}$, and then, equating it to K_c, one gets the propagation condition. This further implies that if, e.g., P_1^* and P_2^* denote two critical loads when acting separately, the interaction diagram (or failure envelope) is linear:

$$\frac{P_1}{P_1^*} + \frac{P_2}{P_2^*} = 1 \qquad (2.80)$$

This is an important practical difference from plasticity, where the interaction diagram is typically a convex curve or convex polygon, while fracture mechanics is at the limit of convexity. Thus the strength for combined loads in fracture mechanics is less than it is in plasticity.

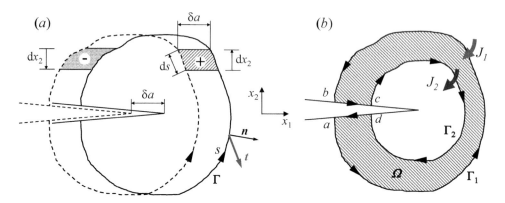

Fig. 2.13 (a) Variation of elastic energy by the J-integral; (b) path independence of J-integral.

and the crack to extend straight ahead. The integration along the contour proceeds, and the path length s is measured, in the counterclockwise direction. The contour is moved forward as a rigid body together with the crack tip as the tip advances to the right by δa (dashed curve). The corresponding transformation of the coordinates can be written as $X_1 = x_1 - a$, $X_2 = x_2$ where (X_1, X_2) are the fixed, or Lagrangian, coordinates attached to the material points in the initial configuration.

The interest here is to calculate the change of the energy supply ΔW to the crack tip as the contour moves with the crack tip. It is clear that ΔW can be written as

$$\Delta W = \left(\frac{\partial W}{\partial a} - \frac{\partial W}{\partial X_1}\right)\delta a = \left(\frac{\partial W}{\partial a} - \frac{\partial W}{\partial x_1}\right)\delta a \tag{2.81}$$

As discussed in Sec. 2.1.1, the energy that flows into the crack tip can be calculated by Eq. 2.3, that is,

$$W = \int_\Gamma t_i u_i \mathrm{d}S - \int_\Omega \bar{U}\mathrm{d}A \tag{2.82}$$

where $\bar{U} = \int \sigma_{ij}\mathrm{d}\varepsilon_{ij}$ = strain energy density, and t_i, u_i = traction and displacement vectors on the contour Γ, respectively. Under quasi-static condition, the condition of stationarity of potential energy in equilibrium states dictates that (Rice, 1968a)

$$\int_\Gamma t_i \frac{\partial u_i}{\partial a}\delta a\,\mathrm{d}s = \int_\Omega \frac{\partial \bar{U}}{\partial a}\delta a\,\mathrm{d}A \tag{2.83}$$

Therefore, Eq. 2.81 becomes

$$\Delta W = -\frac{\partial W}{\partial x_1}\delta a = \frac{\partial(-W)}{\partial x_1}\delta a \tag{2.84}$$

All the purpose of the foregoing four equations ending with Eq. 2.84 has been to demonstrate that, for the purpose of calculating ΔW, we can freeze the original stress and displacement fields during the translation of the contour.

Now consider an element $\mathrm{d}s$ of the contour which sweeps during crack advance δa the area $\delta A = \delta a \mathrm{d}x_2$ (cross-hatched in Fig. 2.13a). This causes the strain energy, $\bar{U}\delta A$, to flow into the moving contour interior if the vertical projection, $\mathrm{d}x_2$, of the vector of the path increment $\mathrm{d}s$ lies on the front side of the contour, and out the contour if on the back side (the distinction between into and out of the contour is automatically accounted for by the sign of $\mathrm{d}x_2$, which is positive in front of the contour and negative in the back, because coordinate s runs counterclockwise). Therefore, we have

$$\Delta U = \int_\Gamma \bar{U} \mathrm{d}x_2 \delta a \tag{2.85}$$

Further we need to include the work that the traction $t_i = \sigma_{ij} n_j$, acting on the contour, does on the incremental displacement, δu_i, of each contour point (here n_i is the unit vector of the outward normal of the contour). Putting together these two kinds of energy supply to contour Γ, the total energy supply to the contour interior during crack advance δa is

$$\Delta W = - \left[\int_\Gamma [t_i \delta u_i] \mathrm{d}s - \int_\Gamma \bar{U} \, \delta a \, \mathrm{d}x_2 \right] \tag{2.86}$$

Now the J-integral is defined as

$$J = \lim_{\delta a \to 0} \frac{\delta W}{\delta a} \tag{2.87}$$

By noting that $\delta u_i = u_{i,1}\delta a$, where $u_{i,1} = \partial u_i / \partial x$ (and $x \equiv x_1$), we thus obtain Rice's famous J-Integral derived in 1968 (Rice, 1968a,b):

$$J = \int_\Gamma \left(\bar{U} \mathrm{d}x_2 - \sigma_{ij} n_j u_{i,1} \mathrm{d}s \right) \tag{2.88}$$

The foregoing analysis indicates that the J-integral can be derived directly from the energy analysis of the contour translation by considering the work done by the domain Ω. In this case, the energy flux is equal to the difference between the strain energy stored in the domain and the external work done, i.e. $\Delta W = \int_\Gamma \bar{U} \, \delta a \, \mathrm{d}x_2 - \int_\Gamma [t_i \delta u_i] \mathrm{d}s$. The essential step of the analysis is that we can freeze the stress and strain fields while translating the contour, as allowed by the stationarity of potential energy.

The J-integral was adopted as the basis of the American Society for Testing Materials (ASTM) standard for measuring the fracture energy of metals ASTM (2017). This integral can be computed whenever all the points on contour Γ behave elastically. However, the J-integral is exactly equal to the Linear Elastic Fracture Mechanics (LEFM) energy release rate, $\mathcal{G}$, only if: (a) the non-elastic zone reduces to a point in the interior of Γ; (b) the crack faces are traction-free, and (c) the crack is planar and extends in its own plane.

Another form of this integral is obtained by setting $\mathrm{d}x_2 = n_i \mathrm{d}s = \delta_{1j} n_j \mathrm{d}s$, where δ_{ij} = the Kronecker delta. This yields:

$$J = \int P_{1j} n_j \mathrm{d}s \tag{2.89}$$

$$\text{where} \quad P_{ij} = \bar{U}\delta_{ij} - \sigma_{jk} u_{k,i} \tag{2.90}$$

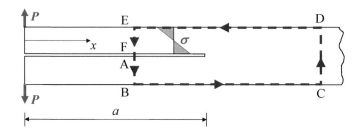

Fig. 2.14 Integration path (dashed line) used to compute the J-integral.

This integral was in fact first formulated as a path-independent integral by Eshelby (Eshelby, 1956), but without relation to fracture mechanics. Tensor P_{ij} is called Eshelby energy momentum tensor.

The salient property of the $J-$integral is that it is path independent, as proven already in 1956 by Eshelby (Eshelby, 1956) (albeit with no relation to fracture mechanics). A purely mathematical proof is more involved (see Appendix A) but the path-independence is easily proven physically. We consider two closed contours Γ_1 and Γ_2, enclosing between them an annular domain A* in which the material is purely elastic (see Fig. 2.13b). If the energy fluxes, J_1 and J_2, through these two contours were different, it would mean that energy would be dissipated in domain A*. But this is impossible if the material is elastic or, more broadly, nonlinear elastic. Hence the fluxes of energy through both contours must be identical, $J_1 = J_2$.

Note that in the case of a finite fracture process zone (FPZ), to be studied later, the J-integral gives the energy flux into the whole FPZ, not just into the crack tip within the FPZ. Also note that the J-integral applies only to straight propagation along the crack plane, i.e., not to any sideways crack extension. In the case of metals, a hardening yielding zone can be (and has been (Hutchinson, 1968; Rice and Rosengren, 1968)) treated as nonlinear elastic provided that the J contour is chosen such that no point within it has suffered unloading. Softening damage is, obviously, excluded.

In geotechnical and some other applications, body forces f_i working on displacements u_i must be included in the J-integral. Obviously, one must use $\bar{U} = \int \sigma_{ij} \mathrm{d}\epsilon_{ij} - f_i u_i$ (Palmer and Rice, 1973).

Example 2.3: *J-integral for a double cantilever beam specimen*

Now consider again the double cantilever beam specimen discussed in Example 2.1. We will now determine the energy release rate by using the J-integral. Because the J-integral is path independent, we should choose the path cleverly to simplify the calculation. For this specimen, we consider the rectangular path ABCDEF shown in Fig. 2.14. Segments BC and ED are at the surface and contribute nothing because, on them, $\mathrm{d}x_2 = 0$ and $t_i = 0$. Segment CD also does not contribute since it lies so far

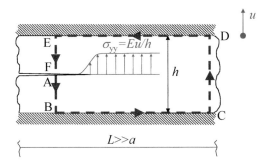

Fig. 2.15 Integration path (dashed line) used to compute the J-integral for a crack within a layer of solidified glue bonded to two rigid plates subjected to uniform displacement.

from the crack tip that that the stresses are zero. Segments AB and EF make equal contributions. So it suffices to consider segment EF, for which

$$J_{EF} = \int_{h/2}^{-h/2} \left(\bar{U} - t_1 u_{1,1} - t_2 u_{2,1} \right) \mathrm{d}x_2 \tag{2.91}$$

According to the engineering beam theory, we have

$$\int_{h/2}^{-h/2} \bar{U} \mathrm{d}x_2 = -\frac{(Px_1)^2}{2EI} \tag{2.92}$$

$$\int_{h/2}^{-h/2} t_1 u_{1,1} \mathrm{d}x_2 = \int_{h/2}^{-h/2} \frac{My}{I} \frac{My}{EI} \mathrm{d}y = -\frac{(Px_1)^2}{EI} \tag{2.93}$$

$$\int_{h/2}^{-h/2} t_2 u_{2,1} \mathrm{d}x_2 = \int_{h/2}^{-h/2} \frac{t_2 P}{EI} \left(\frac{a^2}{2} - \frac{x_1^2}{2} \right) \mathrm{d}y = \frac{P^2}{EI} \left(\frac{x_1^2}{2} - \frac{a^2}{2} \right) \tag{2.94}$$

By substituting Eqs. 2.92, 2.93 and 2.94 into Eq. 2.91, we obtain $J_{EF} = P^2 a^2 / 2EI$. Therefore, the J-integral for the entire path ABCDEF is

$$J = 2J_{EF} = \frac{P^2 a^2}{EI} = \frac{12 P^2 a^2}{Ebh^3} \tag{2.95}$$

which gives the same result as Eq. 2.22 in Example 2.1.

Example 2.4: *Semi-infinite crack in a solidified glue bonded to rigid plates*

Consider a semi-infinite crack in a layer of solidified glue of small thickness h bonded to two parallel rigid plates (Rice, 1968b). The rigid plates are moved apart by distance u. To calculate the J-integral, we choose the rectangular path ABCDEF shown in Fig. 2.15. It is seen that segments BC and DE do not contribute to the J-integral because $\mathrm{d}x_2 = 0$ and $u_{i,1} = 0$. Neither do the transverse segments AB and

EF located far behind the crack tip where the stress is zero. So, the only contribution comes from segment CD. Since it lies far ahead of the crack tip, its stress state is not affected by the presence of the crack. Noting that $u_{i,1} = 0$ along segment CD, we have

$$J = J_{CD} = \bar{U}h = \frac{E'u^2}{2h} \qquad (2.96)$$

where $E' = E/(1 - \nu^2)$.

2.9   Numerical Calculation of Stress Intensity Factors

In previous sections, we showed that the stress intensity factors can be determined either from elasticity solutions or from the energy release rate based on Irwin's relation. However, in general, a closed-form solution for the stress intensity factors may not be available for a given practical problem. So one must often resort to numerical approximations. Most of the strategies available can be grouped into (a) *incremental stiffness methods*, (b) *near-tip field fitting*, and (c) *J-integral evaluation*.

2.9.1   Incremental stiffness method

The incremental stiffness method is based on Eq. 2.20 relating the energy release rate and the complementary potential energy function (Fig. 2.16a–d). The method involves the calculation of the elastic compliance of the specimen for two different, but very close, crack lengths $a - \Delta a$ and $a + \Delta a$. The energy release rate $\mathcal{G}$ can then be estimated as:

$$\mathcal{G}(P, a) = \frac{1}{b}\left[\frac{\partial \Pi^*}{\partial a}\right]_P = \frac{P^2}{2b}\frac{dC(a)}{da} \approx \frac{P^2}{2b}\frac{C(a + \Delta a) - C(a - \Delta a)}{2\Delta a} \qquad (2.97)$$

where $P = $ applied load, $b = $ specimen thickness, $a = $ crack length and $C(a) = $ compliance. The compliances needed in Eq. 2.97 can be easily calculated by finite elements, either (a) by using one mesh and simulating the crack extension by freeing one node so that the crack extends by one element (Fig. 2.16b), or (b) by modifying the mesh for the second calculation in which the node at the crack tip is displaced by Δa (Fig. 2.16c). The first approach is easy to use and does not require any modification of the global stiffness matrix, but requires a fine mesh for the accurate numerical differentiation in Eq. 2.97. The second method decouples the crack extension from the mesh size, but requires a partial recalculation of the stiffness matrix.

2.9.2   Near-tip field fitting

The near-tip field fitting exploits the known near-tip fields of the stress, displacement or crack opening fields to estimate K_I. One can use the stress field $\sigma_{yy}\sqrt{2\pi r}$ close to the crack-tip, where σ_{yy} is the stress normal to the crack plane and r is the distance to the crack tip. In fact, the plot of $\sigma_{yy}\sqrt{2\pi r}$ vs. r should tend to K_I as r approaches zero (Eq. 2.70).

Alternatively, it is possible to use the crack opening which is related to the stress intensity factor as $w = 8K_I\sqrt{r}/E'\sqrt{2\pi}$. Therefore, the limit of $wE'\sqrt{2\pi}/8\sqrt{r}$ as $r \to 0$ will also yield the value of K_I.

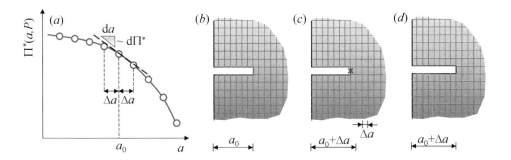

Fig. 2.16 (a) Initial crack a_0 meshed with quadrilateral elements; (b) Crack of length $a_0 + \Delta a$ obtained by freeing one node and (c) by re-meshing; (d) calculation of the energy release rate from the numerical results.

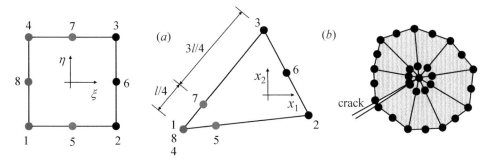

Fig. 2.17 (a) Collapsing of an 8-node quadratic isoparametric element into a singular quarter-node element; (b) Rosette of singular finite elements at the crack tip.

It is worth mentioning that obtaining accurate results (less than 5% error) with near-tip analysis requires extremely fine meshes. This is due to the difficulties in representing the crack tip singularity with ordinary finite elements. Careful convergence studies by Wilson (Wilson, 1971) and Oglesby and Lamackey (Oglesby and Lamackey, 1972) showed that the near-tip approximate solution may not converge to the analytical solution even for extremely fine meshes.

To overcome this issue, it was proposed to include directly a $r^{-1/2}$ singular term in the shape function (see e.g. Aliabadi and Rooke (1991)). These methods, generally incorporate special shape functions and require developing a special dedicated element.

A notable example, extensively used today by almost all the commercial FE codes, is to use the quarter-node isoparametric quadrilateral element (Barsoum, 1975, 1976;

Henshell and Shaw, 1975). In this formulation, the standard 8-node isoparametric quadrilateral is collapsed to a triangular quarter-point element (Fig. 2.17a). The vertex corresponding to the collapsed nodes $1 - 8 - 4$ becomes the singular point, and a $r^{-1/2}$ singularity is achieved by placing nodes 5 and 7 at the quarter-point (from the singular vertex) of the radial sides of the triangle. These elements are placed around the crack tip as shown in Fig. 2.17b.

Note that the aforementioned elements cannot reproduce a homogeneous stress field. Consequently, they are unrealistically stiff with respect to the normal stresses parallel to the crack line, which in LEFM have no effect on K_I but distort the overall stress field (note that they have a large effect on the critical K_{1c}). Therefore, one must avoid situations where the loading makes such stresses large.

It is also possible to surround the crack tip with elements that can reproduce a homogeneous stress field but are enriched with an additional degree of freedom corresponding to the singular strain field (Byskov, 1970; Tracey, 1971, e.g.)

As mentioned earlier, one can also determine K_I by least-square fitting of the near-tip displacement field to the displacements in a group of nodes surrounding the crack front. This approach is used in mesh-free methods, and has the advantage that one does not need to know beforehand where the crack tip is located (Dolbow and Belytschko, 1999; Duarte *et al.*, 2001; Rabczuk *et al.*, 2007).

J-Integral Evaluation

Alternatively, one can numerically evaluate the *J*-integral to determine the energy release rate, from which we can determine the stress intensity factors. Compared to the incremental stiffness method, numerical differentiation is avoided and a single calculation is sufficient. This approach is embedded in most commercial finite element software, for example, ABAQUS. In finite element simulations, instead of using the line integral (Eq. 2.88), the *J*-integral is often evaluated using the domain integral method (Li *et al.*, 1985; Shih *et al.*, 1986), which is able to deliver accurate results even with rather coarse meshes.

2.10   Stress Intensity Factors for Typical Simple Geometries

Up to the 1980s, extensive efforts have been devoted to calculating the stress intensity factors for hundreds of specimen geometries. The results are summarized in *The Stress Analysis of Cracks Handbook* by Tada, Paris, and Irwin (Tada *et al.*, 2000), and also in the handbook by Murakami (Murakami, 1986).

Here we present the expressions of stress intensity factors for some simple geometries under mode I loading. Recall from Eq. 2.55 that the stress intensity factor can always be written as $K_I = \sigma\sqrt{D}k(\alpha)$ and the corresponding energy rate function can be written as $\mathcal{G} = \sigma^2 D g(\alpha)/E$, where σ = nominal stress, D = characteristic specimen size, $\alpha = a/D$ = relative crack size, $k(\alpha)$ = dimensionless stress intensity factor, and $g(\alpha) = k^2(\alpha)$ = dimensionless energy release rate function. In what follows, we will use this general form of the stress intensity factor and present $k(\alpha)$ and the definitions of σ and D for each specimen geometry.

1. For a line crack in infinite plane:

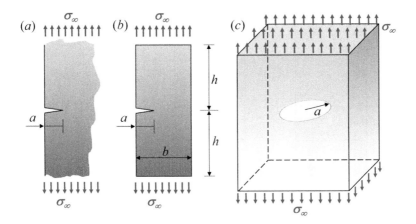

Fig. 2.18 (a) Crack of length a in an half-plane; (b) Single Edge Notch Tension Specimen; (c) Penny-shape crack in infinite medium.

$$\sigma = \sigma_\infty, \quad D = a, \quad k(\alpha) = k_0 = \sqrt{\pi} \tag{2.98}$$

2. For a normal surface crack of depth a in a half-plane (Fig. 2.18a):

$$\sigma = \sigma_\infty, \quad D = a, \quad k(\alpha) = k_0 = 1.12\sqrt{\pi} \tag{2.99}$$

3. For Single Edge Notch Test (SENT) specimen with a crack of length a, witdh b and height $2h$ (Fig. 2.18b):

$$\sigma = \sigma_\infty, \quad D = a, \quad k(\alpha) = \sqrt{\frac{2b}{a}\tan\frac{\pi a}{2b}} \, \frac{0.752 + 2.02\,(a/b) + 0.37\left(1 - \sin\frac{\pi a}{2b}\right)^3}{\cos\frac{\pi a}{2b}} \tag{2.100}$$

4. For a circular (penny-shaped) crack of radius a in a 3D infinite body subjected to remote stress σ normal to crack plane (Fig.2.18b):

$$\sigma = \sigma_\infty, \quad D = a, \quad k(\alpha) = k_0 = \frac{2}{\sqrt{\pi}} \tag{2.101}$$

5. For a long planar strip of width D with a centric normal crack of length $2a$, subjected to remote longitudinal stress σ (Fig. 2.19a) (Feddersen, 1966; Tada *et al.*, 1973),

$$\sigma = \sigma_\infty, \quad k(\alpha) = k_1(\alpha)[\cos(\pi\alpha)]^{-1/2}\sqrt{\pi a} \tag{2.102}$$

where $\alpha = a/D$ and $k_1(\alpha) = 1 - 0.1\alpha^2 + 0.96\alpha^4$ (error $< 0.1\%$).

6. For a strip of width D containing two edge cracks of depth $a = D/2 - r$ and with a short ligament $2r \ll D$ under load P (Fig. 2.19b),

$$\sigma = P/bD; \quad k(\alpha) = \left(\frac{\pi}{2} - \pi\alpha\right)^{-1/2} \tag{2.103}$$

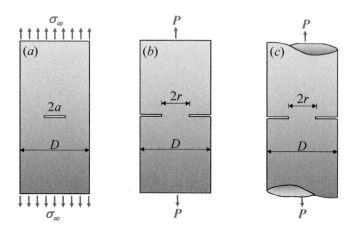

Fig. 2.19 (a) Central crack of length a in infinite strip of width D; (b) Edge crack in infinite strip; (c) Circumferential edge crack in circular bar with circular ligament.

7. For a long circular cylinder of diameter D weakened by circumferential crack of depth $a = D/2 - r$ leaving a small ligament of radius $r \ll D$, under remote axial load P (Fig. 2.19c) (Tada *et al.*, 1973),

$$\sigma = 4P/\pi D^2; \quad k(\alpha) = \frac{\sqrt{\pi}}{8} \left(\frac{1}{2} - \alpha \right)^{-3/2} \tag{2.104}$$

8. For a row of co-linear identical equidistant cracks of length $2a$ and center-to-center spacing D, in an infinite plane remotely loaded by stress σ normal to the cracks (Fig. 2.20a),

$$\sigma = \sigma_\infty, \quad k(\alpha) = \sqrt{\tan(\pi\alpha)} \tag{2.105}$$

9. For a three-point-bend beam under load P, with depth D, span $L = 4D$, and crack depth a (Pastor *et al.*, 1995; Bažant and Planas, 1998, p. 50) (error $< 0.5\%$ for all α, which is better than for the classical formula of (Srawley, 1976)):

$$\sigma = P/bD \tag{2.106}$$

$$k(\alpha) = p(\alpha)\sqrt{\alpha}(1 + 2\alpha)^{-1}(1 - \alpha)^{-3/2}, \quad \alpha = a/D \tag{2.107}$$

$$p(\alpha) = 1.900 + \alpha(0.089 - 0.603\gamma + 0.441\gamma^2 - 1.223\gamma^3) \tag{2.108}$$

where $\gamma = 1 - \alpha$.

10. For a panel of width D containing a crack of length $2a$, where a pair of concentrated loads P is applied at the face centers of the crack (Fig. 2.20c) (Newman Jr., 1971; Tada *et al.*, 1973),

$$\sigma = P/bD \tag{2.109}$$

$$k(\alpha) = (1 - 0.5\alpha + 0.957\alpha^2 - 0.16\alpha^3)[\pi\alpha(1 - 2\alpha)]^{-1/2} \tag{2.110}$$

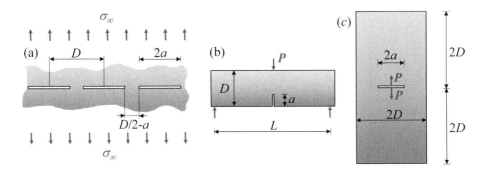

Fig. 2.20 (a) System of collinear cracks; (b) Notched beam in three-point bending; and (c) Crack surface subjected to a pair of concentrated loads.

(error $< 0.3\%$). Note that in this case K_I decreases with a at constant P. This is known as the case of *negative geometry*. In the other aforementioned cases 1–8, K_I increases with a at constant P, which is the case of *positive geometry*.

2.11 Calculation of Elastic Compliance and Deflection from Stress Intensity Factors

If the energy release rate or the stress intensity factor is known for every crack length a, the deflection increase due to fracture can be calculated. By substituting Irwin's relation (Eq. 2.72) into Eq. 2.97 and setting $K_I = Pk(\alpha)/b\sqrt{D}$, we have

$$\left[\frac{\mathrm{d}C(a)}{\mathrm{d}a} \right]_P = \frac{2k^2(\alpha)}{bDE'} \tag{2.111}$$

The compliance now follows by direct integration:

$$C(\alpha) = C_0 + \frac{2\phi(\alpha)}{bE'}, \qquad \phi(\alpha) = \int_0^\alpha k^2(\alpha')\mathrm{d}\alpha' \tag{2.112}$$

where C_0 is the compliance of the structure for the case of no crack ($\alpha = 0$). The corresponding deflection can simply be calculated as $u(a) = C(a)P$ (with regard to the aforementioned crack-parallel stress effect (Nguyen *et al.*, 2020a), note that here we deal with the theory of LEFM, in which there is no such effect)..

Consider now that there are many loads, P_i ($i = 1, 2, ...n$), all of which produce mode I fracture. Based on Eq. 2.55, and because the principle of superposition applies to stresses (though not energy release rates), the total stress intensity factor can be written as

$$K_I = \frac{1}{b\sqrt{D}} \sum_{i=1}^n P_i k_i(\alpha) \tag{2.113}$$

where $k_i(\alpha)$ is the dimensionless stress intensity factor for load P_i. Since the compliances are $C_{ij} = \partial^2 \Pi^* / \partial P_i \partial P_j$, we have

$$
\begin{aligned}
\frac{\partial C_{ij}(\alpha)}{\partial a} &= \frac{\partial}{\partial a} \frac{\partial^2 \Pi^*}{\partial P_i \partial P_j} \\
&= \frac{b \partial^2 \mathcal{G}}{\partial P_i \partial P_j} \\
&= \frac{b \partial^2}{\partial P_i \partial P_j} \left(\frac{K_I{}^2}{E'} \right) \\
&= \frac{\partial^2}{\partial P_i \partial P_j} \left[\frac{1}{bD} \left(\sum_m P_m k_m(\alpha) \right)^2 \right] \\
&= \frac{2}{bDE'} k_i(\alpha) k_j(\alpha)
\end{aligned}
\tag{2.114}
$$

The total compliance matrix is obtained as

$$
C_{ij}(\alpha) = C_{ij}^0 + \frac{2}{bE'} \int_0^\alpha k_i(\alpha') k_j(\alpha') \mathrm{d}\alpha'
\tag{2.115}
$$

where C_{ij}^0 denotes the specimen compliance before any crack was created. Note that Eq. (2.112) is a special case for $n = 1$. The deflection in the sense of load P_i is $u_i(\alpha) = \sum_j C_{ij}(\alpha) P_j$.

Example 2.5: *Calculation of crack mouth opening displacement*
Eq. 2.115 is useful to calculate the crack mouth opening displacement $w_M(a)$ (Fig. 2.21), which needs to be controlled to stabilize fracture tests during the post-peak or snapback regime. In that case $P_1 = P$ = actual load, and $P_2 = P_M$ = imaginary pair of loads acting in the sense of the relative displacement measured by the crack mouth gage (Fig. 2.21). We may write the displacements as

$$
u = C(a)P + C_M(a)P_M
\tag{2.116}
$$
$$
w_M = C_M(a)P + C_{MM}(a)P_M
\tag{2.117}
$$

where u = load-point displacement, w_M = crack mouth opening displacement, $C(a)$, $C_M(a)$, $C_{MM}(a)$ = compliance functions. We denote the dimensionless stress intensity factors of loads P and P_M as $k(\alpha)$ and $k_M(\alpha)$, respectively. Based on Eq. 2.115, the cross-compliance $C_M(a)$ can be calculated as

$$
C_M(a) = \frac{2}{bE'} \int_0^\alpha k(\alpha) k_M(\alpha) \mathrm{d}\alpha
\tag{2.118}
$$

The crack mouth opening displacement when the structure is loaded by P alone is

$$
w_M = \frac{2P}{bE'} \int_0^\alpha k(\alpha) k_M(\alpha) \mathrm{d}\alpha
\tag{2.119}
$$

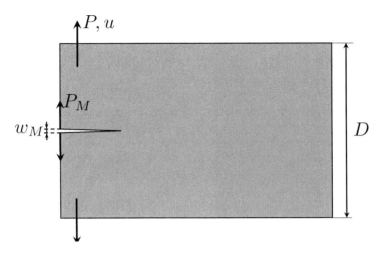

Fig. 2.21 Calculation of crack mouth opening displacement using the compliance method.

A more effective control of fracture test is obtained by crack tip opening displacement (CTOD), measured by a finite-length gauge crossing the crack line at the tip. This displacement can be calculated similarly to Eq. 2.119.

Example 2.6: *Calculation of load-displacement curve of an infinite planar strip with collinear cracks*

Another important application of Eqs. 2.112–2.115 is to calculate the load-deflection curve of a specimen. As a demonstration, we calculate the deflection u_c due to formation of centric crack of length $2a$ in an infinite planar strip under remote stress σ (Fig. 2.20a). According to Eq. 2.102, $K_I^2 \approx \sigma^2 D \tan(\pi a/D)$, and so

$$\frac{1}{b}\frac{\partial \Pi^*}{\partial a} = 2\mathcal{G} = \frac{2K_I^2}{E'} = \frac{2\sigma^2}{E'} D \tan \frac{\pi a}{D} \qquad (2.120)$$

$$\Pi^*(a) = \frac{2\sigma^2}{E'} bD \int_0^a \tan \frac{\pi a'}{D} da' = -\frac{2bD^2\sigma^2}{\pi E} \ln\left(\cos \frac{\pi a}{D}\right) \qquad (2.121)$$

According to the first Castigliano theorem, the deflection due to crack formation is

$$u_c(a) = \left[\frac{\partial \Pi^*}{\partial P}\right]_a = \frac{1}{bD}\frac{\partial \Pi^*}{\partial \sigma} \qquad (2.122)$$

$$= -\frac{4\sigma D}{\pi E} \ln\left(\cos \frac{\pi a}{D}\right) \qquad (2.123)$$

Meanwhile, we can impose the fracture propagation criterion $K_I = K_{Ic}$, which leads to

$$\sigma = \frac{K_{Ic}}{D}[\tan(\pi\alpha)]^{-1/2} \tag{2.124}$$

Substituting Eq. 2.124 into Eq. 2.123 yields

$$u_c = -\frac{4K_{Ic}\ln[\cos(\pi\alpha)]}{\pi E \sqrt{\tan(\pi\alpha)}} \tag{2.125}$$

Eqs. 2.124 and 2.125 form a set of parametric equations, from which we can determine the complete load-displacement response. To plot it, one chooses a sequence of α values, and calculates the corresponding σ and u_c. Note that this resulting load-deflection curve exhibits a snapback behavior starting at the point where $du_c/d\sigma = 0$.

This example, however, is of academic nature. In reality, only one of the two crack tips can propagate. The reason is a bifurcation of equilibrium path; see (Bažant and Cedolin, 1991, Sec. 12.4).

It should also be pointed out that the termination of the load-deflection diagram by a snapback has been shown to be a general property of all fracture situations in which there is a nonzero force resultant transmitted across the ligament (i.e. the uncracked part of cross-section) as it shrinks to zero. If there is only a moment transmitted across the ligament, as in three-point-bend specimens, the load-deflection diagram terminates by post-peak softening, whose slope is negative (Bažant and Cedolin, 1991, Sec. 12.4).

2.12 Bimaterial Interfacial Cracks

In previous sections, our attention was limited to a crack in a homogenous elastic body. Over the past several decades, extensive efforts have also been devoted to cracks lying along an interface between two elastic bodies e.g. (Cherepanov, 1962; England, 1965; Erdogan, 1965; Rice and Sih, 1965; Rice, 1988; Hutchinson and Suo, 1992). Understanding the behavior of bimaterial interfacial crack has important implications for the analysis of layered materials, hybrid lap and scarf joints, composite structures, etc.

The main difference from homogeneous bodies is that the singularity exponent is typically complex, that is,

$$\lambda = \tfrac{1}{2} \pm i\varepsilon \tag{2.126}$$

where $i = \sqrt{-1}$. The imaginary part of the singularity ε is governed by the mismatch of the elastic properties of the two materials. For isotropic materials, the mismatch is characterized by the dimensionless Dundurs number (Dundurs, 1969), to which ε is related.

Here we limit our attention to homogenous isotropic materials. The complex singularity indicates that the stress intensity factor K needs to be expressed also in a complex form. For an interfacial crack of length a subjected to remote stresses, we have

$$K = (\sigma_{22}^{\infty} + i\sigma_{12}^{\infty})(1 + 2i\varepsilon)(\pi a)^{1/2}(2a)^{-i\varepsilon} \tag{2.127}$$

where $\sigma_{22}^{\infty}, \sigma_{12}^{\infty}$ = far-field applied normal and shear stresses, $\sigma = \sqrt{\sigma_{22}^{\infty}{}^2 + \sigma_{12}^{\infty}{}^2}$ = resultant far-field stress, $\psi = \tan^{-1}(\sigma_{12}^{\infty}/\sigma_{22}^{\infty})$ = phase angle of the loading, and $k(\alpha)$ = dimensionless complex stress intensity factor.

The traction on the interface ahead the crack tip and the crack surface displacement behind the crack tip can be expressed by

$$\sigma_{22} + i\sigma_{12} = K(2\pi r)^{-1/2+i\varepsilon} \tag{2.128}$$

$$u_2 + iu_1 = \frac{8}{(1+2i\varepsilon)\cosh(\pi\varepsilon)} \frac{K}{E^*} \left(\frac{r}{2\pi}\right)^{1/2} r^{i\varepsilon} \tag{2.129}$$

$$\text{where:} \quad \frac{1}{E^*} = \frac{1}{2}\left(\frac{1}{E_1'} + \frac{1}{E_2'}\right) \tag{2.130}$$

Eq. 2.129 indicates that the opening displacement is oscillatory. Therefore, in some regions the top and bottom crack surfaces would overlap, which is physically impossible. This problem has been discussed for decades, and the following consensus has been reached.

Based on Eqs. 2.127 and 2.129, the opening displacement u_2 is equal to zero when $\cos[\psi - \varepsilon \ln(a/r)] = 0$. Since the phase angle ψ lies in the range $-\pi/2 < \psi < \pi/2$, the location of the contact points are given by

$$r_n = ae^{-[\psi+(n-\frac{1}{2})\pi]/\varepsilon} \quad (n = 1, 2, ...) \tag{2.131}$$

Note that we can always formulate the problem in such a way that $\varepsilon > 0$. Therefore, the reach of the zone of overlaps is given by Eq. 2.131 with $n = 1$. The ratio of the subsequent contact point coordinates is $r_{n+1}/r_n = e^{-\pi/\varepsilon}$. This indicates that, towards the crack tip, the contact points form a geometric progression with decreasing intervals.

As shown by Rice (Rice, 1988), for most cases where tensile loading is non-negligible, the contact zone behind the crack is very short compared to the crack length, typically less than 1% of the total crack length (and, with regard to the nonlinear fracture process zone discussed in Sec. 3.1, is likely to be shorter than the size of that zone, and thus irrelevant).

More importantly, the oscillatory stress and displacement fields behind the crack tip lead to a well-behaved energy release rate function, that is:

$$\mathcal{G} = \frac{1}{E^* \cosh^2(\pi\varepsilon)} K\overline{K} \tag{2.132}$$

where $\overline{K}$ is the complex conjugate of K. Eq. 2.132 indicates that the energy release rate at the bimaterial interfacial crack depends only on the modulus of the complex stress intensity factor K.

Similar to cracks in homogenous solids, we may consider that an interfacial crack begins to propagate as the energy release rate reaches a critical value G_c. The fracture energy G_c generally depends on the level of mode mixity. However, in many cases, mode I and mode II cannot be decoupled for a bimaterial crack. Rice (Rice, 1988) proposed a mode mixity angle φ:

$$\varphi = \tan^{-1}\left[\frac{\text{Im}(Kl^{i\varepsilon})}{\text{Re}(Kl^{i\varepsilon})}\right] \tag{2.133}$$

where l = reference length. As discussed in Hutchinson and Suo (1992), it is preferable to choose some material length scale, such as the size of the fracture process zone, or

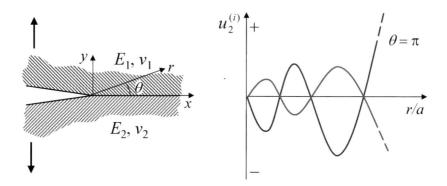

Fig. 2.22 (left) Crack at bi-material interface; (right) Surface overlapping within the Fracture Process Zone (FPZ).

---

the grain size, to represent l. For a given specimen, the mode mixity angle depends on the choice of l, and therefore the fracture energy G_c can generally be expressed as

$$G_c = G(\varphi, l) \tag{2.134}$$

2.13 Comments on Anisotropic Materials and Three-Dimensional Singularities

In such cases, there exist no general analytical solutions of LEFM for the stress intensity factors and near-tip fields. Bao *et al.* (Bao *et al.*, 1992) presented approximate expressions for the stress intensity factor K_I and the energy release rate $\mathcal{G}$ for orthotropic materials, such as fiber composites in the state of plane stress. For a long single-edge cracked strip of width D under remotely applied longitudinal tension σ, the energy release rate is

$$\mathcal{G}(\alpha) = \frac{K_I^2}{E^*} = [Y(\rho)]^2 \sqrt{\frac{1+\rho}{2E_x E_y \sqrt{\lambda}}}\, K_I^2 \tag{2.135}$$

$$K_I = \sigma\sqrt{\pi D \alpha}\, F(\alpha) \tag{2.136}$$

where $\rho = \sqrt{E_x E_y}/2G_{xy} - \sqrt{\nu_{xy}\nu_{yx}}$, $\lambda = E_y/E_x$ and functions $F(\alpha)$ and $Y(\rho)$ are given by approximate equations (10) and (14) in Bao *et al.* (1992). $E_x, E_y, G_{xy}, \nu_{xy}, \nu_{yx}$ are orthotropic elastic constants for plane stress; x, y are the transverse and longitudinal cartesian axes. In Bažant *et al.* (1996), this formulation was used to study the size effect in fracture of orthotropic composites.

The complex exponent of radial coordinate r and the angular distribution of the near-tip stress field is nowadays easy to obtain numerically, for any elastic material.

The most general case of a crack propagating along the interface between two orthotropic materials with orthotropy axes parallel to crack was solved in Achenbach *et al.* (1976*a*). The solution was dynamic, including inertial forces. The special case of statics is obtained by setting the crack propagation velocity $c = 0$. The differential equations of equilibrium in terms of displacements u_x, u_y are transformed to polar coordinates and, as expected, the radial dependence can be separated. This yields a system of two homogeneous linear ordinary differential equations for the dependence of u_x and u_y factors on the polar angle, θ. The radial power-law exponent λ is a parameter in these equations. The interface conditions at $\theta = 0$ and the boundary conditions at $\theta = -\pi$ and π are also homogenous and linear. Introducing finite difference approximations, one gets a matrix eigenvalue problem, with a banded matrix. The eigenvalues are generally complex, and the real part of the lowest one is again -0.5. The eigenvectors define the angular dependence of the near-tip field. A highly accurate computer solution is today virtually instantaneous.

The technique just described has been extended to solve various three-dimensional (3D) singularities in LEFM (Bažant and Estenssoro, 1979). One interesting case is the singular stress field at the surface point of a crack, i.e., a point of intersection of the 3D crack front edge with the surface. The problem is reduced to a homogenous linear partial differential equation on the surface of a sphere surrounding the surface point. The singularity exponent λ is found to depend on the angle φ of the crack front edge with the surface. It is interesting that for the case where $\varphi = \pi/2$ (which was solved analytically by Benthem (Benthem, 1977)), $\mathrm{Re}\lambda$ is not -0.5. In that case, as explained before, the flux of energy into the surface point of a propagating crack is either zero infinite, which makes propagation impossible. Consequently, the angle, β, of the front edge of a propagating LEFM crack with the surface must differ from $\pi/2$. This means that the crack front edge must curve near the surface (unless the Poisson ratio is zero). Depending on the Poisson ratio ν, angle β of a propagating LEFM crack must have a certain value for which $\mathrm{Re}\lambda = -0.5$. If the crack plane is not orthogonal to the surface, its angle affects β, too (Bažant and Estenssoro, 1979).

Exercises

E2.1. To test the fracture behavior of rock, a very large panel of thickness b with a relatively small crack of length $2a$ is tested by injecting fluid into the crack as shown in Fig. 2.23. From Inglis (1913), it is known that under pressure p the straight crack adopts an elliptical shape, with minor axis $c = 2pa/E'$, where the effective Young's modulus $E' = E$ (plane stress) and $E' = E/(1 - \nu^2)$ (plane strain), $\nu =$ Poisson's ratio. 1) Find the complementary energy as a function of p and a, and 2) find the energy release rate function g.

E2.2. Following Exercise E2.1, let $b = 50$ mm, the initial crack length $2a_0 = 100$ mm, $p =$ fluid pressure, $V =$ volume expansion of the crack. Under the assumption of linear elastic behavior with $G_f = 20$ N/m, find and plot p–V curve and g–a curves the panel experiences when it is subjected to a controlled-volume injection until the crack propagates up to 1000 mm, after which it is unloaded to zero pressure. The effective Young's modulus is $E' = 60$ MPa.

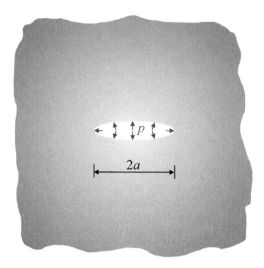

Fig. 2.23 Large panel featuring a central crack of length $2a$ under a fluid pressure p.

E2.3. Consider the complex variable formulation of two-dimensional plane elasticity problems. The displacement field is written as $v = u_1 + iu_2$ and the spatial coordinate can be written as $z = x_1 + ix_2$. Show that the Hooke's law can be written in the following form:

$$\sigma_{11} + \sigma_{22} = 2\left(\lambda + G\right)\left[\frac{\partial v}{\partial z} + \frac{\partial v*}{\partial z*}\right]$$

$$\sigma_{11} - \sigma_{22} + 2i\sigma_{12} = 4G\frac{\partial v}{\partial z*}$$

E2.4. Consider the Westergaard's solution for a central crack in an infinite plate. Show that in the case of a center-cracked panel subjected to remote constant shear stress τ_∞, the Westergaard function to be used is:

$$Z\left(z\right) = \frac{\tau_\infty iz}{\sqrt{z^2 - a^2}}$$

E2.5. Estimate the strength of a large plate under uniaxial tensile stress if it contains an edge crack of length 10 mm. The plate is made of brittle steel with $K_{Ic} = 60$ MPa$\sqrt{\text{m}}$.

E2.6. Consider the plate discussed in problem E2.1. Determine the stress intensity factor using (a) Irwin's relationship, (b) the near-tip expansion for the crack opening.

E2.7. In most handbooks on stress intensity factors, K_I is written in the form $K_I = Y\sigma\sqrt{\pi a}$ where a is the crack length, σ a characteristic stress, and Y a dimensionless

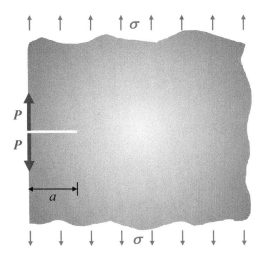

Fig. 2.24 Large panel featuring an edge crack subjected to a remote stress σ and a pair of loads P applied at the crack mouth.

factor depending only on geometrical ratios, in particular on the relative notch depth a/D where D is a characteristic linear dimension of the body. Find the relationship between Y and the dimensionless stress intensity factor $k(\alpha)$ defined in Eq. 2.55.

E2.8. To analyze the behavior of a large structure with cracks, which is assumed to behave in an essentially linear elastic manner, a reduced scale model is built at a $1/10$ scale. The model is tested so that the stresses at homologous points are identical in both model and reality, and we require the full scale and reduced models to break at the same stress level. Determine the scale factors for (a) loads, (b) stress intensity factors, and (c) energy release rates.

E2.9. Calculate the load-deflection curve of the tensioned infinite planar strip with a centric normal crack shown in Fig. 2.19a. For this geometry, K_1 can be calculated using Eq. 2.102. Also, find the critical displacement u at the onset of snapback instability.

E2.10. Consider a plate with a very short edge crack ($a \ll D$) subjected to a remote stress σ as shown in Fig. 2.24. The corresponding stress intensity factor is $K_1 = 1.12\sigma\sqrt{\pi a}$. For the same specimen, if a pair of loads P is applied at the crack mouth of the same plate, the corresponding stress intensity factor is $K_p = 2.594P/\sqrt{\pi a}$ (assume that the width of the specimen is 1). Use LEFM to calculate the crack mouth opening displacement of the plate for the loading configuration as shown in the figure just before the crack starts to propagate (fracture energy G_F).

E2.11. The sandwich composite panel represented in Fig. 2.25 is affected by a man-ufacturing defect in the form of a delamination crack of length $2a$. The air inside the crack is at atmospheric pressure at sea level and there is no leak. The panel is installed in a large rocket reaching an altitude of 56.8 miles, where the atmospheric pressure is $1.09 \cdot 10^{-6}$ kPa.

(a) Assuming LEFM and Euler–Bernoulli beam theories to apply (and the skin to be isotropic), calculate the energy release rate (neglect the rotations at the crack tips);

(b) Assuming an interfacial fracture energy $G_f = 0.1$ N/mm, a thickness of the skin of 0.1 mm, and a longitudinal elastic modulus of the skin of 250 GPa, find the maximum crack length that can be tolerated to guarantee the structural integrity.

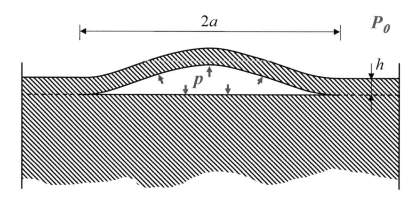

Fig. 2.25 Sandwich composite structure featuring a delamination crack of length $2a$.

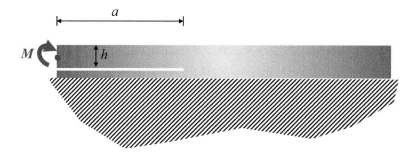

Fig. 2.26 Schematic representation of the configuration investigated in problem E2.12.

E2.12. Use the J-integral to calculate the stress intensity factor for the crack represented in Fig. 2.26. Assume $a \gg h$.

E2.13. Consider LEFM and the Euler–Bernoulli beam theory. Calculate the stress intensity factor for the structures represented in Fig. 2.27 using the J-integral.

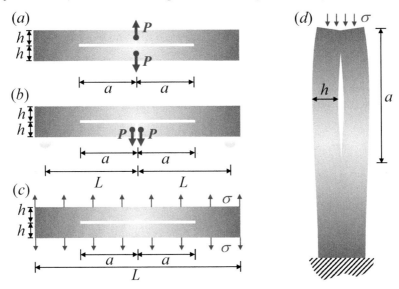

Fig. 2.27 Schematic representation of the structures investigated in problem E2.13.

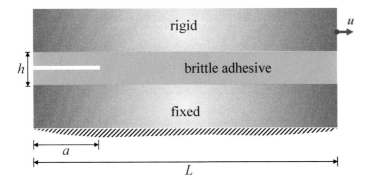

Fig. 2.28 Schematic representation of the structure investigated in problem E2.16.

E2.14 Consider LEFM and the Euler–Bernoulli beam theory. Calculate the stress intensity factor for the structures represented in Fig. 2.27 using the relation between the energy release rate and the elastic potential.

E2.15. A double cantilever beam specimen with arm depths $h = 10$ mm, thickness $b = 10$ mm, and initial crack length $a_0 = 50$ mm, is made of a material with a fracture energy $G_f = 180$ J/m^2 and an elastic modulus GPa. The specimen si tested at controlled displacement rate so that the load goes through the maximum and then decreases, at still increasing displacement, down to 25% of the peak load. When this point is reached, the specimen is completely unloaded. Assuming that LEFM and the beam theory apply, find the $P(u)$ and $\mathcal{G}(a)$ curves. Give the equations of the different arcs and the coordinates of the characteristic points.

E2.16. Consider the structure shown in Fig. 2.28. Assume that $a \gg h$ and $L - 2a \gg h$. Use the J-integral to calculate $\mathcal{G}$ and K_{II} for the layer of adhesive. Consider the bottom layer fixed while the top layer is subjected to a horizontal displacement u.

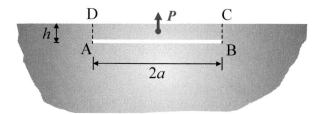

Fig. 2.29 Schematic representation of the structure investigated in problem E2.17.

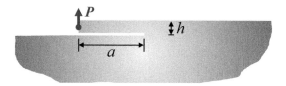

Fig. 2.30 Schematic representation of the structure investigated in problem E2.18.

E2.17. Consider the delamination crack of length $2a$ shown in Fig. 2.29. Treat $ABCD$ as a slender beam of thickness h, loaded by P. Assumed the ends AD and BC to be fixed.

    (a) Formulate the J-integral.
    (b) Using either J or the potential energy calculate the load P_c at which the crack propagates if K_c is known.

E2.18. Consider the structure shown in Fig. 2.30.

    (a) Formulate J-integral.
    (b) Calculate K_I for the crack, assuming that $a \gg h$ and using the Euler–Bernoulli beam theory.

3
Nonlinear Fracture Mechanics—Line Crack Idealization

Fracture is a loss of cohesion, but only up to a point.

Chapter 2 presented the essential concepts of Linear Elastic Fracture Mechanics (LEFM), which applies to structures where the inelastic zone at the crack tip is negligibly small compared to characteristic structure size. However, this is not the case for many engineering structures made of quasibrittle materials, such as concrete, tough ceramics, composites, nanocomposites, and many geomaterials or biomaterials. During the loading process in these structures, a large inelastic zone is formed ahead of the crack tip and, therefore LEFM is inapplicable (Bažant, 1984*b*, 2004; Bažant and Grassl, 2007; Cusatis and Schauffert, 2009; Schauffert and Cusatis, 2011; Schauffert *et al.*, 2011; Salviato *et al.*, 2016*c,b*; Mefford *et al.*, 2017; Salviato *et al.*, 2019; Qiao and Salviato, 2019*b,a*). This calls for nonlinear fracture models, which take the inelastic behavior at the crack tip into account.

This chapter discusses the nonlinear models for quasibrittle materials, which include the equivalent LEFM, the *R*-curve, and the cohesive crack model. These models are largely developed from the concept of LEFM, which renders simple yet reasonable predictions of the overall structural behavior. We mostly leave aside limitations due the recently highlighted effect of crack-parallel stresses, which are better discussed separately.

3.1 Types of Fracture Behavior and Nonlinear Zone

It is evident that the elastic solution of stress distribution at the crack tip is unrealistic because no materials can sustain infinite stress. Therefore there must exist a nonlinear zone at the crack tip. We can classify two types of nonlinear zone: (1) the fracture process zone (FPZ), which is a nonlinear zone characterized by extensive material damage that leads to progressive softening (i.e., by stress decreasing at increasing deformation), and (2) the plastic zone (PZ), which is a non-softening nonlinear zone

Quasibrittle Fracture Mechanics and Size Effect: A First Course. Zdeněk P. Bažant, Jia-Liang Le and Marco Salviato, Oxford University Press. © Zdeněk P. Bažant, Jia-Liang Le, Marco Salviato 2022. DOI: 10.1093/oso/9780192846242.003.0003

surrounding the FPZ and is characterized by hardening plasticity or perfect plasticity (i.e. by stress increasing at increasing deformation or remaining constant).

3.1.1 Three Types of Fracture Behaviors

The sizes of these nonlinear zones relative to the overall structure size lead to three different types of fracture behaviors (Fig. 1.1). In the *first* type (Fig. 1.1a), both the FPZ and PZ are small compared to the structure size. In this case, the whole body except for a negligibly small region surrounding the crack tip behaves as elastic. LEFM then provides a good approximation for very brittle materials such as, for example, glass, plexiglass, brittle ceramics, and brittle metals at conventional length-scales. However, it should be noted that even for these materials there could be cases, e.g. small-scale structures, for which the nonlinear zone cannot be neglected (Bažant, 1984*b*, 2004).

In the *second* type of fracture (Fig. 1.1b), ahead of the crack tip there exists a large nonlinear zone exhibiting elastoplastic hardening or perfect yielding, and the size of the FPZ in which the material breaking occurs is still negligibly small. An archetypal example is provided by ductile metals which, due to dislocation motion, are capable of irreversible deformation without softening ahead of the crack tip. However, for typical structures, the size of the crystal grains at which inter-granular fracturing process takes place is still negligibly small compared to the PZ and the structure. Various ductile metals, especially some tough alloys, belong to this category which is generally defined as *ductile*. This type of behavior is accurately captured by Elasto-Plastic Fracture Mechanics, which is beyond the scope of this book.

The *third* type of behavior (Fig. 1.1c), which is of main interest in this book, includes situations in which the major part of the nonlinear zone undergoes progressive damage with strain softening. The damage is induced by several mechanisms, including the microcracking, void formation, interface breakages, and frictional slips as well as phenomena such as crystallographic transformation in ceramics or fiber debonding and pullout in fiber composites. For this type of behavior, the zone of plastic hardening or perfect yielding is usually negligible, which means that there is a rather abrupt transition from elastic response to damage. This happens in concrete, rock, coarse-grained ceramics, ice, stiff clays, composites and nanocomposites, wood, and various other materials for which the inhomogeneity size is not negligible compared to the structure size. These materials are referred to as *quasibrittle* because, even if no appreciable plastic deformation takes place, the FPZ size is large enough to have to be taken into account calculations.

It is worth mentioning here, however, that in some cases the distinction between the second and third type of behavior can be blurred. For instance, when concrete is subject to a large compression parallel to the crack plane (as in splitting fracture), the specimen could contain both PZ and FPZ of considerably large size.

3.1.2 Effect of Crack-Parallel Stresses

A new type of fracture experiment, called the *gap test*, recently revealed that, in quasibrittle materials, normal stresses parallel to the crack have a large effect on the resistance to fracture–(see Sec. 4.1). While they have no effect on the LEFM, presented

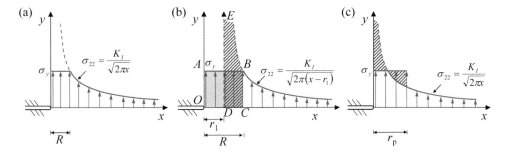

Fig. 3.1 Irwin's estimate of effective crack extension.

in the preceding chapter, they affect its use. Especially, they affect in a significant way the criteria for crack propagation based on the LEFM.

3.2 Irwin's Estimate of the Size of the Inelastic Zone

The length, R, of the inelastic zone at the crack tip was first estimated by Irwin (Irwin, 1958), for plastic (or ductile) materials. A rather crude estimate can be obtained by retaining the singularity of the elastic stress distribution along the crack plane, and cutting off the singularity peak where the stress exceeds the yield strength σ_y, as shown in Fig. 3.1a. For mode I fracture and small x relative to the structure size, the elastic stress distribution is $\sigma_{22} = K_I/\sqrt{2\pi x}$, where K_I = mode I stress intensity factor (SIF). Setting $\sigma_{22} = \sigma_y$ at $x = R$ leads to the following estimate:

$$R = \frac{1}{2\pi}\left(\frac{K_I}{\sigma_y}\right)^2 \tag{3.1}$$

The foregoing expression, however, underestimates the size of the nonlinear zone since it does not account for the fact that cutting off the part of the singular elastic solution reduces the stress resultant and thus disturbs equilibrium. An improved approximation can be achieved by requiring that the stress resultant from an inelastic zone of length R be equal to the stress resultant of the elastic stresses, $\sigma_{22} = K_I/\sqrt{2\pi x}$. Meanwhile, the stress field far from the nonlinear process zone would be unaffected, due to Saint-Venant's principle. To preserve the total stress resultant, we may extend the elastic crack by a distance r_1. With reference to Fig. 3.1b, this means that the area $BCDE$ under the elastic stress solution must be equal to the area $OABCO$. In other words:

$$\int_{r_1}^{R} \frac{K_I}{\sqrt{2\pi\left(x - r_1\right)}}\,dx = R\sigma_y \tag{3.2}$$

while the condition $\sigma_{22} = \sigma_y$ for $x = R$, now with $\sigma_{22} = K_I/\sqrt{2\pi\left(x - r_1\right)}$, leads to:

$$R - r_1 = \frac{1}{2\pi}\left(\frac{K_I}{\sigma_y}\right)^2 \tag{3.3}$$

Solving for R and leveraging the foregoing equations, Irwin Irwin (1958) obtained the following improved estimate:

$$R = \frac{1}{\pi} \left(\frac{K_I}{\sigma_y} \right)^2 \tag{3.4}$$

Note that even this improved estimate is very rough. In fact, the stress distribution used to obtain Eq. 3.4 is mostly a guess. Nevertheless, the solution still captures the dependence of the size of the nonlinear zone on the SIF, or K_I, and on the material strength, σ_y.

Considering that the crack propagates at $K_I = K_{1c}$, where K_{1c} = fracture toughness, we see that Eq. 3.4 implies a characteristic material length $l_{ch} = K_{1c}^2/\sigma_y^2 = E'G_F/\sigma_y^2$, where G_F = total fracture energy, and E' = Young's modulus E (for plane stress), and $= E/(1-\nu^2)$ (plane strain); l_{ch} is generally referred to as the *Irwin characteristic length*. As will be discussed in the following sections, the length of the FPZ of quasibrittle materials is proportional to l_{ch}, and the proportionality constant depends on the material type and the structure geometry.

Eqs. 3.1 and 3.4 give the estimates of the length of the zone in which $\sigma_{22} = \sigma_y$. Globally, more important is the effective length r_p of the plastic yielding zone. Such length should ensure global equilibrium, in the sense that the misfit stresses $\tau = \sigma_{22}(x) - \sigma_y$ are globally in balance (the local inbalance τ decays with distance exponentially, according to St. Venant principle). This means that the virtual work δW of τ on virtual uniform strain $\delta \varepsilon_{22}$ over length r_p should vanish, i.e., $\delta W = \int_0^{r_p} \tau dx \delta \varepsilon_{22} = 0$ for any $\delta \varepsilon_{22}$. This yields the global equilibrium condition $\int_0^{r_p} \tau dx = 0$. Substituting for τ with the near tip elastic field $\sigma_{22} = K_I/\sqrt{2\pi x}$ and integrating, we get the estimate of the global effective yielding zone length for the case of plastic yielding (Fig. 3.1c):

$$r_p = \frac{2K_I^2}{\pi \sigma_y^2} \quad \text{or} \quad r_p = \frac{2}{\pi} l_{ch} \tag{3.5}$$

3.3   Estimation of FPZ Size for Quasibrittle Materials

The FPZ can be estimated more realistically by considering a cohesive stress profile ahead the crack tip (Barenblatt, 1959, 1962). Here we approximate the FPZ by a cohesive (line) crack, which is considered to transmit tensile stresses between the opposite crack faces. The cohesive zone spans from the cohesive crack tip, where the cohesive stress equals material strength, to the point where the cohesive stress is reduced to zero, which is the end of the real (stress-free) crack.

Since the cohesive crack is a line crack, all of the volume of the structure is elastic. So the LEFM concepts can be used, but the stress intensity factor now consists of two parts—K_I^{load} (positive) due to externally applied load, and K_I^{coh} (negative) due to cohesive stresses acting on the crack faces. The basic hypothesis of the cohesive crack model is that the total stress intensity factor for the tip of the cohesive crack must vanish, i.e.,

$$K_I = K_I^{load} + K_I^{coh} = 0 \tag{3.6}$$

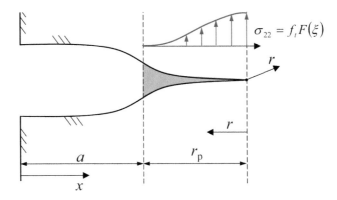

Fig. 3.2 Cohesive stresses in the Fracture Process Zone (FPZ) following $\sigma = f_t F(\xi)$.

The reason is that infinite strains at the cohesive crack tip are avoided only for $K_I = 0$. This further means that, at the tip, the opposite crack faces have a *smooth closing*, as illustrated in Fig. 3.1. The condition of smooth crack closing (proposed earlier, in 1955, in (Zheltov and Kristianovich, 1955) in the context of hydraulic fracture of oil-bearing strata) was the starting point in Barenblatt's 1959 formulation of the cohesive crack models. The condition of smooth closing, i.e., total $K_I = 0$, was also introduced in 1960 by Dugdale (Dugdale, 1960) in his analysis of a slit, or crack, in elastic-perfectly plastic material. His model later served as the basis for the development of a widely used fracture criterion for tough steels British Standards Institute (1979), which postulates a critical value of crack-tip opening displacement (CTOD) at a certain distance behind the actual crack tip.[1]

In the following sections, Eq. 3.6 is used to estimate the size of the FPZ in quasibrittle materials. To demonstrate the use of Eq. 3.6, let us consider the cohesive crack in an infinite plate represented in Fig. 3.2. The FPZ is approximated by a line of length r_p and the cohesive stresses are assumed to follow a generic distribution described by $\sigma(\xi) = f_t F(\xi)$ where f_t = material strength, $F(\xi)$ = dimensionless function describing the stress profile in the FPZ and $\xi = r/r_p$ = dimensionless distance from the tip of the cohesive crack (Fig. 3.2).

To find the FPZ size r_p, we need to calculate the SIF due to the cohesive stresses. Consider a central crack of length $2a$ in an infinite plate subjected to a pair of point loads P per unit thickness as shown in Fig. 3.3. The corresponding SIF can be calcu-

---

[1]Perfunctory readers of Barenblatt's 1962 paper, unaware of his 1959 paper in Russian, often erroneously attributed the origin of the cohesive crack model to Dugdale (Dugdale, 1960). But Dugdale's model, called the "strip yield model", dealt with a different problem—the plastic deformation near the tip of a sharp slit, with no softening. His plastic yielding zone could extend indefinitely with no actual crack or slit growth. It was not a fracture model, and was not presented as such by Dugdale. Nevertheless, as already pointed out, Dugdale's model was later extended into an important fracture criterion, the critical CTOD (or COD) criterion for ductile steels, which has been related to the critical J-integral criterion, generalized to modes II and III (Bilby *et al.*, 1963), and eventually incorporated into British and other standards for fracture of ductile steel.

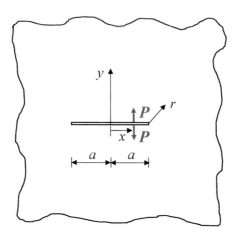

Fig. 3.3 Central crack in infinite plate subject to a point load per unit thickness.

lated as (Tada *et al.*, 2000)

$$K_I^P = \frac{P}{\sqrt{\pi a}} \sqrt{\frac{a+x}{a-x}} \tag{3.7}$$

Let the FPZ be sufficiently small compared to the crack length (i.e. $r << a$). Eq. 3.7 can then be rewritten as

$$K_I^P \approx \frac{P}{\sqrt{\pi a}} \sqrt{\frac{2a}{r}} = P\sqrt{\frac{2}{\pi r}} \tag{3.8}$$

Based on this equation and the superposition principle, we obtain the SIF due to the cohesive stresses:

$$K_I^{coh} = \int_0^{r_p} \sqrt{\frac{2}{\pi r}} \sigma_{22} \mathrm{d}r \tag{3.9}$$

where we used $\mathrm{d}P = \sigma_{22}\mathrm{d}r$. Since $\sigma_{22}\ \mathrm{d}r = f_t F\left(\xi\right) r_p \mathrm{d}\xi$, we obtain

$$K_I^{coh} = \sqrt{\frac{2r_p}{\pi}}\ f_t \int_0^1 \frac{F\left(\xi\right)}{\sqrt{\xi}} \tag{3.10}$$

For the incipient fracture condition $K_I^{load} = K_{Ic}$ and with the smooth closing condition, Eq. 3.6, gives $K_I^{coh} = -K_{Ic}$. Substituting Eq. 3.10 and solving for r_p leads to the following expression:

$$r_p = \frac{K_{Ic}^2}{f_t^2} \frac{\pi/2}{\left[\int_0^1 F\left(\xi\right)/\sqrt{\xi}\mathrm{d}\xi\right]^2} \tag{3.11}$$

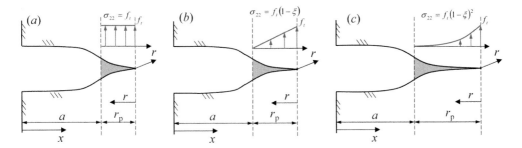

Fig. 3.4 Fracture Process Zone for various stress profiles: (a) constant, (b) linear and (c) quadratic.

which provides an estimate of the FPZ size for any arbitrary cohesive stress profile.

We may now apply Eq. 3.11 to some commonly used cohesive stress profiles, described in general by the function:

$$F\left(\xi\right) = -\left(1 - \xi\right)^{n}, \quad n \in \Re \tag{3.12}$$

3.3.1 Uniform cohesive stress profile

First consider $n = 0$ for Eq. 3.12. The cohesive stresses are uniform and equal to the material strength, and a ductile behavior with a sudden stress drop as in Fig. 3.4a is assumed. For this case,

$$\int_0^1 \frac{F\left(\xi\right)}{\sqrt{\xi}} d\xi = -\int_0^1 \frac{1}{\sqrt{\xi}} d\xi = -2 \tag{3.13}$$

and, substituting into Eq. (3.11):

$$r_p = \frac{K_{Ic}^2}{f_t^2} \frac{\pi}{8} \approx 0.393 l_{ch} \tag{3.14}$$

3.3.2 Linear cohesive stress profile

When $n = 1$ the cohesive stresses follow a linear profile (Fig. 3.4b), we have:

$$\int_0^1 \frac{F\left(\xi\right)}{\sqrt{\xi}} d\xi = \int_0^1 \frac{\xi - 1}{\sqrt{\xi}} d\xi = -\frac{4}{3} \tag{3.15}$$

and, substituting into Eq. (3.11):

$$r_p = \frac{K_{Ic}^2}{f_t^2} \frac{9\pi}{32} \doteq 0.884 \, l_{ch} \tag{3.16}$$

3.3.3   Quadratic cohesive stress profile

When $n = 2$ the cohesive stresses follow a quadratic profile (Fig. 3.4c), we get:

$$\int_0^1 \frac{F(\xi)}{\sqrt{\xi}} \, \mathrm{d}\xi = -\int_0^1 \frac{(1-\xi)^2}{\sqrt{\xi}} \, \mathrm{d}\xi = -\frac{16}{15} \tag{3.17}$$

and, substituting into Eq. (3.11):

$$r_p = \frac{K_{Ic}^2}{f_t^2} \frac{225\pi}{512} \doteq 1.381 l_{ch} \tag{3.18}$$

Generally we see that, for all the stress profiles, the size of the FPZ is always proportional to the Irwin characteristic length, i.e.:

$$r_p = \eta \, \frac{K_{Ic}^2}{f_t^2} = \eta \, l_{ch} \tag{3.19}$$

where coefficient η depends on the cohesive stress profile. This result is not surprising because the basic cohesive crack model involves two material constants K_{Ic} and f_t. A simple dimensional analysis suffices to reach the conclusion that $r_p \propto K_{Ic}^2/f_t^2$.

It is worth highlighting that, as Eq. 3.18 shows, the FPZ of quasibrittle materials can be considerably larger than the size of the PZ of ductile materials. This aspect, often overlooked by the literature and the standards on fracture tests, needs to be taken in consideration when the size of the nonlinear zone is used to determine if a structure is large enough to apply LEFM.

Although the cohesive stress analysis provide a more physical estimation of the FPZ size, the cohesive stress profile must be known a priori, which is normally not the case. As discussed in Sec. 3.12, the profile of the cohesive stresses is governed by the structure geometry and external loading. In general, the cohesive stress profile can be calculated through a set of equilibrium and compatibility equations and a cohesive law, which describes the mechanical behavior of the cohesive crack.

3.4   Equivalent Linear Elastic Crack Model

The failure of quasibrittle materials, in which the FPZ is not negligible relative to the dimensions of the specimen, is a nonlinear problem. When the FPZ is very large, so that it extends over a significant portion of the unbroken ligament, relatively complex models are required to capture the material behavior of the FPZ. On the other hand, if the FPZ occupies only a small but non-negligible part of the ligament, then it is possible to leverage simplified, partly linearized models. These models simulate the overall response of the specimen far from the crack tip by an equivalent elastic crack whose tip is located somewhere within the fracture process zone. This is called *far field equivalence* since the fields of stresses, strains, displacements, etc., are identical within second-order terms, of the order of $(R_c/D)^2$, where R_c is the size of the process zone and D the size of the specimen. As described in Sec. 3.2, a similar approach was used by Irwin 1958 to describe the fracturing behavior in an elastic-perfectly plastic material.

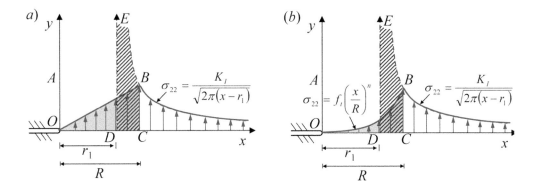

Fig. 3.5 Equivalent crack for a progressively softening material: (a) linear stress distribution and (b) parabolic distribution of degree n.

The problem of defining the *far field equivalence* is to determine where the tip of the equivalent crack, defining the equivalent crack extension

$$\Delta a_e = a_e - a_0 \tag{3.20}$$

should be located for a particular load; a_e = equivalent crack length and a_0 = initial crack length.

Following Sec. 3.2, the simplest way to define the equivalent elastic crack is to impose the remote fields of the equivalent crack to match the ones of the real structure for the same loading. Consider the case represented in Fig. (3.5)a, in which it is assumed that when the critical condition $K_I = K_{Ic}$ is reached, the point at the crack tip has already completely softened to zero stress and the points ahead of it are in intermediate states of fracture such that the stress distribution is linear. We seek an equivalent crack extension $\Delta a_e = r_1$ for which the stress resultant of the elastic stresses equals the resultant of the cohesive stresses ahead of the crack tip. Graphically, this is equivalent to requiring the area of the region $BCDEB$ to be equal to that of the region $OBCO$ in Fig. (3.5)a, i.e.:

$$\int_{r_1}^{R} \frac{K_I}{\sqrt{2\pi(x - r_1)}} dx = \frac{1}{2} R f_t \tag{3.21}$$

We note that $\sigma_{22} = f_t$ for $x = R$ and $\sigma_{22} = K_I/\sqrt{2\pi(x - r_1)}$. By further setting $K_I = K_{IC}$, we have

$$R - r_1 = \frac{1}{2\pi} \left(\frac{K_{Ic}}{f_t} \right)^2 \tag{3.22}$$

Based on Eq. 3.21 and 3.22, we can solve for r_1:

$$r_1 = \Delta a_e = \frac{3}{2\pi} \left(\frac{K_{Ic}}{f_t} \right)^2 \tag{3.23}$$

The foregoing derivations can be generalized for the case shown in Fig. (3.5)b, in which the cohesive stress profile along the fracture process zone is described by a parabola of degree n. Again, the *far field equivalence* can be induced by imposing the areas $OBCDO$ and $DEBCD$ to be equal. This condition leads to the following equation:

$$\int_{r_1}^{R} \frac{K_I}{\sqrt{2\pi\,(x - r_1)}}\,\mathrm{d}x = \int_{0}^{R} f_t \left(\frac{x}{R}\right)^n \mathrm{d}x = \frac{1}{1+n} R f_t \tag{3.24}$$

which, together with Eq.(3.22), gives:

$$r_1 = \Delta a_e = \frac{2n+1}{2\pi} \left(\frac{K_{Ic}}{f_t}\right)^2 \tag{3.25}$$

Meanwhile, using Irwin's approach (Sec. 3.2), the size of the FPZ can be estimated as $R = (n+1)/\pi l_{ch}$. Therefore, the critical equivalent crack extension Δa_e for a very large structure can be written as:

$$\Delta a_{ec} = \beta l_{ch} \quad \text{with} \quad \beta = \eta - \frac{1}{2\pi} \tag{3.26}$$

The foregoing analysis considered a virtually infinite specimen, because the stress distribution is completely dominated by the singular term. So it is convenient to adopt a notation clearly indicating the value for infinite size D. Accordingly, the critical effective crack extension for a semi-infinite crack in an infinite body is defined as:

$$c_f = \lim_{D \to \infty} \Delta a_{ec} \tag{3.27}$$

Accordingly, $c_f = \beta l_{ch}$ where $\beta = \eta - 1/2\pi$ is a material property, and c_f is a material characteristic length, different from Irwin's, although it can be related to it.

3.5 $\mathcal{R}$-**Curves**

The equivalent crack model described in the foregoing section is applicable only to the condition of incipient failure (peak load). It is of interest to develop a simple model that can capture the nonlinear fracture behavior over the entire loading process. The $\mathcal{R}$-curve in such a model relies on the concept of equivalent elastic crack but also allows approximate description of the complete crack growth process and yields the load-displacement curves.

The idea of the $\mathcal{R}$-curve is to preserve the basic formalism of LEFM but, instead of the fracture criterion $G = G_f$ with constant G_f, assume the crack to propagate when $\mathcal{G} = \mathcal{R}$ where $\mathcal{R}$ is a given variable crack growth resistance. In most of $\mathcal{R}$-curve modeling, $\mathcal{R}$ is assumed to depend on the equivalent crack extension, Δa_e only. However, other choices are, in principle possible, for the control variable. One is, e.g., the crack tip opening displacement (CTOD).

The $\mathcal{R}$-curves are suitable for structures with one crack. They are not effective for structure in interacting cracks.

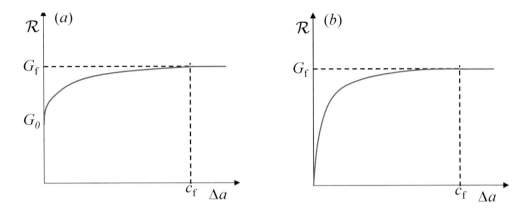

Fig. 3.6 Two typical *R*-curves: (a) with threshold, and (b) without threshold.

3.5.1 Definition of an $\mathcal{R}$-Curve

In an $\mathcal{R}$-curve model, the crack growth resistance $\mathcal{R}$ is expressed as a function of the effective crack extension Δa_e, that is,

$$\mathcal{R} = \mathcal{R}\left(\Delta a_e\right) \tag{3.28}$$

Initially, function $\mathcal{R}\left(\Delta a_e\right)$ was treated as a material property (Irwin, 1960; Krafft *et al.*, 1961). Later, though, it was found that for cracks not too short compared to structural dimensions, as is typical in concrete and other quasibrittle materials, the shape of the $\mathcal{R}$-curve depends significantly on the specimen or structure geometry and the structure size (Bažant and Kazemi, 1990; Bažant *et al.*, 1991; Bažant and Planas, 1998) (for this reason, $\mathcal{R}$-curves are used for concrete structures far less than in materials science). So, if the crack in not relatively very short, the $\mathcal{R}$-curve can be assumed to be unique only for a narrow range of structure geometries, and one needs to determine the $\mathcal{R}$-curve for the given geometry prior to undertaking fracture analysis.

The $\mathcal{R}$-curves generally terminate with an LEFM plateau, $\mathcal{R} = G_f$. The plateau is assumed to be reached when $\Delta a_e = c_f$ (equivalent crack extension). The initial rise of $\mathcal{R}$-curve is sometimes considered to start with a finite value, G_0 (Fig. 3.6a), or with a zero value (Fig. 3.6b). Using a dimensionless function $\bar{\mathcal{R}}(...)$ terminating with 1, one may write

$$R = G_f \bar{R}\left(\Delta a_e / c_f\right) \tag{3.29}$$

The energy dissipated per unit crack advance δa is $R\delta a$. Therefore, in a planar specimen of thickness b, the fracture work per unit thickness needed to extend the crack by a distance Δa:

$$\frac{W_f}{b} = \int_0^{\Delta a / c_f} R(\alpha) D\mathrm{d}\alpha \quad (D\mathrm{d}\alpha = \delta a) \tag{3.30}$$

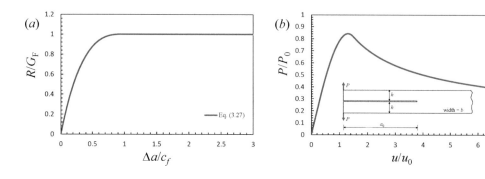

Fig. 3.7 (a) Example of $\mathcal{R}$-curve for a DCB specimen and (b) related P-u curve as constructed from the $\mathcal{R}$-curve.

To exemplify how to describe structural behavior by an $\mathcal{R}$-curve, let us consider a DCB specimen (Fig. 2.4) made of a material with the following $\mathcal{R}$-curve:

$$R = \begin{cases} G_f \left[1 - (1 - \Delta a/c_f)^3 \right] & \text{for } \Delta a \le c_f \\ G_f & \text{for } \Delta a > c_f \end{cases} \tag{3.31}$$

The initial crack length is assumed to be $a_0 = 8h$. The expression for the energy release rate $\mathcal{G}$ is given in Sec. 2.1.2, Eqs. 2.22 and 2.21. Imposing the equilibrium condition $\mathcal{G}(P, a) = \mathcal{R}(a - a_0)$ and using Eq. 3.31, we obtain the following parametric equation for P:

$$P = \begin{cases} (P_0 a_0/a) \left\{ 1 - [1 - (a - a_0)/c_f]^3 \right\}^{1/2} & \text{for } a - a_0 \le c_f \\ P_0 a_0/a & \text{for } a - a_0 > c_f \end{cases} \tag{3.32}$$

where

$$P_0 = \frac{bh}{6a_0} \sqrt{3EG_f h} \tag{3.33}$$

which is the load required to ensure that $\mathcal{G} = G_f$ at $a = a_0$. Now by making use of Eq. 2.21, we can calculate the load-point displacement:

$$u = \begin{cases} u_0 \, (a/a_0)^2 \left\{ 1 - [1 - (a - a_0)/c_f]^3 \right\}^{1/2} & \text{for } a - a_0 \le c_f \\ u_0 \, (a/a_0)^2 & \text{for } a - a_0 > c_f \end{cases} \tag{3.34}$$

where u_0 is the displacement that one would have for $P = P_0$ and $a = a_0$:

$$u_0 = \frac{4a_0^2}{3h^2} \sqrt{\frac{3EG_f h}{E}} \tag{3.35}$$

By combining Eqs. 3.32 and 3.34, we can plot the entire load-displacement curve as shown in Fig. 3.7b.

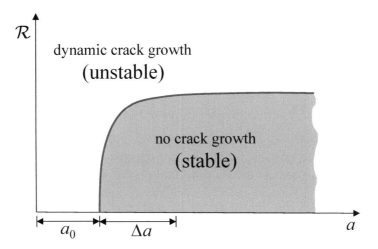

Fig. 3.8 The $\mathcal{R}$-curve divides the space into a stability region (region where no crack growth can take place, shaded area), and an instability region (region of dynamic crack growth.

3.5.2 Stability of Fracture and Critical States

The foregoing analysis dealt with equilibrium and fracture, but gave no information on the stability of equilibrium states. To that end, we note that the energy that must be supplied to an elastic structure under isothermal conditions in order to produce one equivalent crack of length a is:

$$\mathcal{F} = \int_0^c bR(c')\,\mathrm{d}c' + \Pi(a_e) \qquad (c = a_e - a_0) \tag{3.36}$$

where $\mathcal{F}$ represents the Helmholtz free energy, Π is the total potential energy of the structure with a crack of length $c = a_e - a_0$, and $b =$ thickness of the structure in the transverse direction. An equilibrium state of fracture occurs when $\delta\mathcal{F} = 0$, in which case energy is neither supplied nor released when the crack length changes from a_e to $a_e + \delta a_e$. Now, since $\delta\mathcal{F} = (\partial\mathcal{F}/\partial a_e)\,\delta a_e = 0$ and, from Eq. 3.36, $\partial\mathcal{F}/\partial a_e = bR(c) + \partial\Pi/\partial a_e$, the equilibrium condition may be written as:

$$\mathcal{G}(a_e) = \mathcal{R}(c) \quad \text{with} \quad \mathcal{G}(a_e) = -\frac{1}{b}\frac{\partial\Pi(a_e)}{\partial a_e} \tag{3.37}$$

In the special case of LEFM, we have $\mathcal{R}(c) = G_f = $ constant, and then Eq. 3.37 becomes $-\partial\Pi/\partial a_e = bG_f$ as already discussed in Chapter 2.

   If the fracture equilibrium state is *stable*, the crack cannot propagate without any change of loading (applied force or displacement). If the fracture equilibrium state is *unstable*, the crack can propagate without any change of applied load or boundary displacement. Mathematically, the fracture equilibrium is stable if the second variation of Helmholtz free energy is positive, i.e., $\delta^2 \mathcal{F} > 0$. Then, since $\delta^2 \mathcal{F} = \frac{1}{2} \left(\partial^2 \mathcal{F}/\partial a^2 \right) \delta a^2$ and $\partial^2 \mathcal{F}/\partial a^2 = b \left(\mathrm{d}R/\mathrm{d}c \right) + \partial^2 \Pi/\partial a_e^2$, one gets the following stability criterion at the equilibrium state (i.e., $\mathcal{G}(a_e) = \mathcal{R}(c)$):

$$
\begin{aligned}
&\text{Stable if} \quad &&\mathcal{R}'(c) - \mathcal{G}'(a_e) > 0 \\
&\text{Critical if} \quad &&\mathcal{R}'(c) - \mathcal{G}'(a_e) = 0 \\
&\text{Unstable if} \quad &&\mathcal{R}'(c) - \mathcal{G}'(a_e) < 0
\end{aligned}
\tag{3.38}
$$

where $\mathcal{R}'(c) = \mathrm{d}\mathcal{R}(c)/\mathrm{d}c$ and $\mathcal{G}'(a_e) = \partial \mathcal{G}/\partial a_e = -(1/b)\,\partial^2 \Pi (a_e)/\partial a^2$. Note that if $0 < \mathcal{G} < \mathcal{R}(c)$, the equivalent crack is stable regardless of the sign of $\mathcal{R}'(c) - \mathcal{G}'(a_e)$ because it can neither extend nor shorten. This condition is represented by the shaded area in Fig. 3.8.

In the case $\mathcal{G} = 0$, the equivalent crack could start to close near the tip since the crack tip opening displacement δ_c is zero in such a case. One has an equilibrium state of crack shortening ($\delta a_e < 0$) and the stability conditions can be written as follows:

$$
\begin{aligned}
&\text{Stable if} \quad &&\mathcal{G}'(a_e) < 0 \\
&\text{Critical if} \quad &&\mathcal{G}'(a_e) = 0 \quad \text{and} \quad \mathcal{G} = 0 \\
&\text{Unstable if} \quad &&\mathcal{G}'(a_e) > 0
\end{aligned}
\tag{3.39}
$$

It should be noted that when the equilibrium state of crack growth or shortening is unstable, the crack starts to propagate or shorten dynamically, and inertia forces must be taken into the picture.

3.5.3 Stability under Load-Control Condition

The $\mathcal{R}$-curves are instructive for understanding fracture stability. For a structure subjected to a single load P or a system of loads proportional to a single parameter P, one can always write the SIF associated with the equivalent crack as $K_I = P/\left(b\sqrt{D} \right) k\left(\alpha \right)$ where D = structure size, b = thickness of the structure, $\alpha = a_e/D$ and $k\left(\alpha \right)$ = dimensionless function accounting for geometry effects. As a consequence, we write the energy release rate function as $\mathcal{G} = P^2 g(\alpha)/\left(E'b^2 D \right)$, where $g(\alpha) = k^2(\alpha)$.

Consider a structure of positive geometry, i.e., a structure for which $g'(\alpha) > 0$. Fig. 3.9 shows a typical set of curves for $\mathcal{G}(a_e, P)$ for a succession of increasing constant values $P = P_1, P_2, P_3, \dots$. According to Eq. 3.36, the equilibrium states of crack propagation for various load values are represented by the intersection of these curves with the $\mathcal{R}$ curve. Based on Eqs. (3.37), these states are stable if $\mathcal{R}'(c) > \mathcal{G}'(a_e)$ and unstable if $\mathcal{R}'(c) < \mathcal{G}'(a_e)$, at the intersection point. For instance, the curve of $\mathcal{G}$ for $P = P_2$ in Fig. 3.9 shows two equilibrium points, A and B, which are, respectively, stable and unstable.

As the load increases and the crack grows, the difference between the slopes of $\mathcal{R}(c)$ and $\mathcal{G}(a_e)$ gradually diminishes until the slopes become equal or, in other words, the curves become tangent to each other. This condition corresponds to the critical state of stability limit, at which the load is the maximum possible and the structure is in a state of incipient failure. Beyond this point, the crack growth is unstable and dynamic under the load-control condition. The crack propagates dynamically with the rate of

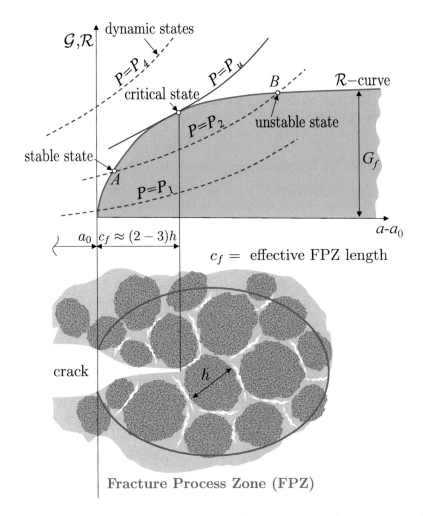

Fig. 3.9 Typical crack resistance and energy release rate curves for structures of positive geometry, $\mathcal{G}'(a_e) > 0$, and the corresponding evolution of FPZ.

kinetic energy supply equal to $\mathcal{G}(a_e) - \mathcal{R}(c)$. Conversely, before the critical state, the equivalent crack (equivalent to the FPZ) grows in a stable manner without inertial forces.

For a structure of negative geometry, that is $g'(\alpha) < 0$, stability is always guaranteed because $\mathcal{R}'(c) > 0$. This case occurs, for example, for the double cantilever beam specimen with a relatively short crack or for specimens with chevron notches. In general, this situation is uncommon since the vast majority of structures feature a positive geometry.

In case $\mathcal{R}(c) = G_f = $ constant, which is a typical case of LEFM, stability requires that $\mathcal{G}'(a_e) < 0$. In other words, there can be no stable crack growth in LEFM except when $\mathcal{G}'(a_e) < 0$. Conversely, if a stable crack propagation is observed and $\mathcal{G}'(a_e) > 0$, the fracture law must be nonlinear.

Finally, it should be noted that comparing structures that are geometrically similar but of different sizes, the curves $\mathcal{G}(a_e)$ are of similar shape but scaled according $1/D$ while the $\mathcal{R}$-curve remains the same. As a consequence, the larger the structure, the larger is the equivalent crack length c at the critical load.

3.5.4 Stability under Displacement-Control Condition

For a structure where the displacement rather than the load is imposed, the stability can be analyzed the same way as in the load-control case. The only difference is that, in this loading scenario, the displacement is kept constant instead of the load and the energy release rate needs to be expressed as $\mathcal{G}(a_e, P)$, rather than $\mathcal{G}(a_e, u)$ as in the previous case. Let us consider, for instance, the double cantilever beam represented in Fig. 2.4, for which the expressions $u = C(a_e)P$ ($C(a_e) = $ elastic compliance) and $G = G(P, a_e)$ have been calculated from Eqs. 2.21 and 2.22, respectively. We can express $\mathcal{G}(a_e, u)$ by

$$\mathcal{G}(a_e, u) = \frac{3}{4} \frac{u^2 E h^3}{a_e^4} \tag{3.40}$$

Fig. 3.10 shows a set of curves of $\mathcal{G}$-a_e at constant displacement, compared with a typical $\mathcal{R}$-curve for this geometry. It is obvious that all the equilibrium states are stable since further crack growth at constant displacements drives the state point into the zone of stability according to Fig. 3.10. It turns out that, for this geometry, the displacement controlled tests are always stable. However, this is not a general property. Some geometries may display instabilities at controlled displacements, a phenomenon usually called *snap-back instability*; (Bažant and Cedolin, 1991).

3.5.5 Experimental Characterization of $\mathcal{R}$-Curve

The characterization of the $\mathcal{R}$-curve requires determining $\mathcal{G}$ for various Δa_e during quasi-static crack growth. The $\mathcal{R} - \Delta a_e$ curve can be obtained by measuring the energy release rate $\mathcal{G} = \mathcal{R}$ during quasi-static crack growth.

There are various ways of determining the $\mathcal{R}$-curve. Usually, the load is measured directly while the equivalent elastic crack length is inferred from a second measurement. While measuring the load is easy, estimation of the crack length is a non-trivial task and various methods differ mainly by the method by which the equivalent crack extension is determined. Five different methods can be distinguished:

- *Direct measurement.* In this method, the length of equivalent crack is measured optically, during a slow fracture test. However, for quasibrittle materials such as concrete or composites this approach has two problems: (a) the identification of the crack tip location is very difficult because of the diffused fracturing damage in the FPZ, and (b) the fracturing crack can be identified only on the surface of the specimen while the effective crack front in the core can be deeper or shallower.

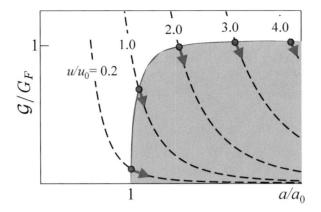

Fig. 3.10 $\mathcal{G}$-a curves at constant displacement (dashed lines), for the DCB specimen of Fig. 2.4. All the equilibrium points are stable.

- *Inference from unloading compliance.* Here the specimen is subjected to stable crack growth up to the required point, and then one or several unloading–reloading cycles are performed. The effective crack length is then computed from the measured unloading–reloading compliance assuming that equivalent LEFM applies. In general, the compliance is measured using the crack mouth opening displacement (CMOD) by means of an extensometer or Digital Image Correlation (DIC). Usually, this method tends to underpredict the crack length due to stiffening effects during unloading, such as hindered crack closure and the contact between crack lips upon load release.

- *Inference from secant compliance.* Here the equivalent crack is defined as that giving the same displacement as observed in virgin loading. Again, the secant compliance can be measured through the CMOD or the load-point displacement.

- *Size effect method.* In this method introduced first for concrete Bažant *et al.* (1991), the $\mathcal{R}$-curve can be also determined by exploiting the fact that, at the peak load, the of $\mathcal{G}$ vs. (a_e) curve for constant load is tangent to the $\mathcal{R}$-curve. By testing geometrically similar specimens of various sizes and recording the peak loads P_u, one obtains a set of the $\mathcal{G}(P_u, a_e)$ curves. One can then determine the $\mathcal{R}$-curve as the envelope of the $\mathcal{G}(P_u, a_e)$ curves. This method will be expounded in Chapter 5.

- *Shape effect method.* Here the shape of the specimen, e.g., the relative notch length $\alpha_0 = a_0/D$, is varied. The $\mathcal{R}$-curve is again the envelope of the $G(P_u, \alpha_0)$ curve for all α_0. But this method gives information for only a narrow part of the $\mathcal{R}$-curve because varying the shape without the size cannot give a sufficient range of a_e. To cover a broader range, the size, D, of the specimen must be varied too, and then the results are even better than the pure size effect method.

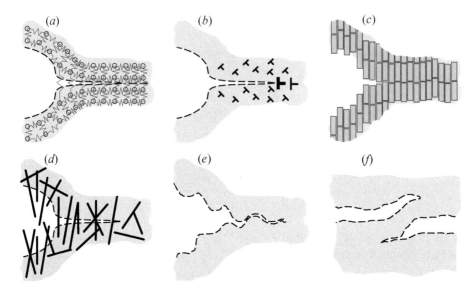

Fig. 3.11 Examples of physical sources of cohesive cracks: (a) atomic bonds, (b) yield (dislocation) strip, (c) grain bridging, (d) fiber bridging, (e) aggregate frictional interlock, and (f) crack overlap (or snubbing).

3.6 Cohesive Crack Model

The cohesive crack model is the simplest model to describe the progressive fracturing damage process ahead of the tip of a preexisting (stress-free) crack, or notch. A cohesive crack is a fictitious crack that is able to transfer stresses from one face to the other. This idea was pioneered in 1959 by Barenblatt (Barenblatt, 1959, 1962) to account for the nonlinear behavior of the atomic bond breaking during crack propagation (many perfunctory readers incorrectly attributed this idea to Dugdale (Dugdale, 1960) who, however, dealt with a different problem—a slit in a metal in which the plastic zone extends indefinitely at no growth of slit length, with no softening stress-displacement relation and no fracture energy as shown in Fig. 3.12; (cf. Bažant (2020)). Dugdale's model, however, served later as the basis of critical CTOD fracture criterion, which was related to the J-integral. It found wide application for tough ductile steels, and has been embodied in the British Standards Institute (1979), Australian, Chinese, Russian, and other standards.

Barenblatt's analysis was limited to cracks very large compared to the cohesive zone itself and at the critical state (i.e. at the onset of steady crack growth). In 1964, Barenblatt extended his model to include micro-defects or microcracking (Barenblatt, 1964); in 1966 to kinetics of quasistatic crack growth and long-time strength (Barenblatt *et al.*, 1966). The formulation of the cohesive crack model was completed in 1968 by Rice (Rice, 1968*b*) who showed, via his path-independent J-integral, that the flux

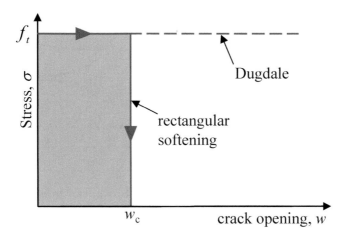

Fig. 3.12 Cohesive stress vs crack opening for the original Dugdale model (dashed line) and for rectangular softening. In the Dugdale model, also called *"strip yield model"*, the stress limit is usually called the plastic flow stress.

of energy into the crack tip, representing the energy release rate (with respect to the crack length) from an elastic structure, is equal to the work of cohesive stresses on the crack opening displacement during crack propagation. This work, equal to the area under the curve of the softening law, represents the fracture energy of the material, G_f.

In 1976, Hillerborg, Modéer and Petersson, under the name *fictitious crack model* (Hillerborg *et al.*, 1976; Modéer, 1979; Petersson, 1981; Gustafsson, 1985; Hillerborg, 1985*b,a*), proposed that a cohesive (or fictitious) crack can form in concrete without any pre-existing crack or notch wherever the tensile strength limit is reached. This recognized the fact that, unlike metals, the concept of crack nucleation is meaningless for concrete since the material is full of densely spaced cracks to begin with, at all scales from nano to near-macro. Subsequently, Petersson (Petersson, 1981), using this version of cohesive crack model, developed an effective fracture simulation algorithm for concrete structures in which the crack and its fracture process zone are not small compared to the structure size.

Since the pioneering studies of Barenblatt, cohesive zone models have often been used to describe the nonlinear behavior near the crack tip in metals and ceramics, and, aside from Dugdale's model, polymers as well. Interestingly, the same theoretical framework has been proven to hold for many distinct micromechanisms, as sketched in Fig. 3.11 with scales ranging from nanometers (Figs. 3.11a–b) to centimeters (Figs. 3.11e–f, for concrete with large aggregates).

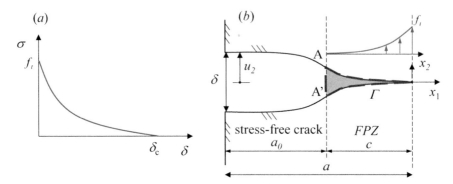

Fig. 3.13 (a) General cohesive law and (b) *J*-integral path used to calculate the fracture energy.

3.6.1 Cohesive Law and Its Relation with Fracture Energy

The essence of the cohesive crack model is the cohesive law $\sigma(\delta)$, which describes the relationship between the stress transmitted through the cohesive crack and the cohesive opening displacement. Fig. 3.13a shows a typical cohesive law for mode I fracture. It is noted that the cohesive law begins with a finite strength value at zero opening displacement, i.e. there is no elastic behavior within the cohesive crack itself. This is because the cohesive crack virtually lumps the FPZ into a line, and the elastic behavior of the FPZ is represented by the elasticity of the bulk material.

It is interesting to use the *J*-integral to determine the relation between the cohesive law and the fracture energy. Let us analyze a cohesive crack in infinite plate and consider the contour represented in Fig. 3.13b that embraces the entire FPZ. The general expression for the *J*-integral is:

$$J = \int_{\Gamma} \left(\bar{u} \mathrm{d}x_2 - t_i \frac{\partial u_i}{\partial x_1} \mathrm{d}s \right) \tag{3.41}$$

Here $\mathrm{d}x_2 = 0$ along the entire contour, no stress is present along crack crossing AA', and the integrals from a_0 to $a_0 + c$ and from $a_0 + c$ to a_0 are equal and of opposite sign. So we have,

$$J = -\int_{\Gamma} t_i \frac{\partial u_i}{\partial x_1} \mathrm{d}s = -\int_{a_0}^{a_0+c} \sigma_{22} \frac{\partial}{\partial x_1} \left(u_2^+ - u_2^- \right) \mathrm{d}x_1 \tag{3.42}$$

where u_2^+ and u_2^- represent the vertical displacement at the top and bottom faces respectively. Now, noting that $\delta = u_2^+ - u_2^-$, denoting $\sigma_{22} = \sigma$, and noting also that $(\partial \delta / \partial x_1) \mathrm{d}x_1 = \mathrm{d}\delta$, one gets:

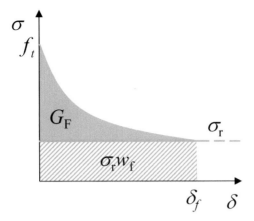

Fig. 3.14 Example of cohesive law with residual stresses.

$$J = \int_{a_0+c}^{a_0} \sigma_{22} \frac{\partial \delta}{\partial x_1}\, \mathrm{d}x_1 \tag{3.43}$$

$$J = \int_{0}^{\delta_c} \sigma\left(\delta\right) \mathrm{d}\delta \tag{3.44}$$

This shows that the fracture energy required to propagate a cohesive crack corresponds to the area under the cohesive law, i.e. $G_F = \int_0^{\delta_c} \sigma_{22}\left(\delta\right)\mathrm{d}\delta$.

We now extend the foregoing analysis to the case in which residual cohesive stresses, σ_r, are present as depicted in Fig. 3.14. In this case, the residual stresses act on the arc AA', and so Eq. 3.43 needs to be complemented by adding the J-integral contribution $J_{AA'}$ on AA':

$$J = \int_{a_0+c}^{a_0} \sigma_{22} \frac{\partial \delta}{\partial x_1}\, \mathrm{d}x_1 + J_{AA'} \tag{3.45}$$

Now, following the same reasoning as before, the first term on the RHS takes the same expression as in Eq. 3.44. Graphically, this corresponds to the total area under the curve in Fig. 3.14. The second term can be easily calculated noting that $J_{AA'} = -J_{A'A}$ and that:

$$J_{A'A} = \lim_{\delta \to 0} \int_{u_2^+}^{u_2^-} t_2 \frac{\partial u_2}{\partial x_2} \mathrm{d}x_2 = -t_2(u_2^- - u_2^+) = \sigma_r(u_2^+ - u_2^-) \tag{3.46}$$

Therefore,

$$J_{A'A} = \sigma_r \delta_c \tag{3.47}$$

which, graphically, corresponds to the rectangular dashed area in Fig. 3.14. Finally, we obtain the following expression:

$$J = \int_0^{\delta_c} \sigma\left(\delta\right) \mathrm{d}\delta - \sigma_r \delta_c \tag{3.48}$$

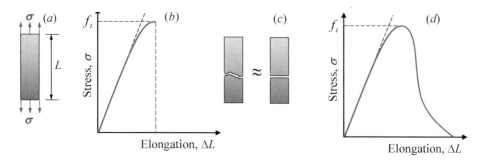

Fig. 3.15 Uniaxial test on a quasibrittle material (a). Typical stress-elongation curve with single crack failure (b). A rough crack is idealized as a flat crack (c). Stress-elongation curve for a stable tension test (d).

---

which corresponds to the solid color area in Fig. 3.14. As can be noted, the J-integral still equals the fracture energy even in the presence of residual stresses.

3.6.2   Relationship between Cohesive Law and Strain Softening Behavior

Consider a uniaxial tension test on quasibrittle specimen (Fig. 3.15a). If the specimen is large compared to the maximum size of material heterogeneities and the test is performed under a load control condition, a nearly linear stress-elongation curve will be recorded up to the peak load. At this condition, catastrophic (dynamic) failure occurs, usually through a single crack (Fig. 3.15b). The crack surface is generally rough due to the material heterogeneity. Nevertheless, we may approximate the final failure by a smooth crack orthogonal to the loading axis (Fig. 3.15c). The irreversible deformation of each of the two pieces the specimen broke into are likely to be very small, at most on the order of the inelastic strain recorded just before peak. Accordingly, one can assume that after the peak load, all the deformation localizes into the final crack. Furthermore, the evolution from no crack at all to a fully broken bar can be assumed to be gradual according to several seminal investigations conducted in late 1960s (Hughes and Chapman, 1966; Evans and Marathe, 1968; Heilmann *et al.*, 1969*a*). It was shown that, if the tensile specimen is short enough and a very stiff testing device is adopted, the crack can evolve in a stable manner allowing the characterization of the complete load-elongation curve, including the post-peak region (Fig. 3.15d). This result proved unambiguously that the structure experiences a gradual softening (i.e. gradual loss of load-carrying capacity) rather than a sudden loss of load-carrying capacity (a similar result on fiber composites was shown recently by the authors thanks to the design of a novel test apparatus (Salviato *et al.*, 2016*b*). The experimental results of (Heilmann *et al.*, 1969*a*) further showed that after the peak load, the strain localizes into a very narrow region while the rest of the specimen unloads.

Leveraging the foregoing observations, a uniaxial tensile test can be idealized in the following manner (Fig. 3.16). Up to the maximum load, the load increases while the strain in the bar remains uniformly distributed in the specimen (arc OP). At the peak

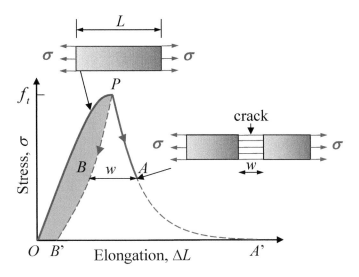

Fig. 3.16 An idealized tensile test.

load, a cohesive crack normal to the loading axis appears at the weakest cross-section somewhere in the specimen. This cohesive crack develops a finite opening w while it is still able to transfer stress. The rest part of the specimen unloads with a uniformly distributed strain, which is represented by arc PB in Fig. 3.16. Accordingly, the total elongation at point A is the sum of a uniform strain corresponding to point B and the crack opening w:

$$\Delta L = L\varepsilon_B + w \tag{3.49}$$

with $L =$ length of the specimen and $\varepsilon_B =$ strain of the portion of specimen that unloads from the peak load.

With the foregoing picture of the fracturing process, the stress transferred through the cohesive crack can be described as a function of the crack opening as originally proposed by Barenblatt (1959, 1962):

$$\sigma = f(w) \tag{3.50}$$

where $f(w)$ is a characteristic function of the material to be determined experimentally. This function is called the *softening curve* and can be extracted from the load-elongation curve following the procedure represented in Fig. 3.17.

To simplify the computations, one can assume that the inelastic strain in the loading–unloading path is negligible or, in other words, that the behavior of the bulk material is linear elastic. Thus, the post-peak elongation can be calculated from Eq. 3.49 as

$$\Delta L = L\frac{\sigma}{E} + w = L\frac{f(w)}{E} + w \tag{3.51}$$

where E is the elastic modulus.

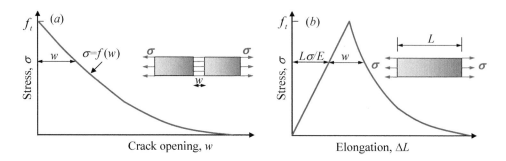

Fig. 3.17 Softening curve (a), and resulting stress-elongation curve when the bulk material behavior is assumed to be linear elastic (b).

Clearly the preceding discussion refers only to the case of pure tension, and an extension to other situations is needed for a general model. Several models to account for, e.g., mixed-mode loading conditions, nonlinear behavior of the bulk and the effect of triaxiality on the crack onset have been proposed in the literature.

3.6.3 Softening Curve, Fracture Energy, and Other Properties

It is evident that the softening curve $\sigma = f(w)$ essentially represents the cohesive law, a basic ingredient of the cohesive crack model. As discussed in Sec. 3.6.1., for mode I fracture, this curve is determined by the material tensile strength f_t, the total fracture energy G_F and other parameters that describe the shape of the curve.

The tensile strength is the stress at which the crack is created and starts to open, i.e. $f(0) = f_t$, and the cohesive fracture energy G_F is the external energy supply required to create and fully break a unit surface area of cohesive crack. This can be calculated by considering a thin element of initial length h centered at the cohesive crack location. The stress that the rest of the specimen transmits upon the faces of this element is σ (Fig.3.18). During the test, the elongation of such an element is δh and the incremental external work is:

$$\delta W = \sigma S \delta h \tag{3.52}$$

where S is the area of the cross-section of the specimen. According to Eq. 3.49, the total elongation of the element is:

$$\delta h = h \delta \epsilon + w \tag{3.53}$$

where $\delta \epsilon$ is the variation of the strain in the bulk. Thus, referring to Fig. 3.16, the total work on an element of length h that includes the crack is:

$$W = hS \int_{OPB'} \sigma d\varepsilon + S \int_{OPA'} \sigma dw \tag{3.54}$$

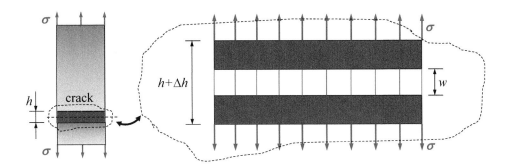

Fig. 3.18 Schematic representation of the strip containing the cohesive crack.

where the first term represents the energy dissipation due to the inelastic deformation of the bulk material and the second term the energy expended to open the cohesive crack. If we consider a vanishingly thin element, $h \to 0$, the bulk energy dissipation becomes negligibly small and only the second term, representing the cohesive energy, remains. By dividing this energy by the crack surface, i.e. the specimen cross-section S, and invoking the definition of fracture energy, we reach the following expression:

$$G_F = \int_0^\infty \sigma \mathrm{d}w = \int_0^\infty f(w)\mathrm{d}w \qquad (3.55)$$

where Eq. 3.50 was used to express the stress in terms of w.

Geometrically, G_F coincides with the area under the softening curve in Fig. (3.17a), which is consistent with the analysis of cohesive law presented in Sec. 3.6.1. Alternatively, G_F can also be represented by the area under the curve σ-ΔL in Fig. (3.17b) (area $OPAA'$, Fig. 3.16). We note that, in a general case in which energy dissipation occurs also in the bulk of the specimen, the energy supply to completely break the specimen is larger than the cohesive energy, and G_F in Fig. 3.16 amounts only to the area $B'BPAA'$, not the total area.

3.7 Integral Equations of Mode I Cohesive Crack Model

Consider now a cohesive crack growing in mode I as shown in Fig. 3.19a and assume proportional loading characterized by a generalized force P. It is assumed that the crack grows monotonically from the initial length a_0 to a so that the crack length a can be used as independent variable to calculate the value of P and all the other variables (e.g., displacements, stresses).

We assume that the bulk of the body is linear elastic, and therefore it is possible to express the overall nonlinear solution as a superposition of known elastic solutions. This can be accomplished by means of the decomposition shown in Fig. 3.19a–c where the overall solution (a) is calculated as the superposition of case (b) in which the external forces act and the cohesive stresses have been set to zero, and the case (c) in which the cohesive stresses alone act on the cohesive zone. The condition that the cohesive stress and the crack opening in the cohesive zone must satisfy the cohesive softening law $\sigma = f(w)$ is expressed as

$$\sigma(x) = f\left[C_x(a)P - \int_{a_0}^{a} C_{xx'}(a)b\sigma(x')\mathrm{d}x'\right] \quad \text{for} \quad a_0 < x < a \qquad (3.56)$$

This equation can be used to find $\sigma(x)$ for any given a and load P.

The crack opening w at any coordinate x can be written as:

$$g\left[\sigma(x)\right] = w(x) = C_x(a)P - \int_{a_0}^{a} C_{xx'}(a)b\sigma(x')\mathrm{d}x' \qquad (3.57)$$

where $w(x) = g\left[\sigma(x)\right]$ is the inverse function of the cohesive law and the function $g\left[\sigma(x)\right]$ is unique since the cohesive law is monotonic. Furthermore, $C_x(a)$ is the cross-compliance function defined as crack opening w at coordinate x produced by unit load P, and $C_{xx'}(a)$ is the cross-compliance function defined as w at x caused by a unit force applied at crack faces at x', both for crack length a. These compliance functions can be calculated analytically or obtained numerically as matrix approximations.

Another condition required to solve the problem is the condition of smooth closing at the tip of the cohesive crack. This requires that the SIF caused jointly by the applied load and the cohesive stresses must vanish. This condition can be written as:

$$\frac{P}{b\sqrt{D}}\hat{k}(\alpha) - \frac{1}{\sqrt{D}}\int_{a_0}^{a} k_G(\alpha, x/D)\,\sigma(x)\,\mathrm{d}x = 0 \qquad (3.58)$$

The first term is the SIF due to the applied external loads and the second term is the SIF due to the cohesive stresses, which can be expressed in terms of Green function.

Solving P from Eq. 3.58 and substituting it into Eq. 3.56, we further get:

$$\sigma(x) = f\left[C_x(a)\frac{b}{\hat{k}(\alpha)}\int_{a_0}^{a} k_G(\alpha, x'/D)\sigma(x')\mathrm{d}x' - \int_{a_0}^{a} C_{xx'}(a)b\sigma(x')\mathrm{d}x'\right] \qquad (3.59)$$

for $a_0 < x < a$. This is a nonlinear integral equation from which the stress distribution $\sigma(x)$ in the cohesive zone can be solved. To this end, the integrals may be approximated by summation over values at nodes along the crack, and $C_{xx'}$, C_x are replaced by influence matrices calculated by finite elements.

It is convenient to rewrite the foregoing equation in a non-dimensional form by introducing the relative coordinates:

$$\xi = \frac{x}{D}, \quad \xi' = \frac{x'}{D} \qquad (3.60)$$

With the new coordinates, we can express the cross-compliance $C_x(a)$ as

$$C_x(a) = \frac{1}{bE'}\,\hat{v}_x(\xi, \alpha) \quad \text{with} \quad \hat{v}_x(\xi, \alpha) = 2\int_{\xi}^{\alpha} \hat{k}(\alpha')k_G(\alpha', \xi)\mathrm{d}\alpha' \qquad (3.61)$$

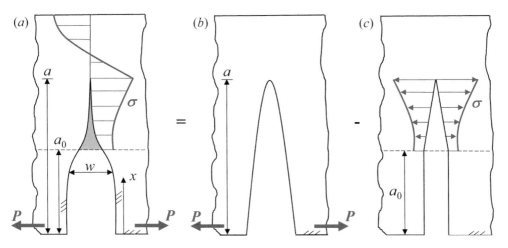

Fig. 3.19 Decomposition of the cohesive crack problem.

and $C_{xx'}(a)$ can be expressed similarly. By replacing the shape factor for the applied load by that for the concentrated load on the crack faces, we have

$$C_{xx'}(a) = \frac{1}{bE'}\,\hat{v}_{xx'}(\xi,\xi',\alpha),$$

$$\text{with} \quad \hat{v}_{xx'}(\xi,\xi',\alpha) = 2\int_{\xi_m}^{\alpha} k_G(\alpha',\xi')k_G(\alpha',\xi)\mathrm{d}\alpha' \quad (3.62)$$

where $\xi_m = \max(\xi,\xi')$. Eq. 3.59 can now be rewritten as

$$\sigma(\xi) = f\left[\frac{D}{E'}\int_{\alpha_0}^{\alpha} N(\xi,\xi',\alpha)\,\sigma(\xi')\mathrm{d}\xi'\right] \quad \text{for} \quad \alpha_0 < \xi < \alpha \quad (3.63)$$

Here the expression in the square brackets is the crack opening $w(\xi)$ and

$$N(\xi,\xi',\alpha) = \frac{\hat{v}_x(\xi,\alpha)}{\hat{k}(\alpha)}k_G(\alpha,\xi') - \hat{v}_{xx'}(\xi,\xi',\alpha) \quad (3.64)$$

3.8   Eigenvalue Analysis of Peak Load and Size Effect

We will now use the result of Sec. 3.7 to calculate the peak load capacity of the structure. The essential idea is to invert the problem—determine the structure size for which the peak load occurs at a given relative crack depth. Thus the problem gets reduced to an eigenvalue problem, which can easily be solved using the available numerical approaches.

We first consider the variation of Eq. 3.57 for an infinitesimal crack growth δa:

$$\delta w(x) = C_x(a)\delta P - \int_{a_0}^{a} bC_{xx'}(a)\delta\sigma(x)\mathrm{d}x' + \delta m \qquad (3.65)$$

where $\delta w(x)$, δP, and $\delta\sigma(x')$ are the variations of the corresponding variables due to δa, and

$$\delta m = \left[-bC_{xa}(a)\sigma(a) + \frac{\partial C_x(a)}{\partial a}P - \int_{a_0}^{a} \frac{\partial C_{xx'}(a)}{\partial a}b\sigma(x')\mathrm{d}x' \right]\delta a \qquad (3.66)$$

It can now be shown that, in fact, $\delta m = 0$ at peak load state. We first note that $C_{xa}(a) = C_{ax}(a)$ due to the reciprocity theorem, and that $C_{ax}(a)$ is the crack opening at the cohesive crack tip, which is zero.

By taking advantage of Eq. 3.61 and recalling that $\partial/\partial a = \partial/\partial(D\alpha)$, we can write

$$\frac{\partial C_x(a)}{\partial a} = \frac{2}{bDE'}\hat{k}(\alpha)k_G(\alpha,\xi) \qquad (3.67)$$

Similarly, the derivative $C_{xx'}(a)$ can be calculated from Eq. 3.62:

$$\frac{\partial C_{xx'}(a)}{\partial a} = \frac{2}{bDE'}k_G(\alpha)k_G(\alpha,\xi') \qquad (3.68)$$

By substituting Eqs. 3.67 and 3.68 into Eq. 3.66, we get

$$\delta m = \frac{2}{bDE'}k_G(\alpha,\xi)\left[k(\alpha)P - \int_{a_0}^{a} k_G(\alpha,\xi')\sigma(x') \right]\delta a \qquad (3.69)$$

Now, invoking Eq. 3.58 and imposing that $K_I = 0$, we note that the term in the square bracket vanishes. Thus, $\delta m = 0$ at peak load, and Eq. 3.65 reduces to:

$$\delta w(x) = C_x(a)\delta P - \int_{a_0}^{a} C_{xx'}(a)b\delta\sigma(x')\mathrm{d}x' \qquad (3.70)$$

The incremental form of Eq. 3.56 can be re-written as

$$\delta\sigma(x) = f'[w(x)]\left[C_x(a)\delta P - \int_{a_0}^{a} C_{xx'}(a)b\delta\sigma(x')\mathrm{d}x' \right] \qquad (3.71)$$

where $f'(w) = \mathrm{d}f(w)/\mathrm{d}w$. Since, at peak load, $\delta P = 0$ by definition, the following condition must be satisfied at the maximum load:

$$\delta\sigma(x) = -f'[w(x)]\int_{a_0}^{a} C_{xx'}(a)b\delta\sigma(x')\mathrm{d}x' \qquad (3.72)$$

This is a homogeneous integral equation for the stress variation $\delta\sigma(x)$ whose nontrivial solutions are the eigenvectors (or eigenfunctions) of the equation. It is convenient to recast the equation in dimensionless form introducing the dimensionless variables $\hat{\sigma}$ and $\hat{w}$. Using Eqs. 3.60 and 3.62, we get:

$$\delta\hat{\sigma}(\xi) = -\hat{f}'\left[w\left(\xi\right)\right]\frac{D}{l_{ch}}\int_{\alpha_0}^{\alpha}\hat{v}_{xx'}(\xi,\xi',\alpha)\delta\sigma(\xi')\mathrm{d}\xi' \tag{3.73}$$

which is seen to take the form of an eigenvalue problem for a homogeneous linear Fredholm integral equation with D/l_{ch} as the eigenvalue. The peak load condition is obtained for the smallest eigenvalue, i.e. for the smallest D.

The solution of the foregoing problem is particularly easy for the case of a linear softening law. For linear softening, $\hat{f}' = -1/2$. Setting $\tau(\xi) = \delta\hat{\sigma}(\xi)/\delta\alpha$, we obtain

$$\tau(\xi) = \frac{D}{2l_{ch}}\int_{\alpha_0}^{\alpha}\hat{v}_{xx'}(\xi,\xi',\alpha)\tau(\xi')\mathrm{d}\xi' \tag{3.74}$$

which can be solved for the smallest eigenvalue $D/2l_{ch}$ and the corresponding eigenvector $\tau(\xi)$.

Next, (Bažant and Li, 1994; Li and Bažant, 1994), the cohesive law can be written as $\hat{\sigma}(\hat{w}) = 1 - \hat{w}/2$ where the opening profile $\hat{w}$ can be calculated from Eq. 3.57 as

$$\hat{w}(\xi) = \frac{D}{l_{ch}}v_x(\xi,\alpha)\hat{\sigma}_N - \frac{D}{l_{ch}}\int_{\alpha_0}^{\alpha}N(\xi,\xi',\alpha)\hat{\sigma}(\xi')\mathrm{d}\xi' \tag{3.75}$$

Now, multiplying the foregoing equation by the eigenvector and integrating over the entire cohesive crack, we get

$$\int_{\alpha_0}^{\alpha}\tau(\xi)\hat{\sigma}(\xi)\mathrm{d}\xi = \int_{\alpha_0}^{\alpha}\tau(\xi)\mathrm{d}\xi - \frac{D}{2l_{ch}}\hat{\sigma}_N\int_{\alpha_0}^{\alpha}\tau(\xi)v_x(\xi,\alpha)\mathrm{d}\xi$$

$$+ \frac{D}{2l_{ch}}\int_{\alpha_0}^{\alpha}\tau(\xi)\int_{\alpha_0}^{\alpha}v_{xx'}(\xi,\xi',\alpha)\hat{\sigma}(\xi')\mathrm{d}\xi'\mathrm{d}\xi \tag{3.76}$$

where $\hat{\sigma}_N = P_m/bDf_t$, and P_m = peak load capacity. By inverting the order of integration in the double integral and using Eq. 3.74, we note that the last term of the RHS and the term on the LHS cancel each other and Eq. 3.76 becomes

$$\hat{\sigma}_N = \frac{D}{2l_{ch}}\frac{\int_{\alpha_0}^{\alpha}\tau(\xi)\mathrm{d}\xi}{\int_{\alpha_0}^{\alpha}\tau(\xi)v_x(\xi,\alpha)\mathrm{d}\xi} \tag{3.77}$$

This provides the solution for the peak load (Bažant and Li, 1994; Li and Bažant, 1994).

Exercises

E3.1. For the metallic materials whose properties are given in Table 3.1 (extracted from Broek (1986)), determine: (a) the size fo the critical inelastic zone using Irwin's estimate; (b) the minimum distance from the crack tip to the specimen surface, D_s, for which LEFM is applicable (according to ASTM E 399, $D_s = 2.5K_{Ic}^2/\sigma_y^2$); (c) the characteristic size $l_{ch} = E'G_f/f_t'^2$ obtained by replacing σ_y with f_t'.

Table 3.1 Yielding stress and fracture toughness of some engineering materials.

Material	σ_y (MPa)	K_{Ic} (MPa$\sqrt{\text{m}}$)
Maraging steel 300	1670	93
D 6 AC steel (heat treated)	1470	96
A 533 B (reactor steel)	343	195
Low strength carbon steel	235	> 220
Titanium, 6Al-4V	1100	38
Titanium, 4Al-4Mo-2Sn-05Si	940	70
Aluminum 7075, T651	540	29
Aluminum 7079, T651	460	33
Aluminum 2024, T3	390	34

E3.2. What should be the minimum distance D_s from the crack tip to the surface for a quasibrittle material? Express D_s as a function of coefficient η in Eq. 3.19 and assume the same ratio D_s/R_c as for metals (ASTM E 399). Give the value of D_s for a concrete with $l_{ch} = 0.25$ m and $\eta = 4$. Find the minimum depth of single-edge notch beams with a notch-to-depth ratio of 0.5 that is needed to make LEFM applicable.

E3.3. For the panel investigated in Exercises 2.1 and 2.2, find and plot the p-V and $\mathcal{G}$-a curves for a test in which the crack extends from 100 mm to 1000 mm under volume expansion control, after which the panel is unloaded. Assume that resistance to crack growth varies with crack extension Δa in the form:

$$\mathcal{R} = 2G_f \left[1 - \frac{\Delta a}{2\lambda} \right] \qquad\qquad \text{for } 0 \le \Delta a \le \lambda$$
$$\mathcal{R} = G_f \qquad\qquad \text{for } \Delta a > \lambda$$

where $G_f = 100$ N/m and $\lambda = 276$ mm. Find the peak pressure and the maximum increase in volume. Use an effective elastic modulus $E' = 60$ GPa.

E3.4. A rectangular panel subjected to tensile stress parallel to one of its edges is made of a material characterized by the following $\mathcal{R}$-curve:

$$\mathcal{R} = G_f \left[1 - (1 - \Delta a/c_f)^n \right] \qquad\qquad \text{for } \Delta a \le c_f$$
$$\mathcal{R} = G_f \qquad\qquad \text{for } \Delta a > c_f$$

Determine the fracture strength of the panel for vanishingly small edge cracks as a function of G_f, E', c_f, and n.

E3.5. (a) Find the external work required to completely break a three-point-bend notched beam of thickness b, depth D, and notch-to-depth ratio α_0, under the assumption that the fracture of the material can be described by a $\mathcal{R}$-curve given by $\mathcal{R} = G_f[1 - \exp\left(-\Delta a/c_0\right)]$ where G_f and c_0 are constants. (b) Find the dependence of the mean fracture energy on the beam depth; the mean fracture energy is the total work supplied, divided by the initial ligament area.

E3.6. Consider a very large panel with a center crack of length $2a_0 = 100$ mm, subjected to remote tension normal to the crack plane. Assuming that the material follows the same $\mathcal{R}$-curve of Exercise 3.4 with $n = 2$, $G_f = 120$ N/m, $c_f = 50$ mm, and that $E' = 600$ GPa, determine (a) the peak stress; (b) the apparent fracture toughness K_{INu}; (c) the ultimate fracture toughness K_{Iu} (value of K_I at the equivalent crack tip at peak load); (d) the true fracture toughness K_{Ic} (value of K_I at peak for an infinitely long crack).

E3.7. Repeat exercise E3.6 for an initial crack length 10 times longer. What is now the difference among K_{INu}, K_{Iu}, and K_{Ic}?

E3.8. Using the $\mathcal{R}$-curve and function $\mathcal{G}(a)$, state the conditions of no propagation, stability limit, and unstable propagation.

E3.9. What would be the effect of a crack-parallel stress σ_{xx} on the $\mathcal{R}$-curve?

E3.10. Consider the function $F(c, D) = \mathcal{G}(\alpha, D) - \mathcal{R}(c)$ where $\alpha = a/D$, $a = a_0 + c$, $a_0 = $ initial stress-free crack length, and $\mathcal{R}(c) = \mathcal{R}$-curve. Sketch $\mathcal{G}(\alpha, D)$ and $\mathcal{R}(c)$ and show stable and unstable equilibrium states. Then, sketch these curves for different sizes D.

E3.11. Show, based on the $\mathcal{R}$-curve analysis, that a positive geometry becomes unstable at controlled load as soon as the peak load is reached.

E3.12. Consider a large panel with a center slit of length 100 mm on which a hydraulic-fracturing simulation is being performed by injecting a liquid inside the crack. Assuming that LEFM applies, with $G_f = 140$ N/m and $E' = 500$ MPa, and that the liquid is incompressible, analyze the test stability if it is carried out (a) at controlled pressure; (b) at controlled volume of injected liquid.

4

Nonlinear Fracture Mechanics—Diffuse Crack Model

What looks like a line has often a finite width, and it matters.

Chapter 3 expounded the nonlinear fracture mechanics based on the line crack idealization. We now proceed to expound the nonlinear fracture mechanics taking into account the existence of fracture process zone (FPZ) of a finite width at crack front. The fact that the fracture process zone must have a finite width has been implied, since the birth of fracture mechanics, by the fact that the surface energy, γ, of homogenous materials, such as unalloyed metal, glass or silicone, is typically several orders of magnitude smaller than the fracture energy, G_f. To explain it, many microcracks and frictional microslips must form on the sides of a growing crack. This approach, which can take into account the great effect of crack-parallel stresses in quasibrittle materials, is a more fundamental and more general approach than the cohesive crack model, although understanding of that model is a prerequisite for the present chapter. As discussed in previous chapters, the failure of quasibrittle structures is often accompanied by the damage localization instability, which give rises to the energetic and statistical size effects on the nominal structural strength.

In this chapter, we discuss how the damage localization instability influences the numerical modeling of quasibrittle fracture. We will particularly focus on the spurious mesh sensitivity in finite element (FE) simulations and present several types of localization limiters, which the crack band model and the nonlocal damage models tackling this issue. In addition to continuum-based FE modeling, we will also discuss the discrete element modeling of quasibrittle fracture, which provides more physical insights into the damage in the FPZ and illuminates the failure mechanisms at the mesoscale.

Quasibrittle Fracture Mechanics and Size Effect: A First Course. Zdeněk P. Bažant, Jia-Liang Le and Marco Salviato, Oxford University Press. © Zdeněk P. Bažant, Jia-Liang Le, Marco Salviato 2022. DOI: 10.1093/oso/9780192846242.003.0004

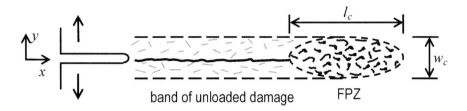

Fig. 4.1 Fracture process zone at the tip of a propagating macrocrack and unloaded damage zone behind the crack.

4.1 Why Crack Band?—Crack-Parallel Stress and Other Evidence

In visual observation, cracks in concrete appear as rugged but sharp lines. However, based on acoustic emissions emanating from forming microcracks, and partly also on microscopic observations and digital image correlation (DIC), it has long been known that, at the front of a propagating cracks, there is fracture process zone (FPZ) that has not only a finite length, l_c, but also a finite width, w_c. The finite FPZ width is also manifested in the minimum possible spacing of stable parallel cracks. The damage process in the FPZ consists of randomly distributed microcrack openings, splitting compression cracks, and frictional dilatant slips. Behind the crack front the microcracks close and leave behind a band of unloaded damaged material (Fig. 4.1), of width w_c, flanking the visible sharp crack.

It thus appears logical and more realistic to model crack propagation in most quasibrittle materials as a crack band rather than a sharp crack (Fig. 4.1). Recently it has been experimentally demonstrated that, in quasibrittle materials, the fracture energy, G_f, depends strongly on crack-parallel stress σ_{xx} (and probably also on σ_{zz} and σ_{xz}). These stresses and their corresponding strains do not represent the basic (i.e. thermodynamic) variables in the cohesive crack model, since it is a line crack model.

These conclusions have been reached upon finding a surprisingly simple, yet novel, adaptation of the three-point-bend test, called the "gap test" (Nguyen *et al.*, 2020*b*,*a*), depicted in Fig. 4.2a. The gap test has four key features:

1. A pair of elasto-plastic pads (Fig. 4.2a) that possess a long plateau of nearly perfect plastic yielding is installed adjacent to the notch mouth, to generate notch-parallel compression of a desired magnitude.

2. Stiff supports at beam ends are installed with a suitable gap so as to engage in contact with the test beam and generate a bending moment only after the pads develop plastic yielding (stage 2 in Fig. 4.2a).

3. In this way, the test beam transits from one statically determinate support system to another, which helps to make evaluation unambiguous and simple.

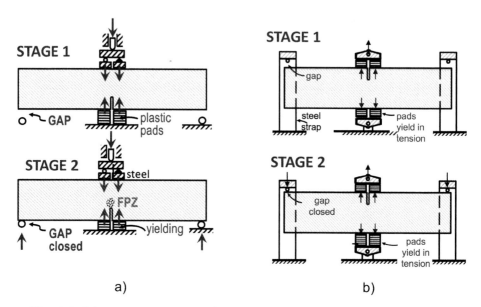

Fig. 4.2 Schematics of gap-test: a) three-point bend test, and b) tension test.

4. The constancy of crack-parallel compression along with the static determinacy of loading allows, for each σ_{xx}, using the size effect method, an easy and unambiguous method, to evaluate the fracture energy G_f and characteristic FPZ size (see Sec. 5.8). Fig. 4.2b shows the set-up of gap-test in tension. For experimental details, see (Nguyen *et al.*, 2020*a*), and for their summary (Nguyen *et al.*, 2020*b*).

The typical test results for concrete are shown in Fig. 4.3. The three experimental data points (solid circles) in Fig. 4.3 are the effective values of fracture energy G_f as a function of three levels of the compression stress σ_{xx} at the crack or notch tip (which is easily calculated from the stress under the yielding pads (Nguyen *et al.*, 2020*b,a*). The ordinate is the ratio σ_{xx}/f_t where f_t denotes the uniaxial compression strength of the material.[1]

As evident from Fig. 4.3, G_f is not constant but strongly depends on σ_{xx}. This result demonstrates that the LEFM and the cohesive crack model do not apply, except for an approximation in which the G_f is made a function of the current value of σ_{xx} (such an approximation, however, ignores very strong dependence on loading history). The curves plotted in Fig. 4.3 are the FE simulations with the crack band model based

---

[1]The fracture energy, G_f, is always understood as the energy per unit area of crack surface, dimension J/m^2 or N/m, same as the surface energy. Thus, although we say "line" crack, taking a two-dimensional viewpoint, we always understand there is also a certain width b in the third dimension or, in axisymmetric crack growth, a radian or circumference.

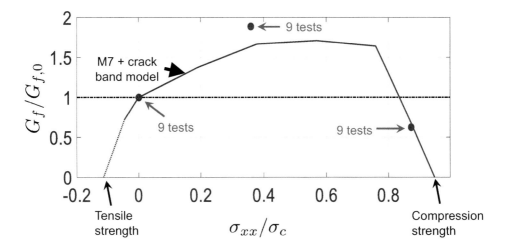

Fig. 4.3 Measured and simulated dependence of effective fracture energy on the crack-parallel stress.

microplane constitutive damage model M7 (Caner and Bažant, 2013) that has been calibrated to match the measured f_c, Young's modulus E, and G_f at zero σ_{xx}.

We can now summarize the arguments for representing quasibrittle fracture with a band of finite width, rather than a line crack. Aside from the fact that the fracture energy is orders of magnitude larger than the surface energy, they are four:

1. The new experimental evidence for crack-parallel stress effects, as just pointed out.

2. The evidence from computational modeling of the FPZ based on a realistic computational damage model, which shows that the line crack models cannot reproduce various tensorial effects in the FPZ seen in experiments (such as the vertex effect) and computations (such as those with the microplane model).

3. Convenience in programming and computations, as recognized by Rashid (Rashid, 1968) already in 1968.

4. The need to set, in a continuum model, the minimum admissible spacing of parallel cracks for situations with no softening localization. Such cracks can be produced, e.g., by cooling, shrinkage, or hydraulic fracturing (as in shale). Since, as will be discussed later, the crack band must have a certain width h that is a material property, placing parallel crack bands next to each other automatically sets minimum possible spacing of parallel fractures.

So it is now clear that the fundamental model for quasibrittle fracture cannot be a line crack. It is a band of tensorially modeled distributed damage, the localization of which eventually leads to a distinct sharp crack. The cohesive crack model and, ultimately, the LEFM, are simplifications, albeit very useful ones. They are essential for: (a) understanding fracture mechanics, (b) for providing accurate solutions of some benchmark tests that must be matched by a realistic tensorial model, and (c) for serving as the basis of generalizations. But the usability of line crack models for quasibrittle materials is quite limited.

On the other hand, the crack band model has the advantage that, in some special cases, the stress analysis problem can be solved analytically and exactly, as in Eqs. 3.6 — 3.18. Such solutions are valuable as benchmark test cases which the numerical solution with the crack band model must match.

In what follows in this chapter, the crack band model is explained, along with the associated aspects of mesh-sensitivity as well as the related nonlocal and discrete models.

4.2   Strain Localization, Mesh Sensitivity, and Localization Limiters

4.2.1   Bifurcation of Equilibrium Path

When a structure contains elements that exhibit a strain softening behavior, strain localization can take place. To demonstrate this fundamental concept, we analyze the behavior of a chain (or series coupling) of two identical elements (Fig. 4.4a). Assume that each element has a softening load-elongation behavior as shown in Fig. 4.4b. What is of interest here is the load-elongation behavior of the entire chain. The equilibrium and compatibility conditions indicate that

$$P(\delta_s) = P_1(\delta_1) = P_2(\delta_2) \tag{4.1}$$
$$\delta_s = \delta_1 + \delta_2 \tag{4.2}$$

where P, P_1, P_2 = force in the chain, element 1 and element 2, respectively, and $\delta_s, \delta_1, \delta_2$ = corresponding elongations.

By using Eqs. 4.1 and 4.2, we can plot the load-elongation curve of the chain. When the element reaches the peak load, there are two possible loading paths for continuing elongation of the chain: (a) both elements undergo softening, i.e. homogenous deformation, and (b) one element undergoes softening and the other experiences unloading; see paths AC and AB in Fig. 4.4c.

The question is which path will occur. In this regard, one must first realize that, since we consider equilibrium states, the principle of virtual work requires that the first-order work of the load or prescribed displacement must be zero. So we need to consider only the second-order work, i.e., the work of load increment on the displacement increment. Based on the Second Law of Thermodynamics, the path that will occur is that for which the entropy increase will be greater (see the detailed discussion of stable path in Sec. 10.2 in Bažant and Cedolin (1991); also Bažant (1976)). This is equivalent to minimizing the potential energy (or Helmholtz free energy), or to maximizing the complementary energy (or Gibbs free energy).

For an imposed small displacement increment, the second-order work done is negative for both paths (i.e., the energy is released by the system) and the path that occurs is that for which it is smaller, i.e. larger in magnitude. For an imposed small load increment, the second-order complementary work done on the system is positive for both paths (i.e., complementary energy is delivered to the system) and, based on

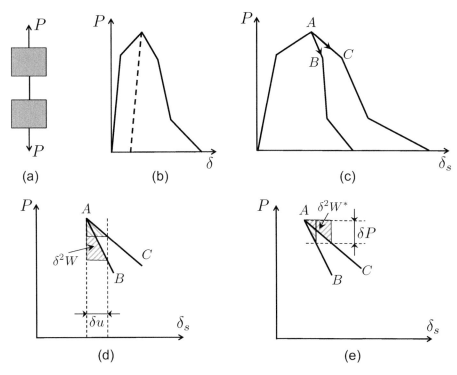

Fig. 4.4 (a) A chain (series coupling) of two softening elements. (b) Load-elongation curve of a single element (the dashed line represents the unloading curve at the peak load). (c) Two possible post-peak load paths of the chain. (d) Analysis of second-order work in the post-peak regime. (e) Analysis of second-order complementary work in the post-peak regime.

---

thermodynamics, the path that occurs is that for which it is larger.[2] Figs. 4.4d and 4.4e show the second-order work and second-order complementary work for the incremental process in the softening regime. It is evident that the nonlocalized path, AC, cannot occur in reality. In other words, the chain system (or series coupling) will exhibit strain localization.

The states on the unbifurcated path (such as C) can nevertheless occur, but only if the structure is restrained from localization. When the restraint is removed, a state

---

[2]The physical proof (simplified from Bažant and Cedolin 1991) is briefly as follows. Let $C^{(1)}$ be the stiffness (or slope) for the path AB, and $C^{(2)}$ for AC in Fig. 4.4a. The second-order works done on displacement increment δu are $\delta^2 W^{(1)} = \frac{1}{2} C^{(1)} \delta u^2$ and $\delta^2 W^{(2)} = \frac{1}{2} C^{(2)} \delta u^2$. Generally, the isothermal increment of entropy is $\Delta S = -\Delta \tilde{W}/T$. So the increments of entropy for paths (1) and (2) departing from the same equilibrium state A are $\mathrm{d}S^{(1)} = -T\delta^2 W^{(1)}$ and $\mathrm{d}S^{(2)} = -T\delta^2 W^{(2)}$. Since the descent of path AB is steeper than that of path AC, we have $C^{(1)} < C^{(2)} < 0$, and so $\mathrm{d}S^{(1)} > \mathrm{d}S^{(2)}$ (in detail, see Sec. 10.2 in Bažant and Cedolin (1991)). Hence AB is the path that maximizes entropy, and thus is the path that will occur.

such as C remains stable only if (a) δ_s is held constant and (b) the stability condition derived by Bažant (1976) is satisfied (see also Sec. 13.2 in Bažant and Cedolin (1991)), or else dynamic motion occurs right after releasing the restraint. Stability is possible only if the material characteristic length is finite, which is a point that revealed the necessity of finite material characteristic length and, consequently, of a nonlocal concept for strain softening damage (Bažant, 1976).

Indeed, the argument in 1976 was that if the material characteristic length acting as a localization limiter did not exist, the gradual strain-softening would have never been observed, and yet it was, on small enough specimens (thanks to the adoption of very stiff testing frames). This was an undeniable proof of nonlocality of softening continuum.

Further it follows that, in experiments, material softening properties can be observed directly only if the specimen size is roughly the same as the size of the fracture process zone, which is naturally the case for concrete (and is why the problem of softening localization was first analyzed for concrete). Otherwise the material softening properties must be deduced indirectly, which requires size effect tests of a broad enough size range.

We can extend the foregoing analysis to a chain of n identical elements (Fig. 4.5a). By using the same argument, we would conclude that, right after the chain attains its peak load, only one element will undergo softening while the remaining $n-1$ elements will undergo unloading. Fig. 4.5b shows a set of nominal stress-strain curves of the chain for n different values.

The essential result of this analysis is that, when the number of elements increases, the strength of the chain remains unchanged but the softening branch gets steeper (i.e. the chain becomes more brittle). As $n \to \infty$, the chain behavior becomes perfectly brittle.

The foregoing analysis is now applied to a homogenous bar of length L subjected to uniaxial tension. The material exhibits a softening stress-strain curve. Imagine that the bar is subdivided into n elements (Fig. 4.6a), and each element has a length $h = L/n$. Clearly the behavior of the bar is same as that of a chain of n elements. Based on the aforementioned analysis, we can conclude that the overall load-deflection curve of the bar would depend on the number of elements used for the subdivision. From a physical viewpoint, this result is absurd because the physical response of the bar cannot depend on the imagined subdivision.

To further demonstrate the implication of strain localization, we consider a simple elastic-softening stress-strain curve as shown in Fig. 4.6b. In the softening regime, the strain can be expressed as

$$\epsilon = \frac{\sigma}{E} + \epsilon_f \tag{4.3}$$

where E = elastic modulus, and ϵ_f = inelastic fracturing strain. We further assume that the unloading is purely elastic and the softening part of the stress-strain curve can be described by a one-to-one relationship between σ and ϵ_f, that is $\sigma = \phi(\epsilon_f)$ (Fig. 4.6c). It is clear that the total work γ_F required to fully break a unit volume of material is given by the area under the stress-strain curve, and so the area under $\sigma - \epsilon_f$ curve, that is:

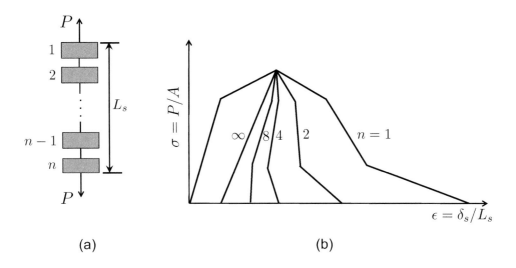

Fig. 4.5 (a) A chain of n softening elements. (b) Nominal stress-strain curve of the chain.

$$\gamma_F = \int_0^\infty \phi(\epsilon_f) \mathrm{d}\epsilon_f \qquad (4.4)$$

Once the bar reaches its peak load capacity, only one element starts to soften, and the remaining elements experience elastic unloading. Therefore, the total elongation of bar can be calculated as

$$\Delta L = \frac{\sigma}{E}(L - h) + \left(\frac{\sigma}{E} + \epsilon_f\right)h = \frac{\sigma}{E}L + \epsilon_f h \qquad (4.5)$$

The first term in the last expression is the elastic elongation while the second term is the elongation due to fracture. The total work required to break the whole bar is just the work needed to break the softening element:

$$\mathcal{W}_F = Ah \int_0^\infty \sigma \mathrm{d}\epsilon_f = Ah\gamma_F \qquad (4.6)$$

where A = cross-sectional area of the bar. Eq. 4.6 indicates that the total energy dissipation of the bar depends on the choice of size h. What would be the preferred value of h? We may determine it by using the condition of the maximum second-order complementary work condition. Evidently, the solution is $h = 0$. This implies that both the total inelastic deformation of the bar and the total energy dissipation are zero, which is obviously unacceptable and contradicts the experimental observations.

The foregoing analysis also has a profound implication for numerical simulations. If we consider the subdivision as the FE meshing, we may immediately conclude that the

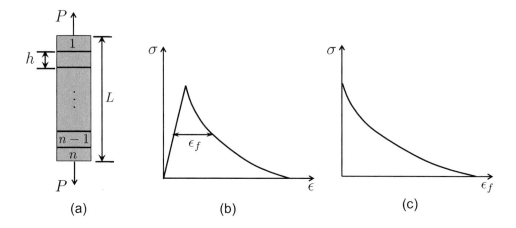

Fig. 4.6 (a) A homogenous bar under uniaxial tension. (b) Elastic-softening stress-strain relationship. (c) Stress-fracturing strain curve.

result of FE simulations would depend on the number of finite elements, or the mesh size, which is a subjective choice of the analyst. Therefore, the FE simulation loses its objectivity, as pointed out by Bažant (Bažant, 1976). This property is commonly referred to as the spurious mesh sensitivity.

To summarize, we show that, under static loading, the simple stress-strain model with strain softening would predict that damage localizes into a zone of zero size, which implies that the fracture work is zero. Meanwhile, it also leads to the issue of spurious mesh sensitivity in FE simulations. Subsequent work by Bažant and Belytschko (Bažant and Belytschko, 1985; Belytschko *et al.*, 1986) showed that similar issues also exist in dynamic analysis. For example, consider that two converging elastic waves propagate from the two ends of a bar towards the center. The analysis showed that the failure is instantaneous and occurs over a zone of zero length indicating zero energy dissipation (Bažant and Belytschko, 1985).

The aforementioned conclusions indicate that, for a continuum formulation using softening stress-strain models, one would need to complement it with additional conditions that would prevent strain localization into a region of measure zero. Such conditions are called localization limiters. In the subsequent sections, we will discuss two classes of localization limiters, namely the crack band models and the nonlocal damage models, which can effectively mitigate the issue of spurious mesh sensitivity in FE simulations of structures made of strain-softening materials.

4.2.2 Stability of Bifurcated Path and Implied Necessity of Localization Limiter

So far our analysis has dealt with equilibrium path bifurcation, as in Figs. 4.4 and 4.5. But at which point of the bifurcated path will the specimen or structure fail? Under the

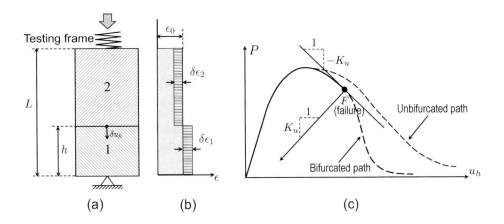

Fig. 4.7 (a) Specimen under uniaxial loading. (b) Profile of strain increment in the specimen. (c) Interpretation of the stability condition using the load-deflection curve.

sole control of load-point displacement, it would fail at the point of vertical tangent or the beginning of snapback of the post-peak curve (which would mean no failure for path AB in Fig. 4.4c). But in reality the specimen under load-point displacement control can fail earlier. To answer this question, we must analyze stability of the equilibrium states with strain softening.

Consider now a bar of length L in a post-peak state of bifurcated path (such as path AB in Fig. 4.4c), loaded through the elastic frame of the testing machine. Within the bar, consider a segment (1 in Fig. 4.7) of length h, assumed to be in a softening state of uniform axial strain. Assume that a small incremental axial displacement δu_h is enforced at the interface of this segment with the rest of the bar, to cause the strain in segment 1 to increase and thus soften, while the deformation in the rest of the bar, of length $L - h$, and in the elastic testing machine frame decreases according to a combined unloading stiffness, K_u (> 0). The incremental stiffness that resists δu_h is $K = K_t + K_u$ with $K_t = E_t A/h$ ($E_t < 0, K_t < 0$) where A is the cross-section area of the bar (uniform within segment 1). Since, in equilibrium, the first-order work is zero, the incremental work done by δu_h is the second-order work:

$$\Delta W = \tfrac{1}{2} K \, \delta u_h{}^2 = \tfrac{1}{2}(E_t A/h + K_u)\delta u_h{}^2 \tag{4.7}$$

Obviously, for $\Delta W > 0$, the equilibrium state will not change if δu_h is not enforced and no work is done. This indicates stable equilibrium. However, if $\Delta W < 0$, energy will be released from the bar spontaneously, which means stability loss.

Mode fundamentally, note that $\Delta W = -T\Delta S$ where T = absolute temperature and ΔS = increment of entropy of the bar-machine system. If the entropy cannot increase, the equilibrium state cannot change,i.e., is stable. If an increase of entropy, $\Delta S > 0$, is possible, it will happen, and so displacement δu_h will occur spontaneously, i.e., stability will be lost. In principle, the concept of stability of mechanical equilibrium is a consequence of the Second Law of Thermodynamics (Bažant and Cedolin, 1991, Chap.10).

So, stability of the postpeak deformation state requires that $K > 0$, that is,

$$-K_t < K_u \quad \text{or} \quad (-E_t)A/h < K_u \tag{4.8}$$

$K_t = E_t A/h$ (< 0) is the stiffness of the softening segment of length h. This condition is interpreted graphically on the curve of load P versus displacement u_h at the softening interface (Fig. 4.7a), which is the curve reached after bifurcations. The specimen will fail at point F at which a line of slope $-K_u$ becomes tangent to the $P(u_h)$ curve, marked in the figure (in the case of sudden slope change, the contact of the line of slope $-K_u$ is at the slope change point).

Eq. 4.8 leads us to observe that if the strain could localize into a line, i.e., if $h \to 0$ were a realistic possibility, then the stability condition (Eq. 4.8) would always be violated during post-peak, which would mean that a stable post-peak softening could never be observed (which is what was generally believed until 1963). Yet, introduction of very stiff testing frames beginning in 1963 demonstrated that stable post-peak states do exist (Rüsch and Hilsdorf, 1963; Hughes and Chapman, 1966; Evans and Marathe, 1968; Heilmann *et al.*, 1969b; Waversik and Fairhurst, 1970; Hudson *et al.*, 1971). .

The inevitable conclusion, made by Bažant in 1976 by a similar argument as here (Bažant, 1976), is that the zone of continued strain-softening must have a certain finite minimum width h_{min}, which must be a material property, representing a material characteristic length. Thus the existence of stable postpeak states is a proof that a material characteristic length exists, and that a continuum model must be nonlocal. In computational mechanics, h_{min} was later called the localization limiter and led in 1983 to the crack band model (Bažant and Oh, 1983) for fracture of quasibrittle materials.

The foregoing argument can also be inverted. By determining experimentally the point, F, of stability loss on the softening curve (Fig. 4.7c), one can deduce the value of h, i.e., the localization limiter (e.g., the proper width of a crack band). To this end one must, of course, correctly determine the post-peak softening curve $P(u_h)$ as modified by the bifurcations. Since E_t depends on $\epsilon = u_h/h$, and the dependence may be unknown a priori, the test for point F would have to be conducted at several scaled sizes (and repeated to counter the random scatter).

4.3   Crack Band Model

4.3.1   Basic Concepts

The simplest localization limiter is the *crack band model*. The general idea of minimum admissible finite element size as a materials property was advanced in 1976. In the late 1970s it was developed for stress-strain models with sudden post-peak stress drop (Bažant and Cedolin, 1979, 1980; Cedolin and Bažant, 1980), and in 1982 for gradual softening stress-strain models (Bažant, 1982; Bažant and Oh, 1983). We will now present the basic features of the crack band model by using a simple uniaxial model while the full details of three-dimensional analysis (except for some later updates) can be found in Chapter 8 of Bažant and Planas (1998).

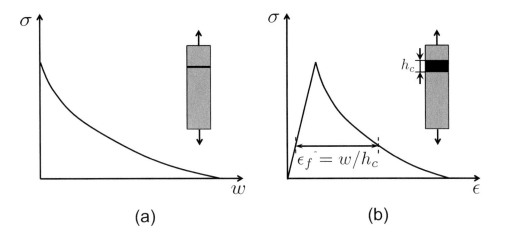

Fig. 4.8 Correspondence between the cohesive crack model and the crack band model: (a) Cohesive stress-separation law, and (b) stress-fracturing strain curve.

Consider again the aforementioned homogenous bar under uniaxial tension. The essential idea of the crack band model is that the element size and the post-peak softening law must be related and, in particular, that there exists a certain minimum finite element size, h_c, that is a material property. By setting $h = h_c$. In the uniaxial setting, the elongation of the element due to fracture can thus be expressed as

$$w = h_c \epsilon_f \tag{4.9}$$

The damage in the element can be approximately represented by the formation of a cohesive line crack (with an FPZ of finite width in front). The fracturing relative displacement w may then be approximated as the opening displacement (or separation) of a cohesive line crack. The curve of stress versus fracturing strain, $\sigma = \phi(\epsilon_f)$, of the crack band model (Fig. 4.6b) can thus be related to the stress-separation law, $\sigma = f(w)$, of the cohesive line crack, i.e.

$$\phi(\epsilon_f) = \phi(w/h_c) = f(w) \tag{4.10}$$

We have thus introduced, for the elastic-softening behavior, a unique relationship between the cohesive line crack and the crack band model defined by a uniaxial softening stress-strain relation (which is, of course, a simplification of a actual tensorial stress-strain relation for the crack band; see Figs. 4.8a and b). These two models must also be made equivalent in terms of energy analysis. For the crack band model, the strain localization indicates that the total work required to break the bar is equal to

$$\mathcal{W}_F = h_c A \bar{W}_F \tag{4.11}$$

where A = cross-sectional area of the bar, and $\bar{W}_F$ is the effective (or average) fracturing work per unit volume or fracturing energy density (dimension J/m^3 or N/m^2). For the cohesive crack model, this energy can now be used to calculate the fracture energy

$$G_F = \mathcal{W}_F/A = h_c \bar{W}_F \tag{4.12}$$

By substituting Eq. 4.4 into Eq. 4.12, we obtain

$$G_F = h_c \int_0^\infty \phi(\epsilon_f) \mathrm{d}\epsilon_f \tag{4.13}$$

Together with Eq. 4.10, we can rewrite Eq. 4.13 as

$$G_F = \int_0^\infty f(w) \mathrm{d}w \tag{4.14}$$

As indicated by Eq. 4.14, the fracture energy is given by the area under the cohesive law, which is a fundamental property of the cohesive crack model discussed in Chapter 3. Using the J-integral, as in Sec. 3.6, one can again show that G_F is equal to the flux of energy into the crack band front (flux with respect to crack band length rather than time).

It is now clear that the cohesive crack model can be translated into an energetically and globally equivalent crack band model, though not vice versa if the crack band is defined tensorially. However, these two models differ locally, near the crack band front, and have different transverse strain and displacement profiles along the crack.

Consider again a uniaxial tensile specimen (Fig. 4.9). The cohesive crack model features a displacement jump across the crack, and thus the transverse strain profile is a Dirac delta function. The crack band model, on the other hand, features a continuous gradual increase of the displacement across the crack band and the strain exhibits sudden jump at the band boundary. This difference becomes unimportant for most engineering practice because usually the structure size is sufficiently larger than the crack band width, and also because the damage in the FPZ has high random scatter whose deterministic approximation is fraught with ambiguity, which is the same for the crack band and line-crack cohesive models.

4.3.2   Rescaling of Softening Law for Increased Finite Element Mesh

As indicated by the foregoing analysis, the crack band model requires h_c to be a material constant, which is of the order of a few inhomogeneity sizes. This imposes a restriction on the mesh size for the FE analysis. In a general case, all the finite elements in the structural domain that does not stay elastic would have to be of size h_c, which could lead to excessive computational burden for the large structures. To overcome this burden, we note that often it is clear in advance that the fracturing damage will localize into one crack or one crack band, and what then matters for failure behavior is the fracture energy of the band. So, for a larger (or smaller) finite element size h', the energy dissipated by fracture in one element, $h_c W_F$, must be preserved, that is,

$$h' W_F' = h_c W_F \quad \text{or} \quad W_F' = \frac{h_c}{h'} W_F \tag{4.15}$$

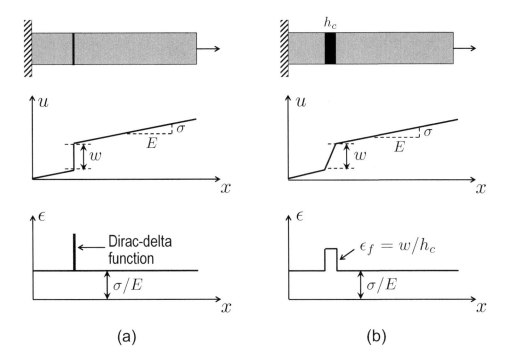

Fig. 4.9 Analysis of uniaxial tension specimen by using (a) cohesive crack model, and (b) crack band model.

---

where W'_F is the mean density of dissipated fracture energy compatible with finite element size h' ("mean"—because the dissipated energy density has a certain random statistical distribution across the element, imagined as bell-shaped). Eq. 4.15 indicates that the elastic-softening stress-strain curve input for the finite element must depend on the element size, and shows that it must be rescaled as

$$\epsilon_{f,e} = \frac{h_c}{h_e}\,\epsilon_f \tag{4.16}$$

Figs. 4.10a and b exemplify the aforementioned rescaling for the simple case of a linear softening stress-strain curve. But we must note that if

$$h_0 > 2\,\frac{EG_F}{f_t^2} = 2l_{ch} \tag{4.17}$$

where l_{ch} is Irwin's characteristic length (Sec. 3.2), the rescaling would lead to a snapback on the effective stress-strain curve (Fig. 4.10c) (Bažant and Cedolin, 1991,

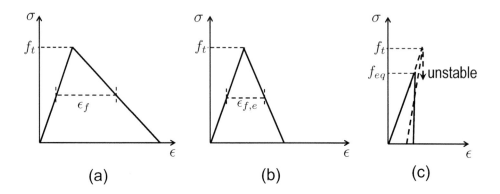

Fig. 4.10 Rescaling of stress-strain curve for crack band model: (a) element size h_c, (b) element size $h_c < h_e < EG_F/f_t^2$, and (c) element size $h_e > EG_F/f_t^2$.

Chap. 4). This would cause stability loss in which the stress would suddenly drop to zero (dynamically, with inertia forces), and the energy dissipation in the element would no longer equal G_F. A simple solution to this problem is to replace the snapback stress-strain diagram by a stress-strain diagram with the same area but a vertical stress drop (Fig. 4.10c) (Bažant and Cedolin, 1979; Bažant, 1985a,b). The only way to do that is to reduce the tensile strength to an equivalent strength f_{eq}, as follows

$$h\frac{f_{eq}^2}{2E} = G_F \quad \text{or} \quad f_{eq} = \sqrt{\frac{2EG_F}{h}} \tag{4.18}$$

It must be emphasized that the rescaling is appropriate only if it known that initial distributed damage will actually localize. This is typical of many structures, e.g., shear failures of reinforced concrete beams, slabs, and shear-walls, or usually cracks in composite airframes. However, there exist problems where the distributed cracking may or may not localize, and then the rescaling may be inappropriate. These problems , for example, the hydraulic fracturing of gas shale, where localization may be prevented by the surrounding field of high triaxial compression. In these cases, a more general energy regularization approach is needed. It has recently been proposed to adjust the rescaling of softening law based on the damage pattern of the structure (Gorgogianni *et al.*, 2020). This approach was successfully used for simulations of static and dynamic failures of quasibrittle structures, which involve both localized and diffused damage patterns.

4.3.3   Compatibility of Energy Density, Softening Law and Element Size

Let us now choose element size h_{ref} as the reference size, different from h' and h_c. An equation of the same form as Eq. 4.15 must again hold, i.e., $h_cW_F = h_{ref}W_{F,ref}$. Its substitution into Eq. 4.15 gives $h_{ref}W_{F,ref} = h'W'_F$ or

$$W_{F,ref} = \frac{h_{ref}}{h'} W'_F \qquad (4.19)$$

This equation reveals an important feature of crack band model—any element size h_{ref} (within a reasonable range) can be chosen as the *reference* size.

Ideally, all the finite elements in the structural domain that undergoes fracturing ought to have this same element size, h_{ref}, in which case the softening law remains unchanged. But again, for computational reasons already mentioned, the element size can be changed to another size h', larger or smaller than h_{ref}. Then, in analogy to Eq. 4.16, the stress-strain curve must be adjusted as

$$\epsilon'_f = \frac{h_{ref}}{h'} \epsilon_{f,ref} \qquad (4.20)$$

This result means that one has considerable freedom in the choice of element size. In one aspect, however, there is no freedom. The softening law for the chosen element size h_{ref} must be energetically compatible with the mean energy dissipation density $W_{F,ref}$ in the finite element, so as the give the correct fracture energy G_F for a localized crack (assuming that the fracturing localizes into one dominant crack).

4.3.4 Calibration of Crack Band Width or Element Size

The effective width, h_c, of the crack band, which is the effective width of the fracture process zone, cannot be obtained from the standard fracture tests because the data from such test can be fitted equally well with a range of finite element sizes h. Elaborate experimental investigations of the damage band width (e.g., Otsuka and Date (2000)) show that the statistical density of damage decreases from the centerline very gradually, which makes extracting some effective width ambiguous. Even if the size effect in fracture specimens is measured, what matters for the size effect is mainly the length of the fracture process zone, not the crack band width. The salient problem is how to determine h_c or h_{ref}. There are diverse choices:

1. One choice, the basic one, is to determine h_c as the minimum possible spacing of parallel cracks. This can be done in several ways.
   (a) Short parallel cracks can be experimentally observed at the start of cooling or drying from the surface. But the interpretation is hard because such cracks localize as soon as their length exceeds about 1.8 times their spacing (Bažant and Ohtsubo, 1977; Bažant and Cedolin, 1991).
   (b) A theoretical idea to calculate the initial spacing appeared in relation to vertical cracks in a pavement. The h_c can be solved from three conditions: (i) The strength limit f_t at the surface has been reached; (ii) distributed cracking coalesces into a distinct crack of a certain crack length a_0 for which the energy release rate per crack $\mathcal{G} = G_F$; and (iii) a crack is imagined to grow from 0 length to a_0 as the FPZ develops, and the total energy per crack released during this growth according to LEFM is set to equal G_F. This leads to a singular integral equation which is easy to solve numerically. But the physical justification of the third condition would need to be clarified by a particle model such as the Lattice Discrete Particle Model.

(c) Long nonlocalized parallel cracks can be obtained by hydraulic fracturing if the fluid injection into the cracks is so slow that the fluid pressure becomes almost uniform along the cracks, or if there is a high compression in front of the cracks, or both. This was demonstrated by simulating the fracking of shale (Rahimi-Aghdam *et al.*, 2019). But such conditions are hard to achieve in a laboratory.

2. Another test can be based on preventing localization in a tensioned concrete layer restrained by bonded steel plates 1989. But it is not clear whether the FPZ can fully develop under such restraint.

3. Still another rough estimate of h is provided by Eq. 4.5 and explained in Fig. 3.1c.

4. The problem of optimal h changes in practical situations in which there may be complex triaxial stress states at the crack front, including crack-parallel stresses (Sec. 4.1), or if the available material tests are on a scale larger than the Representative Volume Element (RVE), while the scale of structural application is still much larger. In that case one need not aim at getting h_c. Rather, one needs to determine the h_{ref} that is *compatible* with the energy density W_{ref} dissipated in calibration tests even if that h_{ref} may be much larger than h_c. For calibration, we assume to possess (a) a sufficient set of uni-, bi-, and tri-axial material tests, which preferably though not necessarily fracture specimen tests, and (b) one or more laboratory-scale tests of structural failure that we want to extrapolate by crack band model to large sizes or other geometries. For the purpose of this extrapolation, the following three-step procedure may be used:

> **Step 1** Using the same constitutive damage law, first run repeated optimum fits of all the calibration tests, which consist of both (a) and (b) mentioned above. Since the material tests correspond to a much smaller scale than the laboratory structure tests, the quality of these fits must depend on the choice of h.
>
> **Step 2** Then find h_{ref} as the element size h for which the fit of *all* these tests is the closest in the sense of minimizing the coefficient of variation of fitting errors.
>
> **Step 3** Use this h_{ref} to predict the response for larger sizes or different geometries.

Note that all these calibration and extrapolation procedures depend on having a good damage constitutive law. If this law is not sufficiently realistic, e.g., if it does not capture the crack-parallel stress effect (Sec. 4.1), even the optimal calibration fits will not be close. Thus the problem of determining the correct h is inseparable from the problem of finding a realistic tensorial damage constitutive law for the crack band model.

An example of the foregoing procedure is the recent prediction of size effect on the strength of multi-story building shear walls (Rasoolinejad and Bažant, 2019). Extensive data on laboratory tests on reinforced concrete single-story shear walls were fitted with different h using microplane model M7 whose constitutive law was previously shown to match a broad range of material test data (Caner and Bažant, 2013). The element size h_{ref} giving overall the best fit was identified and then used to predict the strength of various kinds of multistory shear walls.

An example of necessity of using h_{ref} much larger than the RVE is the analysis of branching of hydraulic crack in a deep stratum of shale. Using finite elements of RVE size, which roughly equals the FPZ size and is about 2 to 3 mm, would lead to a prohibitive computational burden. But finite elements of size $h_{ref} = 100$ m may be used if the strength and softening law of shale is adjusted to give an approximately correct value of the fracture energy, G_F.

There are cases in which material test data sufficient for determining the constitutive material parameters are lacking. In such cases, one can first choose a convenient mesh size h_{ref} for which calibration of the material parameters by optimum fitting of the available test results for laboratory-scale structures leads to the closest fit. This h_{ref} and the constitutive parameters calibrated in this way may then be used in the crack band model to simulate specimens or structures of different sizes and geometries.

Note that adjustment of the softening damage law for an increased element size is only approximate and thus introduces some error. Therefore it is more accurate, except in elastic regions, to use the same element size for all structure sizes. Besides, this also allows handling situations in which closely spaced parallel cracks may be stable.

4.3.5   Implementation in General FE Analysis

Having compared the crack band model and the cohesive crack model, we should emphasize that the main advantage of the crack band model is the possibility of using complex tensorial damage constitutive laws. Compared to the scalar cohesive laws used in the cohesive crack model, the tensorial constitutive models are capable of capturing the influence of all stress components, among which the normal compressive stresses (Sec. 4.1) are most important since they can promote fracture and can alone cause splitting fracture. The shear components, on the other hand, may cause a change in crack band growth direction. These important features cannot be captured by the cohesive crack model.

Since, for a realistic but complex constitutive law, it is quite difficult to identify which material parameters should be adjusted to regularize the energy dissipation in the softening regime, an adjustment external to the constitutive law was formulated. The material properties and parameters are kept unchanged while the crack band is coupled with fictitious springs (or elastic elements) in the principal stress directions as shown in Fig. 4.11 (Cervenka *et al.*, 2005). These springs change the slope of the post-peak. In the FE analysis, the added spring has the material's elastic properties but the effective length of element equivalent to the spring is larger, possibly much larger, than the crack band size. During the fracture process, the energy stored in the elastic spring is transferred to the crack band and gets dissipated. It is with this algorithm that the crack band model has been implemented (together with complex damage constitutive law) in commercial finite element softwares such as ATENA.

The crack band model gives best results when the mesh is aligned with the direction of crack propagation. If the direction of propagation is not known, the crack band may propagate in a zig-zag manner (Fig. 4.12a). In this case, the accuracy is considerably impaired since the development of shear stresses in the direction parallel to the overall zig-zag band causes some degree of interlocking. The error can be sig-

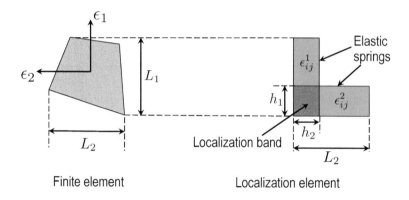

Fig. 4.11 2D representation of elastic spring-crack band arrangement.

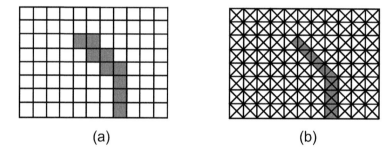

Fig. 4.12 Modeling of crack growth inclined to mesh lines: (a) zig-zag propagation path in a square mesh, and (b) triangular mesh for a better representation of crack path.

nificantly mitigated by using a triangular mesh consisting of horizontal, vertical, and diagonal mesh lines (Fig. 4.12b). The triangular mesh gives a better approximation of arbitrary fracture propagation direction in the case where the actual propagation direction is unknown. An alternative remedy, proposed in (Bažant, 1985c) and adopted in commercial softwares such as ATENA (Cervenka, 1998), is to introduce an empirical correction factor in terms of the angle between the mesh line and crack band direction.

Among various types of nonlinear computational fracture models, the crack band model is what now dominates for concrete in structural and geotechnical engineering industry and for composites in the aircraft industry, and is also an embedded automatic feature of various commercial softwares (e.g. SBETA, ATENA, DIANA) and open-source code OOFEM, or is facilitated as in ABAQUS.

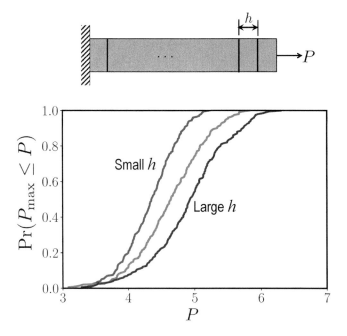

Fig. 4.13 Simulated probability distributions of peak load capacity of a uniaxial tension specimen with different discretizations.

4.3.6 Recent Development for Stochastic Computations

The concept of the crack band model has recently been extended to stochastic FE simulations of quasibrittle fracture (Le and Eliáš, 2016). In stochastic analysis, the material properties, such as strength and fracture energy, are considered to exhibit random spatial variability. Consider again the previous analysis of a uniaxial tension specimen divided into n elements as shown in Fig. 4.6. We now assume that each element has a random strength which follows a certain probability distribution function. It is evident that the strain localization would occur in the element with the smallest strength value, and the peak load capacity of the chain is dictated by the strength of the weakest element. Therefore, the probability distribution of the strength of the chain can be calculated by using the weakest-link model. As will be discussed in Chapter 6, the weakest-link model predicts that the strength distribution of the chain would depend on the number of the elements in the chain, as shown in Fig. 4.13.

Similar to aforementioned dependence of post-peak behavior on the number of elements, the result shown in Fig. 4.13 is physically unacceptable since the strength distribution of the bar cannot depend on the choice of the division. It is noted that this feature occurs only in stochastic analysis. In previous deterministic analysis (Fig. 4.6), the bar strength is independent of the number of elements because every element has the same strength. This simple analysis indicates that stochastic FE simulations of quasibrittle fracture would suffer a more severe issue of spurious mesh dependence as compared to the deterministic case. As discussed in Sec. 4.2, in deterministic analysis the mesh dependence issue arise from the lack of energy regularization. For stochastic analysis, besides the issue of energy regularization, the other issue is the lack of consideration of the mesh dependence of strength distribution when the element size is larger than the crack band width.

In a recent study, Le and Eliáš developed a probabilistic crack band model to mitigate the mesh dependence issue in stochastic simulations (Le and Eliáš, 2016). The essential idea is to adjust the probability distribution functions of the tensile strength and fracture energy density according to the mesh size. In addition to the aforementioned regularization of energy dissipation, which is the essence of the conventional crack band model, this probabilistic model also takes into account the random onset of damage localization inside one finite element. The random location of the crack band is related to the randomness of the tensile strength, which is described by a finite weakest-link model.

4.4 Nonlocal Integral and Gradient Models

The most general method to handle distributed softening damage and control its localization instability is through nonlocal damage models. The nonlocal concept in continuum mechanics was proposed in 1960s. The early nonlocal models were introduced to capture, in a continuous manner, the sizes and spacing of material inhomogeneities in elastic and plastic-hardening materials (Eringen, 1966; Kröner, 1966, 1967; Eringen, 1972; Eringen and Edelen, 1972). These models are not be able to regularize spurious damage localization.

The first and simplest nonlocal model for softening damage was proposed in early 1980s (Bažant *et al.*, 1984). In this model, the stress tensor at point $\boldsymbol{x}$ is calculated from the nonlocal continuum strain tensor $\bar{\boldsymbol{\epsilon}}(\boldsymbol{x})$, which is defined as the weighted average of the strain tensors of a certain neighborhood of point $\boldsymbol{x}$:

$$\bar{\boldsymbol{\epsilon}}(\boldsymbol{x}) = \frac{1}{V_\alpha(\boldsymbol{x})} \int_{V_0} \alpha(\boldsymbol{x} - \boldsymbol{x}')\boldsymbol{\epsilon}(\boldsymbol{x}')\mathrm{d}V(\boldsymbol{x}') \tag{4.21}$$

where $\alpha(\boldsymbol{x} - \boldsymbol{x}') =$ weighting function, $V_\alpha(\boldsymbol{x}) =$ volume under the weighting function, and $V_0 =$ volume of the neighborhood region for which the weighting function takes effect. The simplest weighting function is the uniform distribution function defined over a neighborhood of size l (Fig. 4.14). Since the nonlocal integral is normalized by the volume of the weighting function, the weighting function can be scaled by an arbitrary factor. It is convenient to scale $\alpha(\boldsymbol{x} - \boldsymbol{x}')$ such that $\alpha(0) = 1$ (Fig. 4.14).

It was shown that the numerical convergence can be improved slightly if a smooth bell-shaped weighting function is chosen. An effective choice of such a weighting function is given by Bažant (1990b):

$$\alpha(\boldsymbol{x} - \boldsymbol{x}') = \begin{cases} \left[1 - \left(\dfrac{|\boldsymbol{x} - \boldsymbol{x}'|}{\rho_0 l}\right)^2\right]^2 & (|\boldsymbol{x} - \boldsymbol{x}'| < \rho_0 l) \tag{4.22a} \\ \\ 0 & (|\boldsymbol{x} - \boldsymbol{x}'| \geq \rho_0 l) \tag{4.22b} \end{cases}$$

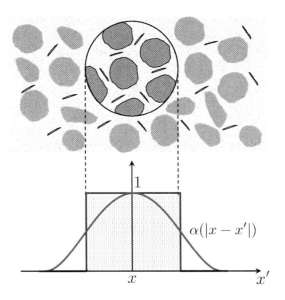

Fig. 4.14 Nonlocal spatial averaging: uniform and smooth decaying weighting functions.

where ρ = a constant chosen such that the volume of weighting function $\alpha(\boldsymbol{x} - \boldsymbol{x}')$ is equal to that of the uniform distribution function (Fig. 4.14). Based on this requirement, we have $\rho = 0.9375$ in 1D, 0.9086 in 2D, and 0.8178 in 3D.

As mentioned earlier, the nonlocal concept was initially developed for elastic heterogenous materials until finite element studies revealed that the spatial averaging of elastic strain could cause spurious instabilities Bažant *et al.* (1984). To demonstrate this, consider a long bar with a nonlocal uniaxial elastic stress-strain relation:

$$\sigma(x) = E\bar{\epsilon}(x); \qquad \bar{\epsilon}(x) = \int_{-\infty}^{\infty} \epsilon(x+r)\alpha(r)\mathrm{d}r \qquad (4.23)$$

where E = Young's modulus, $\epsilon(x)$ = local strain, $\bar{\epsilon}(x)$ = nonlocal strain, $\alpha(r)$ = weighting function scaled so that $\int_{-\infty}^{\infty} \alpha(r)\mathrm{d}r = 1$. Any theory of an elastic continuum must satisfy the following requirements: (a) if the stresses are everywhere zero, the strain in the material must be zero as well, i.e. there is no zero-energy deformation mode; and (b) the wave propagation speed must be real.

Requirement (a) indicates that, when $\sigma(x) = 0$, we have

$$\int_{-\infty}^{\infty} \epsilon(x+r)\alpha(r)\mathrm{d}r = 0 \qquad (4.24)$$

Eq. 4.24 must not have any non-trivial solution of $\epsilon(x)$. We note that a general strain distribution can be approximated by $\epsilon(x) = \sum_{k=1}^{\infty} a_k \exp(i\omega_k x)$, where a_k, ω_k are real numbers and $i = \sqrt{-1}$, and the actual strain is represented by the real part of this series. We want to ensure that no single term of this series would satisfy Eq. 4.24, i.e. $\int_{-\infty}^{\infty} \alpha(r) \exp[i\omega(x+r)] \mathrm{d}r = 0$ must not have any solution. By dividing this equation by $\exp(i\omega x)$, we obtain the following condition Bažant and Chang (1984):

$$\alpha^*(\omega) = \int_{-\infty}^{\infty} \alpha(r) \exp(i\omega r) \mathrm{d}r \neq 0 \quad \text{for all real } \omega \qquad (4.25)$$

In other words, the Fourier transform of the weighting function, $\alpha^*(\omega)$ must be either positive everywhere or negative everywhere.

For requirement (b), we limit our attention to the case of small-deformation, where the local strain field can be expressed by $\epsilon(x) = \partial u(x)/\partial x$ ($u(x) = $ displacement field). The equation of motion can be written as

$$\frac{\partial \sigma}{\partial x} = \rho \frac{\partial^2 u}{\partial t^2} \qquad (4.26)$$

where $\rho = $ mass density. Substituting Eq. 4.23 into Eq. 4.26, we obtain

$$E \frac{\partial}{\partial x} \int_{-\infty}^{\infty} \frac{\partial u(r)}{\partial s} \alpha(x - s) \mathrm{d}s = \rho \frac{\partial^2 u}{\partial t^2} \quad (s = x + r) \qquad (4.27)$$

We note that any wave may be decomposed into harmonic components of the type $u(x) = a \exp[i\omega(x - vt)]$, where $v = $ wave velocity and $\omega, a = $ real constants ($\omega \neq 0$). Substitution of this displacement field into Eq. 4.27 yields

$$\frac{E}{\rho} \frac{\partial}{\partial x} \left[e^{i\omega x} \int_{-\infty}^{\infty} e^{i\omega r} \alpha(-r) \mathrm{d}r \right] = v^2 i \omega e^{i\omega x} \qquad (4.28)$$

Since the weighting function is symmetrical, i.e. $\alpha(-y) = \alpha(y)$, Eq. 4.28 reduces to (Bažant and Cedolin, 1991, Sec. 13.10); (Bažant and Chang, 1984):

$$v^2 = \frac{E}{\rho} \alpha^*(\omega) \qquad (4.29)$$

Therefore, for the wave velocity to be real, we require that $\alpha^*(\omega)$ to be positive for all $\omega \neq 0$. Combining the condition derived for requirement (a), we conclude that $\alpha^*(\omega)$ must be positive for all ω values.

One can show that in fact many weighting functions do not satisfy the requirement that their Fourier transform be positive everywhere. For example, both the aforementioned uniform and triangular weighting functions would violate the aforementioned two requirements. One remedy is to add a spike of the Dirac delta function to the uniform or triangular weighting function at $r = 0$ (Bažant *et al.*, 1984; Bažant, 1986). Such a model physically implies an overlay of a local elastic continuum onto a nonlocal continuum model. However, this overlay could prevent strain-softening from reducing stress to zero (Bažant and Planas, 1998). In the subsequent studies, it was found that the need for using the overlay of local elastic continuum is an unrealistic artifice and it is also unnecessary.

The root of aforementioned spurious instability is that there are multiple elastic strain fields that satisfy Eq. 4.21 since the strain can accept alternating solutions. Therefore, the solution is to use a nonlocal continuum with local elastic strain, in which only the softening damage strain or its parameter is subjected to nonlocal averaging of the type of Eq. 4.21 (Pijaudier-Cabot and Bažant, 1987; Bažant and Pijaudier-Cabot, 1988; Bažant and Jirásek, 2002). Besides the softening damage strain, nonlocal averaging procedure can also be used for other state variables of the constitutive model that characterize the damage process (Bažant and Jirásek, 2002). In these models, the nonlocal variables are positive and ever increasing due to the irreversibility of the damage process and consequently no arbitrary solution could exist. It has been shown that, for quasibrittle materials, the nonlocal averaging of inelastic strain or other damage-related variables can be physically explained by the interaction of microcracks (Bažant and Ožbolt, 1990; Bažant and Planas, 1998; Bažant and Jirásek, 2002). A nonlocal model based on microcrack interactions has also been developed (Bažant, 1994), but its implementation in a nonlocal FE program (Ožbolt and Bažant, 1996) has been too complicated.

In addition to the integral-type nonlocal models (Eq. 4.21), called also strongly nonlocal, there also exist nonlocal gradient models, which are called weakly nonlocal and take into account only the close vicinity of the material point. Early attempts led to explicit-gradient models (Aifantis, 1984, Bažant, 1984a), which can be derived from the Taylor series expansion of the nonlocal integral models. Consider the integral-type nonlocal averaging of some variable κ:

$$\bar{\kappa}(x) = \int_0^\infty \alpha(r)\kappa(x+r)\mathrm{d}r \qquad (4.30)$$

The Taylor series expansion of the local variable $\kappa(x+r)$ is

$$\kappa(x+r) = \kappa(x) + \frac{\partial \kappa}{\partial x}r + \frac{\partial^2 \kappa}{2\partial x^2}r^2 + ... \qquad (4.31)$$

By substituting Eq. 4.31 into Eq. 4.30 and noting that $\int_0^\infty \alpha(r) = 1$ and $\alpha(r)$ must be an even function, we obtain

$$\bar{\kappa}(x) = \kappa(x) + c\nabla^2 \kappa \qquad (4.32)$$

where $c \propto l^2$ and l = material characteristic length defining the size of the nonlocal averaging neighborhood. It is seen from Eq. 4.32 that the nonlocal variables can be calculated from the explicit second-order derivative of the corresponding local variable. Due to the truncation of the second-order derivative, the explicit gradient models take into account only the near-field nonlocal interaction. Besides, the inclusion of the second-order derivative also imposes challenges in numerical implementation; e.g., higher-order finite elements are required.

A more robust approach is to use implicit gradient models, in which the nonlocal variables are calculated by solving a second-order differential equation Peerlings *et al.* (1996, 2001). The implicit gradient models can also be derived from the nonlocal integral model. Consider again the Taylor expansion of the local quantity κ (Eq. 4.31).

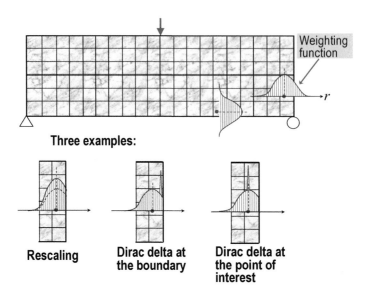

Fig. 4.15 Treatments of weighting function when nonlocal averaging zone protrudes structure boundary.

Instead of truncating at the second term, we now write the full Taylor expansion of the nonlocal variable $\bar{\kappa}$:

$$\bar{\kappa}(x) = \kappa(x) + c_1\nabla^2\kappa(x) + c_2\nabla^4\kappa(x) + \ldots \tag{4.33}$$

where $c_1 \propto l^2$ and $c_2 \propto l^4$. By applying the Laplacian and multiplying with c_1 on both sides of Eq. 4.33, we get

$$c_1\nabla^2\bar{\kappa}(x) = c_1\nabla^2\kappa(x) + c_1^2\nabla^4\kappa(x) + c_1c_2\nabla^6\kappa(x) + \ldots \tag{4.34}$$

By subtracting Eq. 4.34 from Eq. 4.33 and ignoring the terms with high-order Laplacian, we obtain

$$\bar{\kappa}(x) - c_1\nabla^2\bar{\kappa}(x) = \kappa(x) \tag{4.35}$$

This differential equation is supplemented with the Neumann boundary condition $\mathbf{n}\nabla\kappa = 0$, where $\mathbf{n}$ is the unit normal vector of the structure boundary. In FE simulations, Eq. 4.35 can be solved by using the same mesh as that used for the equilibrium equations except that an additional degree of freedom needs to be introduced at each node for solving the nonlocal variable $\bar{\kappa}$ (Peerlings *et al.*, 1996).

For all integral-type nonlocal models, a difficulty arises near the structure boundary where the averaging neighborhood tends to protrude outside the boundary (Fig. 4.15) or overlap across a crack. Various ad hoc convenient assumptions were introduced. For example, one may rescale the weighting function within the body (Bažant and Jirásek,

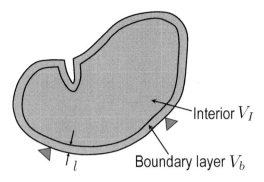

Fig. 4.16 Concept of boundary layer model

2002) or place a Dirac delta function either along the structural boundary, or at the center point of the nonlocal integral (Borino *et al.*, 2003). However, these approaches were not based on physical arguments.

The problem with the boundary can be eliminated by a recently formulated boundary layer model. In this model, a boundary layer of thickness l_1 is excluded from the body (Fig. 4.16) (Bažant *et al.*, 2010; Bažant and Le, 2017). The nonlocal averaging procedure is then performed only for all material points in the interior V_I. In the boundary layer, the average quantities (e.g. strain, stress, damage) are calculated along the thickness direction of the layer, and the nonlocal averaging of these average quantities is then performed only along the boundary layer. This model implies a diminishing nonlocal effect when the point of interest gets closer to the boundary. Similar concept has recently been proposed in several other nonlocal models (Krayani *et al.*, 2009; Havlásek *et al.*, 2016).

Similarly, the Neumann boundary condition for the implicit gradient model has not been physically justified, which is analogous to the aforementioned issue of the treatment of the nonlocal weighting function near the structural boundary. The boundary layer model implies that Eq. 4.35 should be accompanied by a mixed boundary condition.

The gap test has recently injected a new problem into the nonocal and gradient models. These tests suggest that the nolocal characteristic lengths in the direction of damage propagation and transverse to it should be different, and that their magnitude might depend on the crack-parallel stresses. This has emerged as a major problem, to be addressed in future research. These tests also indicate that the band width in phase field models might have to vary with these stresses.

4.5 Discrete Computational Models

With the drastic increase in computer power, discrete computational models have attracted much attention over the past decade. The discrete models, in the form of either lattice or particle models, offer a realistic approach to simulation of softening damage in quasibrittle materials.

A discrete lattice model was proposed by Hrennikoff in 1941 for elasticity problems. Various other forms were later developed to handle quasibrittle materials (Zubelewicz and Bažant, 1987; Bažant *et al.*, 1990; Schlangen and van Mier, 1992; Jirásek and Bažant, 1995*a,b*; Cusatis *et al.*, 2003*a,b*). The discrete particle model was preceded by the so-called distinct element model developed for particulate materials such as cohesionless soils and rock masses (Cundall, 1971; Cundall and Strack, 1979). The model was further extended to concrete materials (Zubelewicz and Bažant, 1987), in which the material domain is discretized by a set of rigid polyhedral elements connected by nonlinear springs of zero length (Kawai, 1978; Bolander and Saito, 1998; Bolander *et al.*, 2000). All these discrete models involve some characteristic length scale (e.g. particle size or lattice size), which effectively serves as a localization limiter regularizing the energy dissipation in the fracture process. In discrete particle models, the particles could either represent the actual material inhomogenieties (e.g. aggregates in concrete materials) or simply be used as an efficient means to introduce a length scale to the model (Jirásek and Bažant, 1995*a,b*).

The constitutive behavior of the lattice or the interaction of the particles plays a central role in discrete element modeling. The simplest model is to consider axial deformation only. However, such a model would predict a fixed Poisson ratio of the homogenized continuum, $\nu = 1/3$ in 2D and $\nu = 1/4$ in 3D, while the simulated damage band also appears to be too narrow. These issues can be overcome by a model in which the lattice or particle interaction features both axial and shear deformation (Zubelewicz and Bažant, 1987; Schlangen and van Mier, 1992; van Mier and Schlangen, 1993). Over the last decade, significant advances have been achieved towards a realistic discrete element model for various concrete materials including both normal concrete (Cusatis *et al.*, 2003*a,b*, 2011) and fiber-reinforced concrete (Schauffert and Cusatis, 2011; Schauffert *et al.*, 2011). The discrete element models have recently been extended to rate-dependent failure of concrete (Smith *et al.*, 2014; Le *et al.*, 2018*a*).

One attractive feature of discrete models is that they provide an explicit representation of the heterogeneities of the material, which makes them well suited to handle the random size distribution of the material heterogeneities. With further incorporation of the randomness of material properties, the discrete model has successfully been used for stochastic modeling of quasibrittle fracture (Grassl and Bažant, 2009; Eliáš *et al.*, 2015; Le *et al.*, 2018*a*).

Discrete element models usually require high computer power, which could be prohibitive, especially for analyzing large-scale structures. Aiming at high computational efficiency and at the same time retaining the ability of capturing the microstructural features, recent efforts have been directed towards the development of multiscale computational frameworks, which bridge the discrete element model and the continuum finite element model, e.g. (Rezakhani and Cusatis, 2016).

Exercises

E4.1. In the gap test (Sec. 4.1), the bending moment and the crack-parallel compression increase non-proportionally. As an alternative, one might consider avoiding the gaps and applying the crack-parallel compression by jacks. In that case the crack-parallel compression could be increased proportionally. However, it would make test evaluation much more difficult. Explain why.

E4.2. As a problem analogous to a chain under tension (Fig. 4.4), consider a long layer containing a set of n parallel shear cracks with a softening stress-slip relation, subjected to shear stress at both surfaces of the layer. The same shear stress is transmitted by each crack. Explain why only one of the cracks will propagate.

E4.3. Eq. 4.7 applies to the case of controlled displacements. What happens if the load is controlled, like a gravity weight load, and the onset of negative K_t is reached?

E4.4. Consider energy release balance in the case of parallel cooling cracks. Would a reduction of G_f by crack-parallel stress lead to smaller or greater crack spacing?

E4.5. Consider the crack band model. Calculate the critical element size for which the sudden vertical stress drop gives the correct fracture energy.

E4.6. When the crack band model is applied to a large element, the adjusted stress-strain curve could exhibit a snap-back behavior. What would be the issue associated with such a snap-back behavior, and what is the remedy?

E4.7. What is the requirement for the weighting function $\alpha(r)$ of total strain $\epsilon(x)$ of the nonlocal model to prevent zero energy modes of deformation?

E4.8. Explain the reason for applying the nonlocal operator only to inelastic strain ϵ^p or its parameters.

E4.9. Explain the difference between the integral-type and gradient-type (strongly or weakly) nonlocal models. How can the latter be derived from the former?

E4.10. Explain problems with boundary conditions in nonlocal models of both types.

E4.11. What is the difference between the explicit and implicit gradient models. What is the type of equation governing the nonlocal strain in the implicit model?

E4.12. Compare the discrete lattice truss models and the particle models and describe their difference in representing the fracture process zone.

E4.13. In presence of significant crack parallel stress σ_{xx}, can the damage law in the crack band consist of isotropic softening as a function of strain invariant I_1^ϵ only? Or of unidirectional softening as a function of ϵ_{yy} only? Can both of these laws fit the test data when $\sigma_{xx} \approx 0$?

5

Energetic Size Effect in Quasibrittle Fracture

Big and small are both similar and dissimilar, and that's the key.

In Chapter 3, we discussed the concept of nonlinear fracture mechanics. A salient feature of nonlinear fracture mechanics is that it requires both the material strength and fracture energy, which have different dimensions (N/m^2 and N/m). As a consequence, the formulation naturally involves a characteristic length, which is Irwin's length. It represents a material property independent of structure geometry and size. Its existence sets a length scale and leads to intricate size effects in the failure of quasibrittle structures.

In this chapter, we first discuss the power-law scaling of physical phenomena in the absence of characteristic length, and then present in detail two types (types 1 and 2) of the energetic deterministic size effects on the so-called nominal strength of quasibrittle structures. The type 1 size effect applies to structures failing at macrocrack initiation, and the type 2 size effect applies to structures with a large stable crack formed prior to the peak load. We will derive the laws of these two size effects by equivalent linear elastic fracture mechanics (LEFM), matching two-sided asymptotic behaviors, and alternatively also by dimensional analysis. We will compare these two size effect laws (SELs) with the existing experimental results of many quasibrittle structures.

5.1 Nominal Structural Strength and Size Effect

For the purpose of structural design as well as measurement of fracture parameters, we are particularly interested in the size effect on the maximum (or peak) load, P_m (or load capacity), of the structure. It is customary to characterize the size effect in terms of the nominal strength, σ_N, which is a load parameter with the dimension of stress. The nominal strength, σ_N, may be defined as the maximum elastic stress within the structure. But such a definition cannot be used when the structure has a sharp crack or notch because then the maximum elastic stress is always infinite.

Quasibrittle Fracture Mechanics and Size Effect: A First Course. Zdeněk P. Bažant, Jia-Liang Le and Marco Salviato, Oxford University Press. © Zdeněk P. Bažant, Jia-Liang Le, Marco Salviato 2022. DOI: 10.1093/oso/9780192846242.003.0005

Therefore, the nominal strength is generally defined as $\sigma_N = P_m/A$ where A = area of of any chosen cross-section of the structure, which must be positioned homologously when geometrically similar structures of different sizes are compared. The geometric similarity can be two-dimensional (2D) or three-dimensional (3D), and then $A \propto bD$ or D^2, respectively. The nominal strength is defined as:

$$\sigma_N = c_n P_m/bD \ \text{ for 2D}, \qquad \sigma_N = c_n P_m/D^2 \ \text{ for 3D} \tag{5.1}$$

where D is the characteristic structure size (or dimension), which must be measured along homologous lines; P_m is the maximum (or peak) load that the structure can carry, and b is the structure thickness in the transverse dimension, which is constant in the case of 2D similarity. Coefficient c_n is a dimensionless (size-independent) constant, chosen for convenience. It may be chosen so as to make σ_N represent, for example, the maximum elastic stress (if there is no singularity), or the average stress in a cross-section, or the stress at any homologous points of similar structures of different sizes. Often we simply set $c_n = 1$ and then σ_N does not represent the stress at any particular point.[1]

As will be shown, in elasticity or plasticity, σ_N is independent of D, which is the case of no size effect. In fracture mechanics, there is *size effect*, which is understood as the dependence of nominal strength σ_N on structure size D. Since only the relative changes of the structure size are considered, D can be defined as any characteristic dimension; for example, in the case of a beam as its depth or its span, or in the case of a circular plate as its diameter or thickness, or in the case of a hollow tube as its diameter or wall thickness, etc.

We should note that, in this book, the term "strength" with no modifiers is understood in the usual sense of a material property. For structures, instead of the term "nominal strength" defined by Eq. 5.1, we also write "structural strength" or "strength of structure."

5.2   Power-Law Scaling in Absence of Characteristic Length

Before delving into the scaling of quasibrittle fracture, we first discuss the power-law scaling. It is the most basic scaling law applicable to all the physical phenomena that do not involve any characteristic length. In this section, we derive the power-law scaling in the context of mechanical response of a structure subjected to external loading, although the same derivation can be applied to other physical phenomena as well.

Consider a set of geometrically similar specimens of different sizes (Fig. 5.1). We are interested in a certain response, Y, of the specimen of size D. The response Y could be the nominal structural strength, the peak load capacity of the specimen, or

---

[1] In testing, we usually prefer the 2D similarity, for two reasons: (1) the large specimens are much lighter and (2), more importantly, we exclude the size effect of thickness b. The thickness effect is generally much weaker than the size effect of D but not negligible. It includes: (1) the so called wall effect in concrete, due to a lower large-aggregate content in the boundary layer whose thickness is constant, (2) the shear-lip effect known from plasticity, and (3) the effect of 3D elastic stress singularity at intersection of crack front edge with wall surface, caused by the fact that the wall surface is almost in plane stress and crack front in the wall core in almost in plane strain (which makes a difference as the Poisson ratio is non-zero); cf. (Bažant and Estenssoro, 1979).

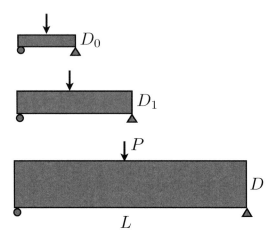

Fig. 5.1 Geometrically similar specimens.

the maximum deflection, etc. We now choose a reference specimen size D_0, and write the response of a specimen of any size D as

$$Y(D) = Y_0 f(D/D_0) \tag{5.2}$$

where Y_0 is the response of the specimen of reference size D_0.

As in important point, when no characteristic length is involved, we may alternatively choose another specimen size D_1 as a reference. The response Y can then be written as

$$Y(D) = Y_1 f(D/D_1) \tag{5.3}$$

where Y_1 is the response of specimen of size D_1. Based on Eqs. 5.2 and 5.3, and noting that $Y_1 = Y_0 f(D_1/D_0)$, we obtain a functional equation

$$f(D/D_1) = \frac{f(D/D_0)}{f(D_1/D_0)} \tag{5.4}$$

To solve Eq. 5.4, we note that if it is valid for every interval $(D, D + \delta D)$, it is valid for all D. So we differentiate Eq. 5.4 with respect to D and set $D_1 = D$. This yields

$$\frac{\mathrm{d}f(\lambda)}{f(\lambda)} = s\frac{\mathrm{d}\lambda}{\lambda} \tag{5.5}$$

where $\lambda = D/D_0$ and $s = \mathrm{d}f(\lambda)/\mathrm{d}\lambda|_{\lambda=1}$. By integrating Eq. 5.5 and imposing the initial condition that $f(1) = 1$, we obtain a power-law scaling function $f(\lambda)$, that is,

$$f(\lambda) = \lambda^s \tag{5.6}$$

This is why the power-law function plays a fundamental role in the mathematical description of scaling behavior of all physical phenomena in which there is no characteristic length.

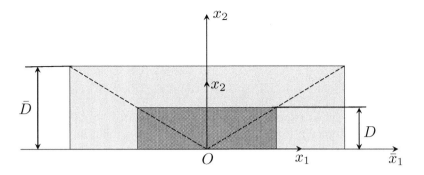

Fig. 5.2 Polar affinity transformation of geometrically similar specimens.

To apply the power-law scaling relation to stress analysis of structures, we note that scaled geometrically similar structures of different sizes can always be obtained from the reference structure (after proper positioning) by means of polar affinity transformation (Fig. 5.2):

$$\bar{x}_i = \lambda x_i \tag{5.7}$$

where x_i are the Cartesian coordinates for the reference structure of size D_0, $\bar{x}_i$ are the coordinates for a geometrically similar scaled structure of size D. For the sake of brevity, we will denote $\partial/\partial x_i = \partial_i$, $\partial/\partial \bar{x}_i = \bar{\partial}_i$. From the chain rule of differentiation, $\partial_i = \lambda \bar{\partial}_i$, $\bar{\partial}_i = \lambda^{-1} \partial_i$.

For the reference structure of size D_0 and the similar scaled structure of size D, the equilibrium equations and the boundary conditions are

For D_0:

$$\partial_j \sigma_{ij} + f_i = 0 \tag{5.8}$$

$$\varepsilon_{ij} = (\partial_j u_i + \partial_i u_j)/2 \tag{5.9}$$

$$\sigma_{ij} n_j = p_i \quad \text{on } \Gamma_1 \tag{5.10}$$

$$u_i = U_i \quad \text{on } \Gamma_2 \tag{5.11}$$

For D:

$$\bar{\partial}_j \bar{\sigma}_{ij} + \bar{f}_i = 0 \tag{5.12}$$

$$\bar{\varepsilon}_{ij} = (\bar{\partial}_j \bar{u}_i + \bar{\partial}_i \bar{u}_j)/2 \tag{5.13}$$

$$\bar{\sigma}_{ij} \bar{n}_j = \bar{p}_i \quad \text{on } \bar{\Gamma}_1 \tag{5.14}$$

$$\bar{u}_i = \bar{U}_i \quad \text{on } \bar{\Gamma}_2 \tag{5.15}$$

in which σ_{ij} and ε_{ij} are the stresses and strains in Cartesian coordinates x_i (the strains being assumed to be small), u_i = displacements of material points, Γ_1 and Γ_2 are the portions of the boundary with prescribed surface tractions p_i and with prescribed displacements U_i; f_i = prescribed body forces; and $n_j = \bar{n}_j$ = direction cosines of unit outward normals on the stress boundary.

Assuming that there is no characteristic length governing the behavior of the structure, we can relate the displacements of the reference and the scaled structures by the scaling law (Eq. 5.6), that is,

$$\bar{u}_i = \lambda^{s+1} u_i \tag{5.16}$$

where s is an unspecified exponent. Substituting this into the differential equations and boundary conditions in Eqs. 5.12–5.15, we obtain the following scaling relations:

$$\bar{\varepsilon}_{ij} = \varepsilon_{ij}\lambda^s, \quad \bar{\sigma}_{ij} = \sigma_{ij}\lambda^s \tag{5.17}$$

$$\bar{p}_i = p_i\lambda^s, \quad \bar{f}_i = f_i\lambda^{s-1}, \quad \bar{u}_i = u_i\lambda^{s+1} \tag{5.18}$$

These relationships indicate how a solution for one size can be transformed to a solution for another size. However, the value of s is not yet known. To determine it, we must rely on the constitutive law and the failure criterion. Next we consider in this regard two basic cases:

1. Elastic, plastic or elastic-plastic behavior. In the theory of plasticity or elasticity with a strength limit, the constitutive relation and the failure condition (either the yield condition or the condition of allowable stress) can be written as

$$\sigma_{ij} = \mathcal{F}_{ij}(\varepsilon_{km}), \qquad \phi(\sigma_{ij}, \varepsilon_{ij}) = \sigma_0 \tag{5.19}$$

where $\mathcal{F}_{ij}$ are tensor-valued functions or functionals of a tensorial argument, ϕ is a nonlinear scalar function of tensorial argument, and σ_0 is the material yield limit or allowable stress limit. After transformation of scale, Eq. 5.19 takes the form $\bar{\sigma}_{ij} = \mathcal{F}_{ij}(\bar{\varepsilon}_{km})$, $\phi(\bar{\sigma}_{ij}, \bar{\varepsilon}_{ij}) = \sigma_0$. In view of the fact that function ϕ (and possibly also function $\mathcal{F}$) is nonlinear, these two equations and Eq. 5.19 hold true simultaneously if and only if $\bar{\sigma}_{ij} = \sigma_{ij}$ and $\bar{\varepsilon}_{km} = \varepsilon_{km}$, which indicates that $s = 0$. The transformation rules from Eqs. 5.17 and 5.18 then become:

$$\bar{u}_i = \lambda u_i, \quad \bar{\varepsilon}_{ij} = \varepsilon_{ij}, \quad \bar{\sigma}_{ij} = \sigma_{ij} \tag{5.20}$$

$$\bar{p}_i = p_i, \quad \bar{f}_i = f_i/\lambda, \quad \bar{u}_i = u_i\lambda \tag{5.21}$$

Also

$$\bar{\sigma}_N = \sigma_N \tag{5.22}$$

So, the nominal structural strength is independent of the structure size. We say in this case that there is no size effect.[2] This is characteristic for all failure analyses according

---

[2]Plasticity is *irreversible* time-independent deformation at *constant* stress and *constant* unloading stiffness. Hardening plasticity is *irreversible* deformation at increasing stress, again at *constant* unloading stiffness. "Softening plasticity" does not exist because the unloading stiffness decreases, which is characteristic of damage or quasibrittleness. However, this established terminology, dating back to von Mises, Hencky, Prager, Drucker, etc., is recently getting confused by a few authors, e.g. (Voyiadjis and Yaghoobi, 2019). In Voyiadjis and Yaghoobi (2019), the authors include size effects due to fracturing damage or quasibrittle fracture, or grain size effects due to dislocation arrest at grain boundaries such as Hall–Petch, etc., as if they were some kind of "plasticity." The readers are warned of this confusion. Here plasticity is *plasticity*.

to elasticity with allowable stress limit, plasticity, and classical continuum damage mechanics (as well as viscoelasticity and viscoplasticity, because time has no effect on this analysis).

2. Linear elastic fracture mechanics. Consider that the failure of the specimen is governed by LEFM. In this case, the constitutive relation and the failure condition can be written as

$$\sigma_{ij} = E_{ijkl}\varepsilon_{kl}, \quad J = G_f \tag{5.23}$$

in which E_{ijkl} is the fourth-order tensor of elastic constants, G_f is the fracture energy (a material property), and J is the J-integral (Rice, 1968a,b), expressed from Eq. 2.88 as

$$J = \oint \left(\frac{1}{2}\sigma_{ij}\varepsilon_{ij}\mathrm{d}y - \sigma_{ij}n_j\partial_1 u_i \mathrm{d}S \right) \tag{5.24}$$

Based on Eqs. 5.17–5.18, we find that the J-integral transforms as

$$\overline{J} = \oint \left[\frac{1}{2}(\lambda^s\sigma_{ij})(\lambda^s\varepsilon_{ij})\lambda\mathrm{d}y - \lambda^s\sigma_{ij}n_j\lambda^{-1}\partial_1(\lambda^{s+1}u_i)\lambda\mathrm{d}S \right]$$

$$= \lambda^{2s+1}\oint \left(\frac{1}{2}\sigma_{ij}\varepsilon_{ij}\mathrm{d}y - \sigma_{ij}n_j\partial_1 u_i \mathrm{d}S \right) = \lambda^{2s+1}J \tag{5.25}$$

Since both $\overline{J}$ and J must satisfy the same fracture criterion, that is, $\overline{J} = G_f$ and $J = G_f$ in all cases, it is obviously necessary and sufficient that $2s + 1 = 0$, that is,

$$s = -1/2 \tag{5.26}$$

So, according to Eqs. 5.17 and 5.18, the transformation laws for LEFM are

$$\overline{u}_i = u_i\sqrt{\lambda}, \quad \overline{\varepsilon}_{ij} = \varepsilon_{ij}/\sqrt{\lambda}, \quad \overline{\sigma}_{ij} = \sigma_{ij}/\sqrt{\lambda} \tag{5.27}$$

$$\overline{p}_i = p_i/\sqrt{\lambda}, \quad \overline{f}_i = f_i\lambda^{-3/2}, \quad \overline{U}_i = U_i\sqrt{\lambda} \tag{5.28}$$

$$\overline{\sigma}_N = \frac{\sigma_N}{\sqrt{\lambda}} \tag{5.29}$$

Therefore, the nominal strength depends on the structure size D as, $\sigma_N \sim 1/\sqrt{D}$, or

$$\log \sigma_N = -\frac{1}{2}\log D + \text{constant} \tag{5.30}$$

So, in the plot of $\log \sigma_N$ versus $\log D$, the LEFM failures are represented by a straight line of slope $-1/2$, while all the stress- or strain-based failure criteria correspond to a horizontal line. The foregoing argument can be generalized to nonlinear elastic behavior, to which the J-integral is also applicable.

5.3 Dimensional Analysis of Size Effect

In Sec. 5.2, we used the argument of self-similarity to derive the size effect on the nominal strength of structures governed by either an elastoplastic or an LEFM failure criterion (Eqs. 5.22 and 5.29). A simpler way to derive these two size effect equations is to use dimensional analysis, which is based on the Vashy–Buckingham Π-theorem (Vashy, 1892; Buckingham, 1914, 1915; Barenblatt, 1979, 1987). The theorem states that any physical law can be written in a dimensionless form and the number of dimensionless variables (say, n) can be determined as

$$n = n_{tot} - n_{ind} \tag{5.31}$$

where n_{tot} = total number of governing variables, and n_{ind} = number of parameters with independent dimensions.

1. Plastic and elastic-plastic materials. First consider the failure load of plastic and elastic-plastic materials, which is governed by the strength limit or the yield stress, σ_0 (or the failure load considered as the maximum load obtained by elastic analysis with maximum allowable stress σ_0). The applied load at failure can be expressed in terms of the nominal strength σ_N. Both σ_0 and σ_N have the metric dimension of N/m^2. Meanwhile, the failure state also depends on the characteristic structure size D and other dimensions such as span L, notch length a and various other geometric characteristics, all of which have the metric dimension of m. According to the Vashy–Buckingham Π-theorem, the equation governing failure state must have the functional form:

$$\Phi \left(\frac{\sigma_N}{\sigma_0}, \frac{L}{D}, \frac{a}{D}, \ldots \right) = 0 \tag{5.32}$$

σ_0 is a material constant. For geometrically similar structures, $L/D, a/D, \ldots$ are constants, and so is σ_0 while function Φ is fixed. According to Eq. 5.32, this can be true only if the nominal strength σ_N is a constant, too. In other words, when the failure criterion is defined in terms of only the material strength (or yield limit), we have in Eq. 5.6:

$$s = 0, \quad \sigma_N \propto D^0 = \text{constant} \tag{5.33}$$

The same scaling must apply to stresses and strains at any homologous points of geometrically similar structures of different sizes. Accordingly, the displacements u at homologous points scale as $u \propto D$.

2. Linear elastic fracture mechanics. Within the framework of LEFM, the failure is governed by the fracture toughness (the critical stress intensity factor) K_{1c}, whose metric dimension is Nm$^{-3/2}$. The other quantities determining failure are the same as before, including σ_N, D, L, a, etc. The Vashy–Buckingham Π-theorem states that the failure condition can be expressed as

$$\Phi \left(\frac{\sigma_N \sqrt{D}}{K_{Ic}}, \frac{L}{D}, \frac{a}{D}, \ldots \right) = 0 \tag{5.34}$$

where K_{Ic} is a material constant. Since the ratios $L/D, a/D, \dots$ are all constant, for geometrically similar structures, it follows that the ratio $\sigma_N \sqrt{D}$ must be a constant, too. Hence, we have

$$s = -1/2, \quad \sigma_N \propto D^{-1/2} \tag{5.35}$$

provided that not only the structure dimensions but also the crack lengths are geometrically scaled. If there are crack-parallel stresses σ_{xx}, varying with D, K_{Ic} varies, too, and the size effect in Eq. 5.35 gets modified.

5.4 Second-Order Asymptotic Scaling Behavior at Small Size Limit

In the previous sections, we derived the power-law scaling relations for structures whose failure criterion is governed by the theories of plasticity and LEFM. For quasibrittle structures, the fracture process zone (FPZ) is not negligibly small as compared to the overall structure size. Therefore, it requires nonlinear fracture mechanics models, such as the cohesive crack model or the equivalent LEFM.

Nevertheless, if the structure size is much larger than the size of FPZ, the peak load can be calculated by means of the LEFM theory, and consequently the scaling of nominal strength follows Eq. 5.35. On the other hand, if the structure size is comparable to the FPZ size, the cohesive crack model would predict that, at the peak load, the cohesive stress is, asymptotically, uniform and is equal to the tensile strength. Therefore, the peak load can be determined by using a plastic analysis, and so the size effect on the nominal strength is absent (i.e. Eq. 5.33).

The foregoing two asymptotic behaviors indicate that the size effect plot of $\log \sigma_N$ versus $\log D$ must approach on the left a horizontal asymptote representing the small size limit, and approach on the right a straight line of $-1/2$ slope representing the large size limit (the second-order asymptotic terms will be given later).

In this section, following the exposition in Bažant (2001, 2004, 2005), we use the cohesive crack model to analyze the boundary value problem to understand how the small-size horizontal asymptote is approached. We consider the static boundary value problem of linear elasticity formulated in Cartesian coordinates x_i $(i = 1, 2, 3)$:

$$\sigma_{ij} = E_{ijkl} \tfrac{1}{2}(u_{k,l} + u_{l,k}), \qquad \sigma_{ij,j} + f_i = 0 \qquad \text{(in } \mathcal{V}) \tag{5.36}$$
$$n_j \sigma_{ij} = p_i \quad \text{(on } \Gamma_s), \qquad u_i = u_{i0} \quad \text{(on } \Gamma_d) \tag{5.37}$$

where σ_{ij} = stress tensor components, $\tfrac{1}{2}(u_{i,j} + u_{j,i}) = \epsilon_{ij}$ = strain tensor components, E_{ijkl} = elastic moduli, f_i = body forces, p_i = surface tractions, prescribed on surface domain Γ_s, and n_i = unit normal of the surface; let Γ_d be the surface domain where the displacements are prescribed. For geometrically similar structures of size D, we introduce the dimensionless coordinates and variables, labeled by an overbar;

$$\bar{x}_i = x_i/D, \qquad \bar{u}_i = u_i/D, \qquad \bar{\sigma}_{ij} = \sigma_{ij}/f_t \tag{5.38}$$
$$\bar{p}_i = p_i/\sigma_N, \quad \bar{f}_i = f_i D/\sigma_N, \quad \bar{E}_{ijkl} = E_{ijkl}/f_t, \quad \bar{u}_{i0} = u_{i0}/D \tag{5.39}$$

The load magnitude is characterized by the nominal strength σ_N as a single parameter, and so $\bar{p}_i$ represents a size-independent (or dimensionless) distribution of the surface

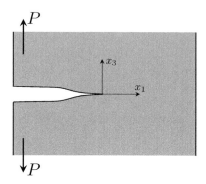

Fig. 5.3 Schematic of a cohesive crack.

tractions on Γ_s, and $\bar{f}_i$ is a size-independent distribution of body forces in volume $\bar{\mathcal{V}}$. The surface normals n_i at homologous points are independent of size D (and thus need no overbar), and f_t denotes the material tensile strength.

Denoting $\partial_i = \partial/\partial\bar{x}_i$ and $\partial/\partial x_i = (1/D)\partial_i$, we can transform Eqs. 5.36 and 5.37 to the following dimensionless form:

$$\bar{\sigma}_{ij} = \bar{E}_{ijkl}\tfrac{1}{2}(\partial_k\bar{u}_l + \partial_l\bar{u}_k), \qquad \partial_j\bar{\sigma}_{ij} + \bar{f}_i\,\sigma_N/\sigma_0 = 0 \quad (\text{in } \bar{\mathcal{V}}) \tag{5.40}$$
$$n_j\bar{\sigma}_{ij} = \bar{p}_i\,\sigma_N/\sigma_0 \quad (\text{on } \bar{\Gamma}_s), \qquad \bar{u}_i = \bar{u}_{i0} \quad (\text{on } \bar{\Gamma}_d) \tag{5.41}$$

where $\bar{\mathcal{V}}$ is the domain of structure volume in the dimensionless coordinates, and $\bar{\Gamma}_s$ and $\bar{\Gamma}_d$ are the surface domains in dimensionless coordinates corresponding to Γ_s and Γ_d, respectively.

We now position coordinates x_i such that the crack would lie in the plane (x_1, x_3) and that the tip of the cohesive crack would be at $x_1 = 0$ (Fig. 5.3). If the structure size is small enough and if the compression strength is unlimited, the cohesive zone of the crack at maximum load will occupy the entire cross-section or, in the case of a notch, the entire ligament. Then the dimensionless crack length $\bar{L} = L/D = \text{constant}$ (if the compression strength is limited and the cross-section is for instance subjected to bending, L/D will not necessarily be size independent but we may assume it to be such, as an approximation for small D).

Eqs. 5.40 and 5.41 must be supplemented by two conditions for the cohesive crack:

1. The dimensionless total stress intensity factor $\bar{K}_t = K_t\sqrt{D}/\sigma_N$ at the cohesive crack tip must vanish in order to ensure the smooth closing condition, that is,

$$\bar{K}_t = 0 \tag{5.42}$$

2. The cohesive (crack-bridging) stresses σ must satisfy the softening law of the cohesive crack, i.e., the curve relating σ to the opening displacement $w = 2u_2$ on the crack plane.

At the small size limit, what matters is the initial part of the cohesive law (i.e. w is small). We will consider the cohesive law with an initial descent as

$$\sigma = f_t[1 - (w/w_0)^p] \qquad \text{(for } -L \le x_1 < 0, x_2 = 0) \qquad (5.43)$$

where $p, w_0 =$ positive constants (the linear initial descent is included for $p = 1$). According to Eq. 5.39, we may now write the dimensionless form of the assumed softening law as follows:

$$\bar{\sigma} = 1 - (\bar{D}\bar{w})^p \qquad \text{(for } -\bar{L} \le \bar{x}_1 < 0, \bar{x}_2 = 0) \qquad (5.44)$$

with $\bar{\sigma} = \sigma/f_t$, $\bar{w} = w/D$, $\bar{D} = D/w_0$.

Now let us focus on the dependence of the solution on the structure size, D, and hypothesize that the dimensionless displacements, stresses and total stress intensity factor approach their small-size limit ($\bar{D} \to 0$) as power functions of $\bar{D}$ with exponent p. Let us try to verify the correctness of this hypothesis. So, for small enough $\bar{D}$, we set:

$$\sigma_N = \sigma_N^0 + \sigma_N' \bar{D}^p, \qquad \bar{\sigma}_{ij} = \bar{\sigma}_{ij}^0 + \bar{\sigma}_{ij}' \bar{D}^p, \qquad \bar{\sigma} = \bar{\sigma}^0 + \bar{\sigma}' \bar{D}^p \qquad (5.45)$$

$$\bar{u}_i = \bar{u}_i^0 + \bar{u}_i' \bar{D}^p, \qquad \bar{w} = \bar{w}^0 + \bar{w}' \bar{D}^p, \qquad \bar{K}_t = \bar{K}_t^0 + \bar{K}_t' \bar{D}^p \qquad (5.46)$$

in which $\sigma_N^0, \sigma_N', \bar{\sigma}^0, \bar{\sigma}', \bar{\sigma}_{ij}^0, ..., \bar{K}_t'$ are size-independent. These expressions may now be substituted into Eqs. 5.40 to 5.44. The resulting equations must be satisfied for not one but all different small sizes $\bar{D}$. For $\bar{D} \to 0$, the dominant terms in these equations are those of the lowest powers of D, which are those with $\bar{D}^0$ and $\bar{D}^p$. By collecting the terms without $\bar{D}$ and those with $\bar{D}^p$, we obtain two independent sets of equations. It so happens that each of these two sets defines a physically meaningful boundary value problem of elasticity for a body with a cohesive crack. This proves the validity of our hypothesis made in Eqs. 5.45 and 5.46.

Elasticity Problem I: By isolating the terms that do not contain $\bar{D}$ (i.e. contain $\bar{D}^0$), we get:

$$\bar{K}_t^0 = 0, \qquad \bar{\sigma}^0 = 1 \quad \text{(for } -\bar{L} \le \bar{x}_1 < 0, \bar{x}_2 = 0) \qquad (5.47)$$

$$\bar{\sigma}_{ij}^0 = \bar{E}_{ijkl} \tfrac{1}{2}(\partial_l \bar{u}_k^0 + \partial_k \bar{u}_l^0), \qquad \partial_j \bar{\sigma}_{ij}^0 + \bar{f}_i \, \sigma_N^0/\sigma_0 = 0, \quad \text{(in } \bar{V}) \qquad (5.48)$$

$$n_j \bar{\sigma}_{ij}^0 = \bar{p}_i \, \sigma_N^0/\sigma_0 \quad \text{(on } \bar{\Gamma}_s), \qquad \bar{u}_i^0 = 0 \quad \text{(on } \bar{\Gamma}_d) \qquad (5.49)$$

Elasticity Problem II: By isolating the terms that do contain $\bar{D}^p$, we get:

$$\bar{K}_t' = 0, \qquad \bar{\sigma}' = -(\bar{w}^0)^p \quad \text{(for } -\bar{L} \le \bar{x}_1 < 0, \bar{x}_2 = 0) \qquad (5.50)$$

$$\bar{\sigma}_{ij}' = \bar{E}_{ijkl} \tfrac{1}{2}(\partial_l \bar{u}_k' + \partial_k \bar{u}_l'), \qquad \partial_j \bar{\sigma}_{ij}' + \bar{f}_i \, \sigma_N'/\sigma_0 = 0, \quad \text{(in } \bar{V}) \qquad (5.51)$$

$$n_j \bar{\sigma}_{ij}' = \bar{p}_i \, \sigma_N'/\sigma_0 \quad \text{(on } \bar{\Gamma}_s), \qquad \bar{u}_i' = 0 \quad \text{(on } \bar{\Gamma}_d) \qquad (5.52)$$

The role of stresses and displacements is played by $\bar{\sigma}_{ij}^0$ and $\bar{u}_i^0$ in problem I, and by $\bar{\sigma}_{ij}'$ and $\bar{u}_i'$ in problem II. In problem I, the crack faces experience uniform tractions whose value is 1. In problem II, in which σ' plays the role of the cohesive stress, the crack faces are subjected to tractions $-(\bar{w}^0)^p$ which vary along the crack faces but can

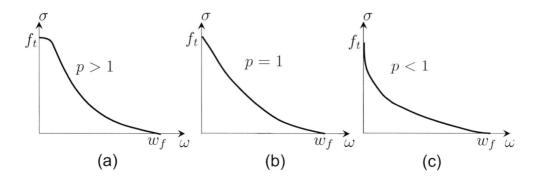

Fig. 5.4 Softening cohesive laws with different initial descending behaviors: (a) $p > 1$, (b) $p = 1$, and (c) $p < 1$.

be determined in advance from the $\bar{w}^0$-values obtained in solving problem I. The fact that isolation of the terms with the zeroth power and the pth power of $\bar{D}$ yield two separate boundary value problems of elasticity is essential for our objective. The rest of the argument is easy and may be stated as follows.

Based on the nature of elasticity, the magnitude of the loads (surface tractions and body forces) is proportional to σ_N^0 in problem I, and to σ_N' in problem II. Elasticity problems are known to have a unique solution, and this applies to both of our problems. If σ_N^0 were zero, that is, if the applied load in problem I vanished, the crack face tractions equal to 1 would cause K_t^0 to be non-zero, which violates Eq. 5.47. Likewise, if σ_N' were zero, that is, if the applied load in problem II vanished, the nonuniform crack face tractions $-(\bar{w}^0)^p$ in problem II would cause $\bar{K}_t'$ to be non-zero, which again violates Eq. 5.47. Meanwhile, it is evident that the loads for problems I and II cannot be infinite. Therefore, both σ_N^0 and σ_N' must be finite.

Based on Eq. 5.45, we can conclude that if the cohesive law of the cohesive crack model has a finite tensile strength and begins its initial post-peak descent as w^p, then the SEL for nominal strength approaches, for $D \to 0$, a finite value and does so as D^p. Most typical cohesive laws for quasibrittle materials have $p = 1$. Thus, according to Eq. 5.45, the SEL must begin near zero size D as a linear function of D, that is,

$$\sigma_N = \sigma_N^0 + (\sigma_N'/w_0)D \quad \text{as } D \to 0 \tag{5.53}$$

This gives an exponential initial SEL curve in the double logarithmic plot, because of the approximation $\ln \sigma_N - \ln \sigma_N^0 = \ln(1 + \sigma_N'\bar{D}/\sigma_N^0) \approx (\sigma_N'/\sigma_N^0) \, e^{\ln \bar{D}}$.

The case $p > 1$ would mean that the softening law would begin its descent from a horizontal initial tangent, which is what might be assumed for ductile fracture of plastic materials (a very short horizontal segment might as well occur for quasibrittle fracture (Fig. 5.4a), although no data exist to answer this question). The case $p < 1$ would mean that the cohesive law would begin its descent with a vertical tangent

(Fig. 5.4c), which would be an unrealistic super-brittle behavior. Many quasibrittle materials such as concrete correspond to the case of $p = 1$ (Fig. 5.4b).

Note that the small-size scaling does not depend on G_f or K_{Ic}, and therefore the crack-parallel stresses have no effect. We should point out that our imposition of the small-size asymptotic properties of the cohesive crack model on the SEL might be regarded as fiction because, for cross-section thicknesses less than several aggregate sizes, the material cannot be approximated as a continuum. Nevertheless, this theoretical asymptotic analysis, describing the average behavior, is useful for validating the subsequent derivation of the approximate size effect equations.

5.5   Derivation of Size Effect Equations Using Equivalent LEFM

As discussed in Chapter 4, the equivalent LEFM provides a useful approximation of the cohesive crack model, where the tip of a sharp LEFM crack is placed roughly into the middle of the FPZ. By considering mode I fracture, we can write the stress intensity factor as

$$K_I = \frac{P}{b\sqrt{D}}\, k(\alpha) \tag{5.54}$$

where P = applied load or loading parameter, b = structure width, and $k(\alpha)$ = dimensionless stress intensity factor, which is expressed as a function of the relative crack or notch length $\alpha = a/D$. We define the nominal strength as $\sigma_N = P_m/bD$ (Eq. 5.1). By recalling Irwin's relation $K_I^2 = E'\mathcal{G}$ and setting $\mathcal{G} = G_f$ = fracture energy of the material, we have

$$\sigma_N = \sqrt{\frac{E'G_f}{Dg(\alpha)}} \tag{5.55}$$

where $E' = E$ = Young's modulus for plane stress and $E' = E/(1 - \nu^2)$ for plane strain (ν = Poisson ratio), and $g(\alpha) = k^2(\alpha)$ = dimensionless energy release function.

Based on the equivalent LEFM, the cohesive crack is approximated by an equivalent LEFM crack with a tip roughly in the middle of the FPZ. This equivalent crack has the length of $a = a_0 + c_f$, where a_0 = original crack length and c_f = additional constant crack length representing about half of the FPZ length. For quasibrittle materials, the FPZ size is about the same for structures of different sizes, and is related to Irwin's characteristic length l_1 corresponding to the initial fracture energy G_f, i.e., $c_f = \gamma_f l_1 = \gamma_f EG_f/f_t^2$. The precise value of γ_f can be calculated for a given specimen geometry from the softening law of the cohesive crack model; for example, for a notched three-point bend specimen, $\gamma_f \approx 0.3$ for a bilinear cohesive law (Bažant and Yu, 2011), and $\gamma_f = 0.44$ for a linear cohesive law (Cusatis and Schauffert, 2009). The dimensionless energy release rate function can be expanded into the Taylor series:

$$g(\alpha) = g_0 + g_0'\theta + \tfrac{1}{2}g_0''\theta^2 + \tfrac{1}{6}g_0'''\theta^3 + \dots \tag{5.56}$$

where $\alpha_0 = a_0/D$, $g_0 = g(\alpha_0)$, $g_0' = \mathrm{d}g(\alpha)/\mathrm{d}\alpha|_{\alpha=\alpha_0}$, etc., and $\theta = c_f/D$.

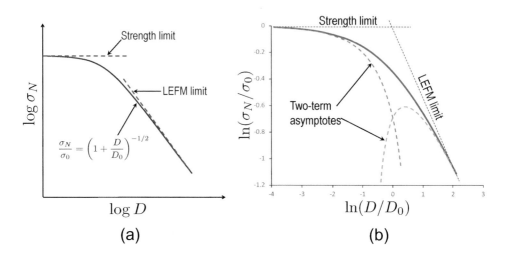

Fig. 5.5 (a) Type 2 size effect curve predicted by Eq. 5.57, and (b) two-term small-size and large-size asymptotes.

5.5.1 Type 2 size effect

We first discuss the type 2 size effect, which applies to structures containing a large notch or a preexisting stress-free crack (i.e. $a_0 \neq 0$). Further we assume that the structure has a positive geometry, that is, $g_0' > 0$), which is applicable in many situations. For geometrically similar structures, the relative crack length α_0 is a constant. By truncating the Taylor series expansion of the dimensionless energy release rate function (Eq. 5.56) after the second term, we obtain a size effect equation of the form (Bažant and Kazemi, 1990, 1991; Bažant and Planas, 1998; Bažant, 2005):

$$\sigma_N = \sqrt{\frac{E'G_f}{g_0'c_f + g_0 D}} = Bf_t \left(1 + \frac{D}{D_0}\right)^{-1/2} \tag{5.57}$$

$$\text{where} \quad D_0 = c_f \frac{g_0'}{g_0}, \quad Bf_t = \sqrt{\frac{E'G_f}{c_f g_0'}} \tag{5.58}$$

where the transitional size D_0 governs the deviation from the LEFM scaling limit. It is noted that D_0 is proportional to the effective size c_f of the FPZ, and the proportionality constant is governed by the dimensionless energy release rate function $g(\alpha)$, which is influenced by the structure geometry. Thus, the SEL in Eq. 5.57 expresses not only the effect of size but also the effect of structure geometry (or shape). It might better be called the "size-shape" effect law because it can be applied even to structures that are not geometrically similar (if their functions $g(\alpha)$ are known).

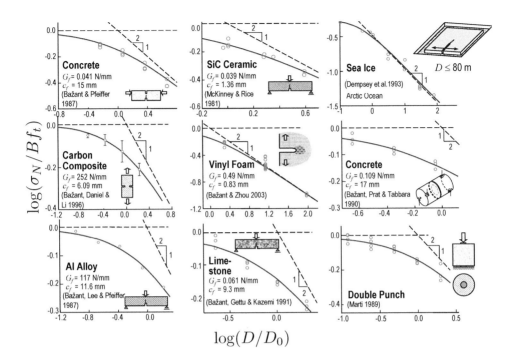

Fig. 5.6 Optimum fitting of size effect data by using Eq. 5.57 (Bažant, 2004).

It should be pointed out that, in the present derivation, we have, in anticipation, truncated the expansion of the energy release rate function $g(\alpha)$ in Eq. 5.56 already after the second term. Although including high-order terms would improve the approximation for large sizes D, it would predict a divergent scaling behavior at the small-size limit giving, $\lim_{D\to 0} \sigma_N = \infty$. This would violate the small-size asymptotic scaling behavior derived in Sec. 5.4.

Fig. 5.5 plots in logarithmic scale the size effect curve predicted by Eq. 5.57. It has been widely documented that Eq. 5.57 agrees well with the experimentally observed size effect for structures made of many quasibrittle materials including concrete, composites, ceramics, wood, rock, sea ice, foam, bone, and limestone (e.g. Fig. 5.6).

Following Eq. 5.57, we may define a brittleness number β,

$$\beta = D/D_0 \qquad (5.59)$$

It is clear that the case $\beta \to \infty$ represents a perfectly brittle behavior, which is characterized by the LEFM limit, whereas the case $\beta \to 0$ represents a quasi-plastic behavior corresponding to the strength limit. The key point is that the brittleness number is influenced by both the structure size relative to the FPZ size and the structure geometry.

It is worthwhile to comment on the effect of crack-parallel stress on the type-2 SEL. When σ_{xx} affects G_f, the derivation leading to Eqs. 5.57 and 5.58 remains valid, but a variable G_f must be used. Note that the transitional size D_0 remains unaffected, and it is only the pre-factor Bf_t that becomes variable. The shape of the size effect curve seen in Fig. 5.5 remains the same only as long as σ_{xx}, and thus G_f, are the same for all sizes.

Type 2 size effect via asymptotic matching

Asymptotic matching has widely been used in fluid mechanics since Prandtl's creation of boundary layer theory for fluid flow in pipes at the dawn of last century, and has recently emerged as an illuminating approach to quasibrittle fracture. It includes various techniques to get good approximate solutions by "interpolating" between easily obtainable asymptotic properties at opposite infinities (in our case at $\log D = -\infty$ and ∞).

Eq. 5.57 can be rewritten as $\sigma_N = \sigma_0(D/D_0)^{-1/2}(1+D_0/D)^{-1/2} \approx \sigma_0(D/D_0)^{-1/2}(1 - D/D_0 + ...)$, which gives the first two terms of the large-size asymptotic expansion of σ_N. Including the first two terms of the small-size asymptotic expansion according to Eq. 5.56 derived from the cohesive crack model (Eq. 5.53), we can summarize the two-term asymptotic requirements for the size effect as follows:

$$\text{for } D \to 0: \qquad \sigma_N \propto 1 - D/D_s \tag{5.60}$$

$$\text{for } D \to \infty: \qquad \sigma_N \propto \frac{1}{\sqrt{D}}\left(1 - \frac{D_0}{2D}\right) \tag{5.61}$$

where D_s is a constant. Although we have not used Eq. 5.60 in the derivation of the type 2 SEL, we find that Eq. 5.58 is, luckily, correct since, for $D \to 0$, $(1+D/D_0)^{-1/2} \approx 1 - D/2D_0$ where $2D_0 = D_s$. To illustrate the nature of asymptotic matching, the two-term small-size and large-size asymptotes are plotted over the full range of D in Fig. 5.5b.

The foregoing analysis applies only if the crack-parallel stress σ_{xx} is asymptotically constant, which is true if σ_{xx} is independent of D and, most generally, only if the second-order terms in the asymptotic expansions of σ_{xx} as functions of D or $1/D$ vanish.

Note that the second-order asymptotic properties exclude other formulas, such as $\sigma_N \propto 1/(1+\sqrt{D/D_0})$ or $1/(1+ce^{\sqrt{D/D_0}})$ ($c = $ constant), which satisfy only the first-order asymptotic properties. Also note that the energy analysis, as will be shown in Sec. 5.5.3, and even in its most primitive 1984 form (Bažant, 1984b), leads to correct two-term two-sided asymptotic properties. The SEL in Eq. 5.57 could be constructed as the simplest formula with these properties.

5.5.2 Type 1 size effect

Type 1 size effect applies to structures in which there is neither a notch nor pre-existing stress-free crack, i.e., $a_0 = 0$. In this case, we have $g_0 = 0$. Therefore, we must take in Eq. 5.56 the second and third terms in the series expansion, that is, $g(\alpha) = g_0'\theta + \frac{1}{2}g_0''\theta^2$. Substitution into Eq. 5.55 yields

$$\sigma_N = \sqrt{\frac{E'G_f}{g'_0 c_f + \frac{1}{2} g''_0 c_f^2 / D}} \tag{5.62}$$

For many structures, we have $g''_0 < 0$. This causes a problem; σ_N becomes imaginary for small enough D. Even though the approximation in the above equation can be valid only for large enough D, we still want a two-sided asymptotic matching law of general applicability, like the type 2 SEL.

To this end, Eq. 5.62 for large D may be approximated as

$$\sigma_N = f_{r\infty} (1 - x)^{-1/2} \tag{5.63}$$

$$\text{where} \quad f_{r\infty} = \sqrt{\frac{E'G_f}{g'_0 c_f}}, \quad x = \frac{-g''_0}{g'_0} \frac{c_f}{2D} \tag{5.64}$$

Eq. 5.63 can further be modified by a binomial expansion that does not change the first two terms of the expansion. Therefore, it is legitimate to replace this equation by

$$\sigma_N = f_{r\infty} (1 + x)^{1/2} \quad \text{or} \quad = f_{r\infty} [1 + r(x/2)]^{1/r} \tag{5.65}$$

where r is an arbitrary positive number determined by optimum fitting of the experimentally measured size effect curve. Upon setting $x/2 = D_b/D$ where

$$D_b = \left\langle \frac{-g''_0}{4g'_0} \right\rangle c_f \tag{5.66}$$

we arrive at the final form of the type 1 SEL (Fig. 5.7a):

$$\sigma_N = f_{r\infty} \left(1 + \frac{rD_b}{D} \right)^{1/r} \tag{5.67}$$

Note that in Eq. 5.66, the signs $\langle .. \rangle$ represent the Macauley brackets, which mean the positive part of the argument; they were inserted because the ratio g''_0/g'_0 can sometimes be positive, in which case there can be no size effect, that is, D_b would vanish.

It must now be emphasized that Eq. 5.67 has not only the correct first and second-order asymptotic properties for $D \to \infty$, like Eq. 5.63, but also a realistic form for small D, such that $\lim_{D \to 0} \sigma_N$ be neither 0 nor imaginary. The feature that $\sigma_N \to \infty$ for $D \to 0$ is shared by the famous, widely used, Hall–Petch formula for the strength of polycrystalline materials (Hall, 1951; Petch, 1954). One might prefer a finite small-size limit but this does not matter in practice because D for concrete cannot be less than about three maximum aggregate sizes, as the material could no longer be treated as a continuum.

The foregoing derivation of type 1 size effect is anchored purely by the energetic (i.e., deterministic) analysis of the fracture process. It predicts a vanishing size effect at the large size limit. However, for very large size structures, the type 1 size effect transits to a statistical size effect, which can be captured by the classical Weibull size

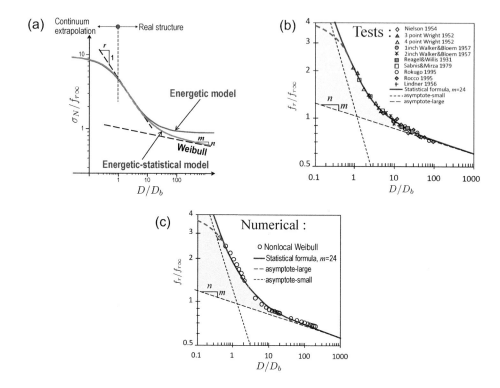

Fig. 5.7 Type 1 size effect: (a) schematic of energetic and energetic-statistical type 1 size effect curves, (b) comparison with classical experimental data for compression strength of different normalized by the mean strength of each concrete, and (c) computer simulations of the mean strength of similar notched three-point bend beams using the nonlocal Weibull theory (Bažant and Novák 2000a).

effect (Fig. 5.7a). We will discuss it in the next chapter. One can combine Eq. 5.67 with the classical Weibull size effect to form an energetic-statistical size effect equation for type 1 failures (Bažant, 2005), which reads

$$\sigma_N = f_{r\infty} \left[\frac{rD_b}{D} + \left(\frac{D_a}{D} \right)^{rn/m} \right]^{1/r} \tag{5.68}$$

where rn/m must be < 1; D_a, D_b = constants, m = Weibull modulus, and n = number of spatial dimensions in which the failure mode of the structure is scaled. Fig. 5.7b shows the comparison between Eq. 5.68 and the available classical experimental data on concrete beams from ten different labs (Bažant and Novák, 2000b,c).

Eq. 5.68 is actually an asymptotic matching approximation supported by two-sided asymptotic properties: (a) For $D/D_a \to \infty$ it gives the Weibull size effect; and (b) for

$m \to \infty$ it matches the deterministic energetic size effect, including both asymptotic terms in Eq. 5.60. Chapter 6 will present an alternative derivation of Eq. 5.68 based on the theory of probability.

In Bažant *et al.* (2007*a*), Eq. 5.68 was generalized to terminate for $D/D_a \to 0$ with a horizontal asymptote. It must be admitted that such an asymptote makes no sense physically because D cannot be smaller that the maximum aggregate size. But it is of theoretical interest since it matches the asymptotic property of the cohesive crack model. This continuum model can provides useful support for the size effect curve in the small size range.

5.5.3   Simple Derivation of Size Effect Law from Dimensional Analysis

The dimensional analysis expounded in Sec. 5.3 may be extended to derive the SEL. In Appendix B this is done in full detail, taking into account the two-sided second-order asymptotic properties (based on Bažant 2004). Here, disregarding these properties and the connection to the energy release function of LEFM, we present a much simpler derivation of the basic SEL of type 2, based merely on: (a) the Vashy–Buckingham Π-theorem, (b) the energy conservation during failure, and (c) the existence of a material characteristic length, c_f, without invoking any fracture mechanics concepts except energy balance (or the First Law of Thermodynamics). Fracture mechanics concepts beyond energy balance need not even be revoked. As before, we assume that (as typical of reinforced concrete) the structural geometry is such that a large stable crack forms before the maximum load. We paraphrase here the derivation given in Yu *et al.* (2016) and slightly refined in Dönmez *et al.* (2020).

Let a be the length of the *effective* crack or (crack band) of length a_0 whose front lies at the *effective* centroid of the fracture process zone (FPZ). Also assume $c_f = a - a_0 =$ material constant. To satisfy geometric similarity, the crack front at peak load of structures of different sizes D must lie at roughly homologous locations. The complementary strain energy, W, of the structure must be proportional to load P, to nominal strain energy density $\sigma_N^2/2E$ ($\sigma_N = P/bD$), and to a certain dimensionless geometry factor $f(\alpha)$ which must depend on dimensionless crack length $\alpha = a/D$. Therefore, W must have the general form:

$$W = \frac{\sigma_N^2}{2E}(bD^2)f(\alpha), \quad \alpha = \frac{a}{D} \tag{5.69}$$

To check that this agrees with the Vashy–Buckingham Π-theorem of dimensional analysis or laws of similitude (Vashy, 1892; Buckingham, 1914, 1915; Barenblatt, 1987), note that there are five governing parameters and two independent dimensions, length and force. Since $5 - 2 = 3$, there must be three and only three governing independent dimensionless parameters, which are

$$\theta_1 = W/bD^2E, \quad \theta_2 = \sigma_N/E \quad \theta_3 = a/D \tag{5.70}$$

Eq. 5.69 must represent an equation relating these dimensionless parameters, and indeed one can check that it is identical to the following equation:

$$F(\theta_1, \theta_2, \theta_2) = \theta_1 - \theta_2 f(\theta_3) = 0 \tag{5.71}$$

The energy conservation (or the First Law of Thermodynamics) requires that the energy release ΔW from the structure during extension Δa of the crack or crack band be equal to the energy dissipated:

$$\Delta W = (\sigma_N^2/2E)bD^2\Delta f(\alpha) = G_f b\Delta a \qquad (5.72)$$

$$\text{where} \quad \Delta f(\alpha) = \frac{\mathrm{d}f(\alpha)}{\mathrm{d}\alpha}\frac{\mathrm{d}\alpha}{\mathrm{d}a}\Delta a = g(\alpha)\frac{\Delta a}{D} \qquad (5.73)$$

Here $g(\alpha) = [\mathrm{d}f(\alpha)/\mathrm{d}\alpha]_{\alpha=\alpha_0}$ and G_f = fracture energy = material constant. Approximating this function by the first two terms of its Taylor series expansion, we have (as in Eq. 5.56)

$$g(\alpha) = g(\alpha_0 + c_f/D) = g_0 + g_0' c_f/D \qquad (5.74)$$

where $g_0 = g(\alpha_0)$ and $g_0' = \mathrm{d}g(\alpha_0)/\mathrm{d}\alpha$. Substituting this into Eq. 5.72 and solving for σ_N yields:

$$\sigma_N = v_u = \sqrt{\frac{2EG_f}{g_0 D + g_0' c_f}} = v_{c0}\lambda_s \qquad (5.75)$$

$$\text{where} \quad \lambda_s = \sqrt{\frac{2}{1 + D/D_0}} \quad D_0 = c_f\frac{g_0'}{g_0}, \quad v_{c0} = \sqrt{EG_f/g_0' c_f} \qquad (5.76)$$

Here λ_s represents the energetic SEL (Bažant, 1984b; Bažant and Kazemi, 1990) (for $d_0 = 10$ in., or 0.252 m, and with the restriction $\lambda_s \leq 1$, λ_s represents the size effect factor embedded into ACI design code provisions for beam shear, slab punching and strut-and-tie model; see ACI-318/2019, articles 22.5.5.1.3, 22.6.5.2, and 23.4.4.1).

The foregoing simple derivation does not provide the relation of g_0 and g_0' to LEFM energy release function, which mean that D_0 and v_{c0} cannot be quantitatively predicted. Yet it is perfectly sufficient for complex situations such as the shear or torsion failure of RC beams and slabs, fracture of sea ice, etc., for which function $g(\alpha)$ is hard to determine.

With regard to the use of λ_s in design code, it is important to note that its expression (Eq. 5.76) remains unaffected by the crack parallel stresses. Only the pre-factor v_{c0} is, but for the code it does not matter since the pre-factor has been determined by fitting of test data (cf. Sec. 7.3.1).

In the case of axisymmetric similarity, as in slab punching, the derivation proceeds similarly. It suffices to replace at the outset bD with $a_0 D$ and bD^2 with $\pi a_0^2 D$.

5.6 Determination of $\mathcal{R}$-Curve from Size Effect Analysis

As discussed in Sec. 5.5.1, the type 2 size effect can be used to determine the fracture energy G_f and the effective FPZ size c_f. In this section, we show that the entire crack-resistance curve ($\mathcal{R}$-curve) can also be obtained from the size effect analysis. The main advantage of size effect analysis is that the $\mathcal{R}$-curve obtained is size independent, and allows easy prediction of the maximum load capacity of the specimen.

As also discussed, the size effect equation (Eq. 5.57) contains the geometrical effect on the nominal structural strength. Therefore, the size effect analysis further allows

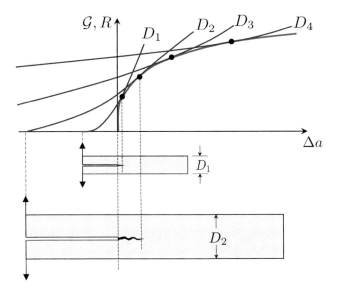

Fig. 5.8 $\mathcal{R}$-curve and the energy release rate curves of geometrically similar specimens of different sizes.

extrapolating the $\mathcal{R}$-curve from one geometry to another, which greatly reduces the experimental efforts required for determining the $\mathcal{R}$-curve.

Consider that size effect experiments have been performed on geometrically similar specimens, from which the size effect curve for the mean structural strength has been obtained. For each specimen size, we can plot the energy release rate function $\mathcal{G}_u(a)$ by setting the applied load to be equal to the peak load. On this energy release rate curve, there is typically one, and only one, point that is tangent to the $\mathcal{R}$-curve (Fig. 5.8). This point represents the critical state.

For different specimens sizes, the $\mathcal{G}_u(a)$ curves will have different tangent points. If we now draw $\mathcal{G}_u(a)$ curves for all the sizes, the $\mathcal{R}$-curve forms the envelope of all these curves (Fig. 5.8).

To describe this envelop mathematically, we first express the nominal strength by using Eq. 5.57:

$$\sigma_{Nu}^2(D) = \frac{E'G_f}{k_0^2(D + D_0)} \tag{5.77}$$

where $k_0 = k(\alpha_0) =$ dimensionless stress intensity factor measured at the original crack length a_0 (note that the dimensionless energy release rate function $g(\alpha)$ is related to $k(\alpha)$ by $g(\alpha) = k^2(\alpha)$). At the peak load, the condition of equilibrium fracture propagation requires $\mathcal{G} = \mathcal{R}$; and so

$$\mathcal{R}(\Delta a) = \mathcal{G}_u(\Delta a, D) = G_f \frac{D}{k_0^2(D + D_0)}[k(\alpha_0 + \Delta a/D)]^2 \qquad (5.78)$$

where Δa = equivalent LEFM crack extension.

Since the $\mathcal{R}$-curve cannot depend on the specimen size D, we must have

$$\frac{\mathrm{d}\mathcal{G}_u(\Delta a, D)}{\mathrm{d}D} = 0 \qquad (5.79)$$

By substituting Eq. 5.78 into Eq. 5.79, we obtain

$$D = \frac{D_0 k(\alpha_e)}{(\alpha_e - \alpha_0)2k'(\alpha_e)} - D_0 \qquad (5.80)$$

where $\alpha_e = \alpha_0 + \Delta a/D$ = relative length of the equivalent LEFM crack and $k'(\alpha_e) = \mathrm{d}k(\alpha)/\mathrm{d}\alpha|_{\alpha=\alpha_e}$.

Eq. 5.80 gives, for specimen size D, the solution for which the peak load is reached at an equivalent LEFM crack of length a_e. Based on Eq. 5.80, we can then express the extension of the equivalent LEFM crack as

$$\Delta a = (\alpha_e - \alpha_0)D = \left[\frac{k(\alpha_e)}{2k'(\alpha_e)} - (\alpha_e - \alpha_0)\right] D_0 \qquad (5.81)$$

Now we insert here the expression of D_0 (Eq. 5.58), and the relations $D_0 = c_f g_0'/g_0 = 2k_0' c_f/k_0$. Then Eq. 5.81 can be rewritten as

$$\frac{\Delta a}{c_f} = f_1(\alpha_e) = \left[\frac{k(\alpha_e)}{2k'(\alpha_e)} - (\alpha_e - \alpha_0)\right] \frac{2k_0'}{k_0} \qquad (5.82)$$

Eq. 5.82 expresses the abscissa of the $\mathcal{R}$-curve as a function of α_e. To obtain the ordinate, we simply substitute Eq. 5.80 into Eq. 5.78, which yields:

$$\frac{\mathcal{R}}{G_f} = f_2(\alpha_e) = \frac{k(\alpha_e)k'(\alpha_e)}{k_0 k_0'} f_1(\alpha_e) \qquad (5.83)$$

Eqs. 5.82 and 5.83 define the $\mathcal{R}$-curve parametrically, through α_e as a parameter (Bažant et al., 1989; Bažant and Kazemi, 1990). To plot the $\mathcal{R}$-curve, we can choose a series of value of α_e values and for each of them evaluate $\Delta a/c_f$ and $\mathcal{R}$. The $\mathcal{R}$-curve defined by this set of two parametric equations has the form:

$$\mathcal{R} = G_f f(\Delta a/c_f) \qquad (5.84)$$

where function f depends solely on the geometry of the specimen or structure.

This formulations is, of course, contingent on G_f being constant. This means that the crack-parallel stress σ_{xx} must be the same (or zero) for all the sizes. Note that the shape of the $\mathcal{R}$-curve, given by Eq. 5.82, is not affected by σ_{xx} but, according to Eq. 5.83, the ordinates of $\mathcal{R}$-curve change in proportion to G_f. If σ_{xx} varies with D, the $\mathcal{R}$-curve does not exist.

5.7 Size Effect Testing of Cohesive Law Parameters

As discussed in Chapter 3, the cohesive crack model is a widely used model for the analysis of fracture of quasibrittle materials. The main advantage of the cohesive crack model is its simplicity. However, determining the cohesive law and the effect of crack-parallel stresses is a challenge. To that end, we need to understand how the size effect curve (e.g. Eq. 5.57) can be related to the shape of the cohesive law and how its parameters can best be determined (Cusatis and Schauffert, 2009; Bažant and Yu, 2011).

It is now widely accepted that the cohesive softening law $\sigma = f(w)$ is not linear, and also not exponential (which was the form used initially for concrete (Hillerborg *et al.*, 1976)). For concrete, various rocks, cold asphalt concrete and probably many other materials, a realistic approximation, proposed by Petersson (Petersson, 1981) and improved by Elices and Planas (Planas and Elices, 1992, 1993; Bažant and Planas, 1998), is a bilinear law consisting of two linear segments—an initial steep descent followed by a long tail. It can be characterized by four parameters: the initial and total fracture energies, G_f and G_F (representing the areas under the initial straight line and under the complete bilinear curve), the tensile strength f_t, and the stress σ_k at the knee point, i.e., the point of slope change.

In view of the recently discovered effect of crack-parallel stresses σ_{xx} (Sec. 4.1), the bilinear law must be scaled as a function of σ_{xx}. Specific test data on this are lacking, but in practice f_t may first be scaled according to the existing data on biaxial or triaxial strength of the material and then the initial steep slope should be scaled to adjust the initial fracture energy G_f. The size effect analysis of the gap test gives no direct information on G_F, but one can probably assume the ratio G_F/G_f to remain constant (typically to 2 to 6).

For quasibrittle many materials such as concrete, rock, composites, bone or cold asphalt mixtures, the size range used in normal laboratory testing is not broad enough for the FPZ to be fully developed at the peak load. Thus all of the cohesive stresses in the FPZ at the peak load lie within the initial steeply descending portion of the cohesive law, and the tail part of the cohesive law does not have yet any influence on the peak load (see the cohesive stress profile on top of Fig. 5.9). The FPZ stress profile reaches the stress values in the tail segment of the cohesive law curve (Fig. 5.9 bottom) only for much larger specimen sizes, which are beyond the range of normal laboratory testing. It is for this reason that, in the previous section, the size effect equations were derived by using only the initial fracture energy, G_f.

According to the type 2 SEL, the initial fracture energy G_f and the effective FPZ size c_f can be determined by rearranging Eq. 5.57 to the following linear equation:

$$Y = \bar{D} + \gamma \quad \text{with} \tag{5.85}$$

$$Y = \frac{f_t^2}{g_0' \, \sigma_N^2}, \quad \bar{D} = \frac{D}{l_1} \frac{g_0}{g_0'}, \quad \gamma = \frac{c_f}{l_1} \tag{5.86}$$

where $l_1 = EG_f/f_t^2$. Parameter γ is a constant for a given cohesive law (Cusatis and Schauffert, 2009; Yu *et al.*, 2010*b*). However, if the cohesive law changes, it depends on the form of the cohesive law only mildly.

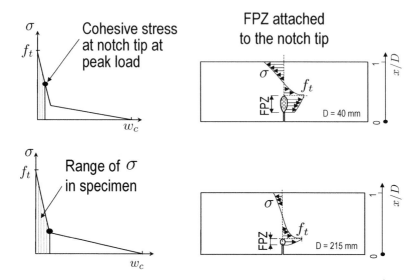

Fig. 5.9 Stress profiles in the FPZ of standard test specimens of different sizes (Yu *et al.*, 2010*b*).

If a bilinear cohesive law is adopted, the dependence Y on $\bar{D}$ exhibits two asymptotes: (a) one asymptote represents what Barenblatt (Barenblatt, 1996, 2003) called the intermediate asymptote and corresponds to Eq. 5.85, and (b) the terminal asymptote, which corresponds to the total fracture energy G_F; see Fig. 5.10 (the intermediate asymptote is rigorously defined by considering the point of transition to the terminal asymptote to recede to infinite size).

For concrete and some other materials, for which the FPZ length is about 0.5 m, the terminal asymptote (Fig. 5.10 left) applies to specimens too large for practical testing. So, normally the FPZ state never reaches the terminal asymptote. However, by virtue of the existence of the second, terminal, asymptote, a long segment of the measured plot of Y vs. $\bar{D}$ runs parallel to intermediate asymptote (Fig. 5.10). This segment allows the most accurate determination of l_1, according to Eq. 5.86, and thus also of G_f (Bažant and Yu, 2011).

So, for optimal determination of the initial fracture energy G_f and tensile strength f_t by Eq. 5.57, one should choose the optimal size range for the size effect tests (see Fig. 5.10). This range depends on the knee point σ_k of the bilinear cohesive law, and on ratio of G_F/G_f. Through extensive cohesive crack simulations, it was shown that for normal concrete, for which the knee point is located lower than 25% of the tensile strength and $G_F/G_f \leq 5$, the optimal size range for the size effect tests is $\bar{D} \in (0.2, 1.8)$ (Bažant and Yu, 2011), and constant γ_f is about 0.30. Note that, by

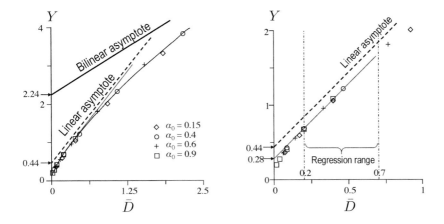

Fig. 5.10 Size effect simulated based on the cohesive crack model with the asymptotes determined by linear and bilinear softening (Bažant and Yu, 2011).

contrast, if we considered a linear softening cohesive law, constant γ_f would be 0.44 (Cusatis and Schauffert, 2009).

It is clear that, in order to determine also the knee point of cohesive law and the total fracture energy G_F, one must test the entire load-deflection curves of normal laboratory-scale test specimens. It has usually been tacitly assumed that the cohesive law can be inferred by testing specimens of one size only. However, this is not true because the cohesive softening softening law is bilinear rather than linear (or exponential) (Yu *et al.*, 2010*b*; Bažant and Yu, 2011; Kim *et al.*, 2013).

As a demonstration, Fig. 5.11 presents the simulated load-crack mouth opening displacement (CMOD) curves of a three-point bend specimen of depth $D = 215$ mm, obtained by considering four very different bilinear softening cohesive laws (Fig. 5.11a). Among these four cohesive laws, the tensile strength f_t is chosen to vary from 4.4 MPa to 7.8 MPa, and the initial fracture energy G_f from 21 to 35 N/m, while the total fracture energy G_F is fixed at 64 N/m. Fig. 5.11b shows that all the four chosen cohesive laws yield virtually the same load-CMOD curve. This reveals that fracture testing of single-size specimens is insufficient to provide an unambiguous prediction of the bilinear cohesive law.

So one must conclude that the test of load-deflection curve of a notched specimen of one size only (known for concrete as the Hillerborg test) gives ambiguous results, such that tensile strength may vary within the ratio 1.77 : 1 and the initial fracture energy within the ratio 1.67 : 1 (Bažant and Yu, 2011) (similar discrepancies are demonstrated for concrete in Bažant and Yu (2011) and for bone in Kim *et al.* (2013)). To avoid this ambiguity, the test of the load-deflection curve of a quasibrittle material must be supplemented by testing the size effect curve of σ_N versus D.

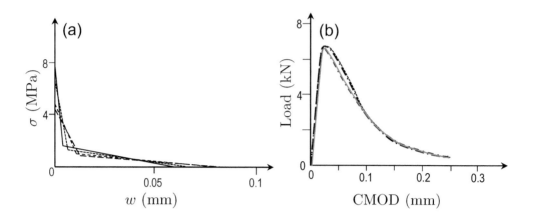

Fig. 5.11 Predictions of the load-CMOD response by using different cohesive laws: (a) cohesive laws, and (b) simulated load-CMOD curves (Bažant and Yu, 2011).

Exercises

E5.1. Restate the proof that any response of geometrically similar structures with no material characteristic length must follow a power law.

E5.2. Prove Pythagoras' theorem by using the dimensional analysis.

E5.3. Use dimensional analysis to derive the scaling equations for the nominal strength of structures by assuming perfect plasticity (i.e. the material yielding is governed by a yield surface).

E5.4. Consider (a) a homogeneous sandstone with grain size 1 mm, and (b) concrete with maximum aggregate size 25 mm. The sizes of geometrically similar specimens are greater than 0.1 m. In which case do you expect the scaling behavior would more likely follow a power law?

E5.5. Logarithmic scale linear regression of one large set of highly scattered experimental data on shear failure of RC beams made of different concretes indicated an optimum fit by power law $D^{-2/3}$. Can that be physically correct, or can it be caused by a wrong hypothesis of linear regression?

E5.6. With a closed book, derive the type 2 energetic SEL.

E5.7. Why must the derivation of the type 2 energetic SEL ignore the third term of the Taylor series expansion of $g(\alpha)$?

E5.8. Why must the derivation of the type 1 SEL take into account the third term of the Taylor series expansion of $g(\alpha)$?

E5.9. The self-weight of notched three-point bend concrete beams deeper than about 2 m causes significant deviations of loading from similarity. What kind of mathematical corrections or physical changes are necessary to apply the type 2 SEL? When the span-to-depth ratio is decreased, are these deviations enhanced or diminished.

E5.10. Does a constant crack-parallel stress σ_{xx} invalidate the type 2 SEL?

E5.11. If the crack-parallel stress increases in proportion to the load applied on a notched three-point bend beam, does the type 2 SEL remain applicable?

E5.12. The type 2 SEL is not, in the mean, affected by material randomness. But the type 1 SEL is. Why?

E5.13. List the five variables entering the type 2 size effect problem. How many independent dimensions do they involve? Recalling the Vashy–Buckingham Π-theorem of dimensional analysis, what is the number of independent governing parameters? State them.

E5.14. Show that axisymmetric geometric similarity leads to the same type 2 SEL.

E5.15. Describe the size effect method of fracture energy test. What is the meaning of the resulting fracture energy relative to the cohesive softening curve with a long tail?

E5.16. A size effect test campaign was conducted to characterize the fracture energy of a novel polymeric material. Geometrically similar SENB specimens as shown in Fig. 5.12 were tested. All the dimensions were scaled except for the thickness t, which was kept constant. For all the specimens, $L/D = 4$, $a_0/D = 0.4$. The nominal strength, $\sigma_N = 3PL/2D^2t$ obtained from the experiments is reported in Table 5.1.

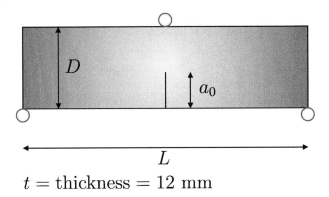

$$t = \text{thickness} = 12 \text{ mm}$$

Fig. 5.12 Single edge notched beam (SENB) specimen.

The Young modulus of the material is $E = 3000$ MPa while Poisson's ratio is $\nu = 0.3$. For all the tests, the initial relative crack length was $\alpha_0 = a_0/D = 0.4$.

1. Calculate the dimensionless energy release rates $g(\alpha_0)$ and $g'(\alpha_0) = dg(\alpha_0)/d\alpha$ starting from the formula for the stress intensity factor provided in Sec. 2.10;

2. Use the following transformation: $X = D$ and $Y = 1/\sigma_N^2$, and plot the experimental data in a Y versus X chart. In such a plot, what type of curve would represent the type 2 SEL?

Table 5.1 Experimental data on geometrically-scaled SENB specimens investigated in problem E5.3.

Specimen width (mm)	Nominal strength (MPa)
$D = 10$	20.34
$D = 10$	19.33
$D = 10$	19.94
$D = 20$	17.47
$D = 20$	18.98
$D = 20$	18.41
$D = 40$	15.85
$D = 40$	16.85
$D = 40$	16.52

3. Perform a linear regression analysis of the data and identify the slope, A, and intercept B of the curve: $Y = AX + B$. How are these parameters linked to the fracture energy and effective length of the FPZ?

4. Construct a plot of $\ln \sigma_N$ versus $\ln D$ with the experimental data and show the fit by means of size effect law.

E5.17. In problem E5.16, the FPZ size is not negligible compared to the specimen size. What would have happened if we used the LEFM model instead of the SEL? For each size D, calculate the fracture energy using LEFM. Are the calculated values the same for every size? Calculate the difference $\Delta = (G_f^{SEL} - G_f^{LEFM})/G_f^{SEL}$ for each size. For what size is the difference the largest? Why?

E5.18. Starting from the SEL derived in problem E5.16, derive the $\mathcal{R}$-curve for the specimen and plot it. Show the energy release rate curves for each size and indicate the related critical points.

E5.19. Following the $\mathcal{R}$-curve derived in problem E5.18, calculate the load-displacement curves for each specimen including the post-peak. The displacement refers to the one of the top pin (Fig. 5.12). If needed, calculate the compliance function $C(u)$ with $C(u)P = u$ analytically.

E5.20. The three-point bend specimen of width $D = 10$ mm shown in Fig. 5.13 failed at a nominal strength of $\sigma_N = 3PL/2tD^2 = 77$ MPa. For this material, a Weibull modulus $m = 12$ was measured. Assuming a 2D scaling, what type of size effect would you expect for this configuration? Using the relevant equation derived in Chapter 5 plot the nominal strength as a function of the structure size D in double-logarithmic scale. Use the following properties: $f_{r\infty} = 300$ MPa, $r = 1.01$, $D_b = 0.0028$ mm, $D_a = 0.0027$ mm. What is the nominal strength when the width is $D = 100$ mm? and when $D = 1$ mm? What would be the nominal strength at $D = 100$ mm if the statistical component of the size effect were neglected?

E5.21. Consider the structure in Fig 2.26. The material used to manufacture the structure is quasibrittle and a finite FPZ develops at the peak load.

1. Introduce your definition of nominal stress in the structure and of relative crack length;

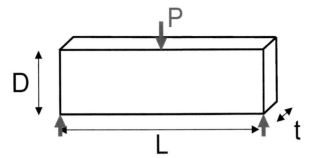

Fig. 5.13 Three-point bend specimen investigated in problem E5.7.

2. Derive the expression of the dimensionless energy release rate and its derivative;

3. Assuming plane strain condition and using an equivalent fracture mechanics approach, write the condition of failure;

4. Assuming the crack front is uniform along the out-of-plane direction, find the expression describing the nominal strength of the structure as function of the structure thickness h;

5. Plot schematically the obtained function in a double-logarithmic plot and describe its asymptotics;

6. Assume the equivalent FPZ length is $0.5l_{ch}$, identify the value of h for which LEFM overestimates the nominal strength of more than 40%.

E5.22. 1. introduce the definition of $\mathcal{R}$-curve and discuss the stability of fracture in quasibrittle media using the structure in problem E5.21 as an example. Sketch the $\mathcal{R}$-curve and energy release curves qualitatively;

2. Consider the geometrically-scaled structures considered in problem E5.21. Discuss the size effect from the viewpoint of the stability of fracture.

E5.23. Show that for a double cantilever beam, $\mathcal{R}$-curve deduced from the type 2 size effect law is independent of the initial crack length.

E5.24. Show that for a crack of length $2a_0$ in an infinite panel subjected to uniform remote tensile stress, the envelope of the $\mathcal{G}_u(\Delta a)$ curves degenerates into a point of coordinates (c_f, G_f). Show that the $\mathcal{R}$-curve can be considered to be defined by the segments of the limiting $\mathcal{G}_u(a)$ curves for $D = 0$ and $D = \infty$ which define the bilinear $\mathcal{R}$-curve: $\mathcal{R} = G_f \Delta a / c_f$ for $0 \leq \Delta a \leq c_f$ and $\mathcal{R} = G_f$ for $\Delta a \geq c_f$.

E5.25. Show that the load-displacement curve expressed as a $\sigma_N - u$ when the behavior is governed by an $\mathcal{R}$-curve deduced from Bažant's SEL always takes the form:

$$\sigma_N = \sigma_1 \phi \left(\frac{u}{u_1}, \frac{c_f}{D} \right), \quad \sigma_1 = \sqrt{\frac{E'G_f}{c_f}}, \quad u_1 = c_f \frac{\sigma_1}{E'}$$

where ϕ is a nondimensional function, which depends implicitly on the structure geometry only.

6

Probabilistic Theory of Quasibrittle Fracture

Abstraction can be perfect, but world is always random.

Since da Vinci's speculation on scaling of structural strength (da Vinci ca 1500s) and Mariotte's demonstration of size effect attributed to randomness of local material strength (Mariotte, 1686), probabilistic theories have played an important role in the long-running efforts to explain the size effect in fracture of solids. The first successful mathematical model was Weibull's statistics of structural strength, which leads to the Weibull size effect. However, it has been shown that the measured strength histograms of quasibrittle structures consistently deviate from the Weibull distribution. The major cause of such deviations is now attributed to two facts: 1) Only the far-out tail of the strength distributions of a material representative volume element (RVE) is a power law, the rest being Gaussian; and 2) The weakest-link model in which the Weibull distribution is anchored cannot be considered *infinite* because the size of the damage localization zone (or the fracture process zone, FPZ) is not negligibly small in comparison to the overall structure size. To match the observed strength statistics of quasibrittle structures, the *finite* weakest-link model of strength statistics, with a finite number of links, has been developed.

In this chapter, we will first derive the classical Weibull statistics of structural strength by using an infinite weakest-link model and discuss the underlying theory of extreme value statistics. We will then present a recently developed finite weakest-link model for the strength distribution of quasibrittle structures. The model is validated by optimum fitting of an extensive database of strength histograms of specimens made of various quasibrittle materials, such as concrete and rocks, engineering and dental ceramics, fiber composites, glass, and asphalt mixtures. The finite weakest-link model predicts a strong size effect on the strength distribution of quasibrittle structures. It is shown that the size effect curve of the mean strength derived from the weakest-link model can be amalgamated, via asymptotic matching, with the deterministic type 1

Quasibrittle Fracture Mechanics and Size Effect: A First Course. Zdeněk P. Bažant, Jia-Liang Le and Marco Salviato, Oxford University Press. © Zdeněk P. Bažant, Jia-Liang Le, Marco Salviato 2022. DOI: 10.1093/oso/9780192846242.003.0006

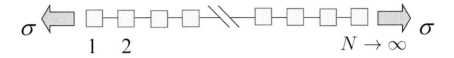

Fig. 6.1 Infinite weakest-link model of strength distribution.

size effect presented in Chapter 4.

6.1   Weibull Statistics of Structural Strength

6.1.1   Infinite Weakest-Link Model

The Weibull distribution (Weibull, 1939) is the most widely adopted probability model for the statistics of structural strength. Due to its wide success in matching observed strength histograms for failure probabilities (for > 0.005, as there are no good data for < 0.005 (Luo and Bažant, 2020)), the Weibull distribution has sometimes been used as an empirical probability distribution for optimum fitting of strength histograms. However, the derivation of the Weibull distribution is anchored in the infinite weakest-link model which, in a broader context, belongs to the class of extreme value distributions (Fisher and Tippet, 1928; Gumbel, 1958).

Consider a structure of positive geometry that fails under load control. For the purpose of calculating the statistics of peak (or maximum) load, the structure can be represented by a chain of links imagined to correspond to the RVEs of the material as shown in Fig. 6.1. Rather, the RVE must be defined as the smallest material element, whose failure can trigger the failure of the whole structure (Bažant and Pang, 2007). This is different from the usual definition of the RVE by means of continuum homogenization of stiffness of heterogeneous elasto-plastic hardening materials (Hill, 1963). The Weibull theory is based on the hypothesis that the number of RVEs is extremely large. This implies that, in comparison to the structure size D, the FPZ is so small that it can be treated as a point. This hypothesis has often gone unmentioned and has usually been considered automatically applicable. But recently it has transpired that, for quasibrittle structures, the number of RVEs potentially failing before the peak load is often far too small to justify infinitely many links in the chain as assumed in the Weibull statistics.

Denote P_k = failure probability of the kth RVE ($k = 1, 2, ..., N$), and P_f = failure probability of the structure as a whole. Since positive geometry is assumed, the structure must fail as soon as one RVE fails. In other words, the structure behaves statistically as a chain, which fails as soon as one link fails (Fig. 6.1).

Equivalently, we may state that the structure will survive if all its RVEs survive. Therefore, the survival probability of the structure, $1-P_f$, must be the joint probability of survival of all the RVEs. If we assume that all P_k are statistically independent (which is acceptable when the autocorrelation length of the random material strength field is not appreciably larger than the spacing of the RVEs), the joint probability theorem gives

$$1 - P_f = (1 - P_1)(1 - P_2) \cdots (1 - P_N) \tag{6.1}$$

$$\text{or} \quad \ln(1 - P_f) = \sum_{k=1}^{N} \ln(1 - P_k) \tag{6.2}$$

Since there are a large number of RVEs in the chain, the failure of the chain must occur at very low stress (or low probability, i.e. $P_i \ll 1$). By introducing the small P_k approximation,

$$\ln(1 - P_k) \approx -P_k \tag{6.3}$$

Eq. 6.2 can now be rewritten as

$$\ln(1 - P_f) = -\sum_{k=1}^{N} P_k \tag{6.4}$$

Based on his extensive experiments, Weibull (Weibull, 1939, 1951) realized that the left (low probability) tail of the cumulative distribution function (cdf) of RVE strength (i.e., failure probability P_k of one RVE) follows a power law, that is,

$$P_k = [\sigma(\boldsymbol{x}_k)/s_0]^m \quad \text{for small } \sigma(\boldsymbol{x}_k) \tag{6.5}$$

Here s_0 and m are material constants; s_0 is called the scaling parameter, m is the Weibull modulus (or shape parameter); and $\sigma(\boldsymbol{x}_k)$ is the positive part of the maximum principal stress at a point of coordinate vector $\boldsymbol{x}_k$. We take the positive part because negative normal stresses do not cause tensile fracture.

By substituting Eq. 6.5 into Eq. 6.4 and replacing the discrete sum by an integral over structure volume V (which is justified if the structure consists of many RVEs), we obtain the well-known Weibull probability integral:

$$-\ln(1 - P_f) = \sum_{k} \left(\frac{\sigma(\boldsymbol{x}_k)}{s_0} \right)^m \approx \int_V \left(\frac{\sigma(\boldsymbol{x})}{s_0} \right)^m \frac{dV(\boldsymbol{x})}{l_0^{n_d}} \tag{6.6}$$

$$\text{or:} \quad P_f = 1 - \exp \left\{ -\int_V \left(\frac{\sigma(\boldsymbol{x})}{s_0} \right)^m \frac{dV(\boldsymbol{x})}{l_0^{n_d}} \right\} \tag{6.7}$$

Here n_d = number of spatial dimensions in which the structure is scaled ($n_d = 1$, 2 or 3), and $l_0^{n_d}$ is the volume of one RVE. The integrand

$$c_f(\boldsymbol{x}) = \frac{[\sigma(\boldsymbol{x})/s_0]^m}{l_0^{n_d}} \tag{6.8}$$

is called the spatial concentration of failure probability. Because the structural strength depends on the minimum strength value in the structure, which is always in the low probability range if the structure is large, Eq. 6.6 is always valid for large structures.

Eq. 6.6 is contingent upon the assumption that the brittle failure of material occurs in tension (rather than shear or a shear-tension combination), and that the random material strength is the same for each spatial direction, i.e., that the strengths in the three principal stress directions are perfectly correlated. Then it is justified to interpret σ in Eq. 6.6 as the positive part of the maximum principal stress at each continuum point.

However, if the random strengths in the principal directions at the same continuum point were statistically independent, then $\sigma^m(x)$ in Eq. 6.6 would have to be replaced by $\sum_{I=1}^{3} \bar{\sigma}_I^m(x)$ where $\bar{\sigma}_I^m(x)$ are the positive parts of the three principal stresses at that point (Freudenthal, 1968; Bažant and Planas, 1998). Nevertheless, considering this kind of statistical independence seems unrealistic.

6.1.2   Size Effect Derived From the Weibull Theory

Consider now geometrically similar structures of different sizes D. In these structures, the dimensionless stress field $\bar{\sigma}(\boldsymbol{\xi})$ is described by a single function of dimensionless coordinate vector $\boldsymbol{\xi} = \boldsymbol{x}/D$, that is, $\bar{\sigma}(\boldsymbol{\xi})$ depends only on structure geometry (or shape) but not on structure size D. In Eq. 6.7, we may then substitute

$$\sigma(\boldsymbol{x}) = \sigma\bar{\sigma}(\boldsymbol{\xi}) \quad (\boldsymbol{\xi} = \boldsymbol{x}/D) \tag{6.9}$$

where σ = nominal stress = P/bD, P = applied load, D = characteristic structure size, b = width of the structure in the transverse direction. Further we may set $\mathrm{d}V(\boldsymbol{x}) = D^{n_d}\mathrm{d}V(\boldsymbol{\xi})$. After rearrangements, Eq. 6.7 yields the standard form of the Weibull strength distribution:[1]

$$P_f(\sigma) = 1 - e^{-(\sigma/S_0)^m} \tag{6.10}$$

where $\qquad S_0 = s_0(l_0/D)^{n_d/m}\Psi^{-1/m}, \quad \Psi = \int_V [\bar{\sigma}(\boldsymbol{\xi})]^m \ \mathrm{d}V(\boldsymbol{\xi}) \tag{6.11}$

Based on the definition of the cdf, we may interpret Eq. 6.10 as the probability for which the nominal strength of the structure is less than the value σ, that is, $P_f = \Pr(\sigma'_N \leq \sigma)$, where $\sigma'_N = P_m/bD$ and P_m = maximum load capacity of the structure, which is intrinsically a random variable. Therefore, σ can also be taken as a prescribed nominal strength level, and Eq. 6.10 essentially represents the strength distribution of the structure. In this sense we may replace σ by σ_N and interpret P_f as the strength cdf.

---

[1]In the special case that all the N links in the chain have the same failure probability P_1, there is a more direct way to derive Eq. 6.10. For an infinite chain,

$$1 - P_f = \lim_{N \to \infty} (1 - P_1)^N = \lim_{N \to \infty} \left(1 - \frac{x}{N}\right)^N, \quad x = NP_1$$

According to the Euler relation, the last limit is equal to e^{-x}. So,

$$P_f = 1 - e^{-x} = 1 - e^{-NP_1} = 1 - e^{-N(\sigma/s_0)^m}$$

where we substituted the tail approximation from Eq. 6.5. Note that, unlike Eq. 6.3, this derivation does not explicitly require assuming P_1 to be small (although, for $N \to \infty$, P_1 of the weakest link must be small).

According to Eq. 6.10, the tail probability of structural failure is a power law:

$$P_f \approx (\sigma_N/S_0)^m \qquad \text{(for } \sigma_N \to 0) \qquad (6.12)$$

For $P_f \leq 0.02$ [or 0.2], its deviation from Eq. 6.10 is $< 1\%$ [or $< 10\ \%$] of P_f.

Eq. 6.10 accounts for the effect of structure geometry (or, equivalently, dimensionless stress field) on the failure probability, which is embedded in integral Ψ. Because the Weibull modulus m in this integral is typically > 20 (and $0.8^{20} = 0.012$), the regions of structure in which the stress is less than about 80% of the maximum stress, $\max[\sigma(\boldsymbol{x})]$, have a negligible effect.

Note that, for a given structure, P_f depends only on the parameter

$$s_0^* = s_0 l_0^{n_d/m} \qquad (6.13)$$

and not on s_0 and l_0 separately. So, the RVE size l_0 is not a material constant and is used here merely for convenience, to serve as a chosen unit of measurement. However, as will be shown in Sec. 6.2, l_0 will play an important role in the subsequent formulation of the finite weakest-link model for quasibrittle structures, and Eq. 6.10 represents an asymptotic case of that model. This is why it is convenient to introduce l_0 here.

Based on Eq. 6.10, we can calculate the nominal strength corresponding to a given failure probability P_f, that is,

$$\sigma_N = C_0 \, (l_0/D)^{n_d/m} \qquad (6.14)$$
$$\text{where} \qquad C_0 = \Psi^{-1/m} s_0 [- \ln(1 - P_f)]^{1/m} \qquad (6.15)$$

Since C_0 and s_0 are independent of D, Eq. 6.14 indicates a power-law scaling of nominal strength of structure when the probability P_f is specified, i.e. $\sigma_N \propto D^{-n_d/m}$. As discussed in Sec. 5.2, the power-law scaling relation implies the absence of characteristic length scale in the problem, which is consistent with the aforementioned discussion on l_0.

Based on Eq. 6.10, we can also calculate the mean nominal strength (Fig. 6.2):

$$\bar{\sigma}_N = \int_0^\infty \sigma p_f(\sigma_N) \mathrm{d}\sigma_N = \int_0^1 \sigma_N \mathrm{d}P_f = \int_0^\infty (1 - P_f) \mathrm{d}\sigma_N \qquad (6.16)$$

where $p_f(\sigma_N) = \mathrm{d}P_f(\sigma_N)/\mathrm{d}\sigma_N =$ probability density function (pdf) of structural strength, while $P_f(\sigma_N) =$ strength cdf (cumulative density function). Note that the last expression of Eq. 6.16 is written based on the fact that σ_N is non-negative.

Substituting Eq. 6.10 into Eq. 6.16 and noting that $\int_0^\infty t^{z-1} e^{-t} \mathrm{d}t = \Gamma(z) =$ gamma function, we obtain, after rearrangements, the well-known Weibull scaling law for the mean nominal strength as a function of structure size D and geometry parameter Ψ;

$$\bar{\sigma}_N(D, \Psi) = s_0 \Gamma(1 + 1/m) = C_s(\Psi) \, D^{-n_d/m} \qquad (6.17)$$
$$\text{where} \qquad C_s(\Psi) = \Gamma(1 + 1/m) \, l_0^{n_d/m} s_0 / \Psi^{1/m} \qquad (6.18)$$

For the gamma function we may use the approximation $\Gamma(1 + 1/m) \approx 0.6366^{1/m}$, which is accurate within the range $5 \leq m \leq 50$ (Eq. 12.1.22 in Bažant and Planas (Bažant and Planas, 1998)).

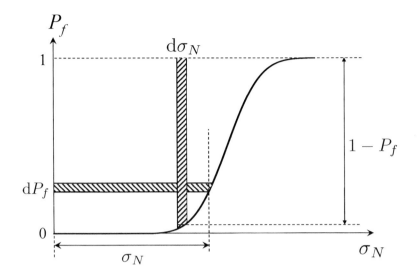

Fig. 6.2 Calculation of mean structural strength.

The coefficient of variation (CoV) of σ_N is calculated as

$$\omega_N{}^2 = \bar{\sigma}_N{}^{-2} \int_0^\infty (\sigma - \bar{\sigma}_N)^2 p_f \mathrm{d}\sigma_N = \bar{\sigma}_N{}^{-2} \int_0^\infty \sigma_N{}^2 \ \mathrm{d}P_f(\sigma_N) - 1 \qquad (6.19)$$

Substitution of Eq. 6.10 gives, after rearrangements, the following well-known expression for the strength CoV:

$$\omega_N = \sqrt{\frac{\Gamma(1 + 2/m)}{\Gamma^2(1 + 1/m)} - 1} \qquad (6.20)$$

This reveals an important property of Weibull strength distribution—ω_N is independent of structure size as well as geometry. Approximately, $\omega_N \approx (0.462 + 0.783m)^{-1}$ for $5 \leq m \leq 50$ (Eq. 12.1.28 in Bažant and Planas (Bažant and Planas, 1998)).

6.1.3 Equivalent number of elements

As discussed earlier, the Weibull distribution (Eq. 6.10) automatically includes the effect of structure geometry (or dimensionless stress field) on the failure probability of the structure. It is useful to introduce the concept of the equivalent number, N_{eq}, of RVEs for which a chain with N_{eq} links gives the same strength distribution. For a chain under the same tensile stress $\sigma = \sigma_N$ in each element, we have

$$P_f = 1 - \mathrm{e}^{-N_{eq}(\sigma_N/s_0)^m} \qquad (6.21)$$

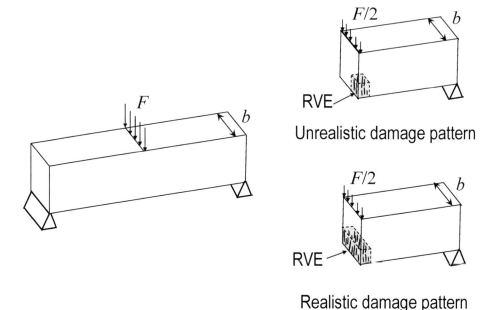

Fig. 6.3 Damage pattern in a three-dimensional structure (Bažant and Le, 2017).

Setting this equal to Eq. 6.10, we can solve for N_{eq}, that is,

$$N_{eq} = (s_0/S_0)^m = (D/l_0)^{n_d} \Psi \tag{6.22}$$

Eq. 6.22 indicates that, in the Weibull statistical theory, the geometrical dependence of the failure probability is equivalent to the size dependence.

It should be pointed out that the equivalent number of RVEs, N_{eq}, is a more convenient alternative to what is called the Weibull stress, σ_W, in the Beremin model (Beremin, 1983) for large crack-tip plastic zone. The Beremin model is defined by setting

$$P_f = 1 - e^{-(\sigma_W/s_0)^m} = 1 - e^{-(V_{eff}/V_0)(\sigma_N/s_0)^m} \tag{6.23}$$

where $V_{eff}/V_0 = (D/l_0)^{n_d} \Psi$. Equating this to Eq. 6.10, we see that

$$\sigma_W = \sigma_N \, \Psi^{1/m} (D/l_0)^{n_d/m} \tag{6.24}$$

that is, the Weibull stress can be regarded as a nominal stress modification taking into account the structure size and geometry.

Eq. 6.23 is the reason why the statistical size effect has often been called the "volume effect." But this term may be misleading. In Eq. 6.22, the integration over volume V of a three-dimensional structure may have to be made two-dimensionally ($n_d = 2$) if the fracture must propagate simultaneously along the whole width of the fracture front, since mechanics prevents the fracture from extending within only a small portion of the width (Fig. 6.3) and not within the rest. According to the "volume effect," the widening of a narrow beam, which increases the structure volume, would reduce the mean nominal strength $\bar{\sigma}_N$ of the beam, while experience shows virtually no effect of the widening on $\bar{\sigma}_N$. For this reason, calling the statistical size effect the volume effect is not quite accurate and can be misleading.

6.1.4 Extreme Value Statistics and Stability Postulate

We will now discuss the Weibull distribution in a broader context of the theory of extreme value statistics. As discussed in Sec. 6.1.1, the Weibull distribution is derived from an infinite weakest-link model. In this model, the strength of the structure (or chain) is determined by the strength of the weakest element. Therefore, the Weibull distribution essentially represents a probability distribution of the minimum value of a large set of independent and identically distributed random variables. This is why the Weibull distribution belongs to the class of extreme value distribution functions (Gumbel, 1958). Here we consider the case of a chain of elements subjected to a uniform stress σ, since generalization to the case of non-uniform stress distribution (e.g., elements of different cross-sectional areas) is straightforward.

The essential idea to derive the extreme value distribution purely mathematically is to subdivide the set of all elements (or links), N in number, into n identical subsets, each of which contains ν elements. To obtain the asymptotic behavior, we must consider each subset size ν to tend to infinity, that is,

$$\nu \to \infty \tag{6.25}$$

whereas the number of subsets, n, can take any finite value. Obviously, the total number of elements in the whole set is also infinite, i.e., $N = n\nu \to \infty$. Let $f(\sigma)$ and $F(\sigma)$ be the cumulative probabilities of survival, under stress σ, of each infinite subset and of the whole infinite set, respectively. Since the whole set will survive if and only if all the subsets survive, the joint probability theorem gives

$$F(\sigma) = f^n(\sigma) \tag{6.26}$$

And since the whole set as well as the subsets are infinite, their survival distributions must be of the same type, though having different mean and standard deviation since the smallest random strength value among n subsets must be equal or smaller than that in one subset. Therefore, the distribution functions $F(\sigma)$ and $f(\sigma)$ must be related by a linear transformation, that is,

$$F(\sigma) = f(a_n\sigma + b_n) \tag{6.27}$$

where a_n and b_n are coefficients depending on number n. Combining Eqs. 6.26 and 6.27, we obtain the following functional equation for the unknown asymptotic cumulative survival distribution $f(\sigma)$:

$$f^n(\sigma) = f(a_n\sigma + b_n) \tag{6.28}$$

This functional equation is called the *stability postulate of extreme value statistics*—'stability', because, in passing from one subset to the whole set, the distribution must remain of the same type, in other words, stable.

In arguably one of the most famous probability papers of the last century, Fisher and Tippett (Fisher and Tippet, 1928) showed that, remarkably, there exist *three and only three* asymptotic distributions that satisfy the stability postulate (Eq. 6.28) (some restrictions were later added by Gnedenko (Gnedenko, 1943)). These distributions are called the extreme value distributions (or stable extreme value distributions). In terms of failure probability under stress σ, $P_f(\sigma) = 1 - f(\sigma)$, they are[2]

$$\text{Gumbel (Type 1) distribution:} \quad P_f(\sigma) = 1 - e^{-e^{\sigma}} \quad (6.29)$$

$$\text{Fréchet (Type 2) distribution:} \quad P_f(\sigma) = 1 - e^{\sigma^{-m}} \quad (6.30)$$

$$\text{Weibull (Type 3) distribution:} \quad P_f(\sigma) = 1 - e^{-\sigma^{m}} \quad (6.31)$$

Difficult though it is to show (Fisher and Tippet, 1928) that the Gumbel, Frechet, and Weibull distributions are the only distributions satisfying the stability postulate (Eq. 6.28) (except for some far-fetched cases of no relevance to strength), it is quite easy to check that they satisfy the postulate. For example, if we substitute the Weibull distribution, Eq. 6.10, we get:

$$(e^{-\sigma^m})^n = e^{-(a_n\sigma + b_n)^m} \quad (6.32)$$

$$n\sigma^m = (a_n\sigma + b_n)^m \quad (6.33)$$

$$(n^{1/m} - a_n)\sigma = b_n \quad (6.34)$$

Since the last equation must be valid for any σ, we conclude that

$$a_n = n^{1/m}, \quad b_n = 0 \quad (6.35)$$

These are the two conditions for satisfying the stability postulate, and they agree with the scaling we derived.

For a given physical problem, the type of the extreme value distribution is determined by the so-called domain of attraction (Gnedenko, 1943; Gumbel, 1958; Ang and Tang, 1984; Bouchaud and Potters, 2000; Vanmarcke, 2010). Gnedenko (Gnedenko, 1943) further derived the conditions for the elementary distribution function to converge to one of the aforementioned three extreme value distributions. The combined results of Fisher, Tippett, and Gnedenko are usually referred to as the Fisher–Tippett–Gnedenko Theorem of extreme value statistics. The theorem states that, if the distribution of the normalized maximum (or minimum) converges to a non-degenerate distribution, then the limiting distribution $f(\sigma)$ would belong to one of the three extreme value distributions (Eqs. 6.29–6.31). It should be mentioned that not all elementary distribution functions would lead to one of these three extreme value distributions

---

[2]Since both Eqs. (6.30) and (6.31) were derived by Fisher and Tippett (Fisher and Tippet, 1928), they should, in fairness, be called the Fisher–Tippett–Weibull and Fisher–Tippett–Gumbel distributions. Why have Fisher and Tippett's names been omitted? — Probably because they demonstrated no applications nor experimental comparisons. Weibull and Gumbel did both, extensively so. Fréchet (Fréchet, 1927) preceded Fisher and Tippett.

(Vanmarcke, 2010). However, the distributions whose extreme value distribution are not Weibull, Gumbel or Fréchet are rather unusual and do not occur in strength statistics.

For the Weibull distribution, the domain of attraction includes all the elementary distributions with a left tail in the form of a power law of non-negative numbers; for Gumbel, all those with an exponential tail decaying to minus infinity; and for Fréchet, all those with an inverse power law infinite tail decaying to plus infinity. Among these three, only the Weibull distribution is suitable for strength of structures following the weakest-link model. This indicates that strength distribution of one RVE should have a power-law tail. As will be discussed in the following section, this power-law tail distribution can be justified physically, by combining the transition rate theory for failure of nanoscale structures and a multiscale transition model of failure statistics.

6.2 Finite Weakest-Link Model of Strength Distribution of Quasibrittle Structures

The foregoing discussion indicates that the validity of the Weibull distribution is anchored by the assumption of the infinite weakest-link model. This assumption is valid for brittle structures failing at macrocrack initiation, in which the FPZ is negligibly small compared to the structure size. This is why the Weibull distribution provides an optimum fitting of strength histograms of fine-grain ceramics specimens. However, the infinite weakest-link model is not applicable to quasibrittle structures, where the FPZ size is not negligible and sets a characteristic length scale.

Evidently, to modify the weakest-link model for strength statistics of quasibrittle structures, we must relax the restriction that there are an infinite number of RVEs in the chain, that is, N in Eq. 6.1 must be finite. Because N is finite, what matters for the probability distribution of the structure is no longer the tail distribution of the RVE strength. To formulate a general finite weakest-link model, we need to know the functional form of the entire distribution function of RVE strength.

In this section, we present a multiscale statistical model, conceived in 2006 (Bažant and Pang, 2006) to determine the functional form of the strength distribution of one RVE, and the consequent probability distribution function of the nominal strength of the structure; for detailed development, see Bažant and Le (2017).

6.2.1 Failure Statistics at Nanoscale as the Basis of Power-Law Probability Tail

Damage and failure of macroscopic structure originates from fracture of its nanoscale structure, for example, a completely disordered system of nanoparticles of the calcium silicate hydrate (C-S-H) in concrete or a regular atomic lattice such as a single crystal grain of ceramics. Therefore, it appears logical to begin with the failure statistics of a nanoscale structure.

Another important reason to study nanoscale fracture is that there exists a well-established analytical model for the interatomic bond breakage. This analytical model is anchored by the atomistic rate process theory (Eyring, 1936; Kramers, 1940; Glasstone *et al.*, 1941; Tobolsky and Erying, 1943; Krausz and Krausz, 1988; Kaxiras, 2003), which theoretically justifies the Arrhenius thermal factor and has long been used to transit from the atomic scale to the material scale, to obtain the temperature and stress dependence of the rates of chemical reactions, phase changes, creep, diffusion, adsorption, etc.

The third reason is that the frequency of breakage of interatomic bonds is almost always orders of magnitude smaller then the frequency of thermal vibrations of atoms. This implies that the breakage process is quasi-stationary, in which case the probability is equivalent to the frequency. The concept of stationarity should seen in the atomic scale perspective. For most dynamic loading (e.g. projectile impact, blast), the crack velocity is at most 30–50% of the Rayleigh wave speed (Buehler and Gao, 2006; Bhat *et al.*, 2012). Consider a nanocrack propagating at a velocity of 10^3 m/s. The average frequency of bond breakage is about 10^{12}/s, while the frequency of random thermal vibrations of atoms is about 10^{14}. We may conclude that, even under impact, an atom vibrates about 100 times on average before a jump over the activation barrier, i.e., bond break. So the atomic breakage process can be treated as quasi-stationary at the nano-level. Obviously, this includes all mechanical processes (though not the nuclear chain reaction). The main point of this argument is that, for a stationary process, the frequency is equivalent to probability.

Consider now that a nanoscale structure, either a disordered system of nano-particles (e.g., C-S-H particles (Allen *et al.*, 2007)) or an atomic lattice block (Figs. 6.4a and b) is subjected to applied remote stress τ (or its resulting load P). The process of crack propagation in the nanoscale structure involves a sequence of breakages of atomic bonds or disordered nanoparticle connections. As the crack propagates, the separation δ between the opposite atoms or nanoparticles across the nanocrack cannot be sudden but increases gradually, by small jumps. Fig. 6.4c shows such a crack-tip deformation predicted by MD simulation (Buehler and Keten, 2010).

The force F_b transmitted between the opposite atoms or particles is obtained by differentiating the interatomic or inter-particle potential function with respect to the separation. So, $F_b = \partial \Pi_b(\delta)/\partial \delta$ where $\Pi_b(\delta)$ is the corresponding interatomic potential, which could, for example, be approximated as the Morse potential (Morse, 1929), Lennard-Jones potential (Lennard-Jones, 1924), Tersoff potential (Tersoff, 1988), and Stillinger-Weber potential (Stillinger and Weber, 1985). However, for the current purpose, the precise form of the interatomic potential function is not necessary because the aim is to obtain only the basic *functional form* of the failure probability of a nanoscale structure.

Fig. 6.5 represents schematically the behavior of different pairs of atoms and nano-particles along the crack path. The pair with the peak bond force defines the front of the nanoscale cohesive crack. At the front, the slope of the potential curve $\Pi_b(\delta)$ reaches its maximum slope (state 3 in Fig. 6.5b–c). The real nanocrack terminates at the pair at which the bond force F_b vanishes (state 5 in Fig. 6.5b–c) and the adjacent atom pairs undergo a large and irreversible separation. The atomic pairs between states 3 and 5 represent the cohesive zone (or the FPZ), in which the bond force decreases with an increasing bond separation. In fracture mechanics, the decreasing bond force is the idea originally due to Barenblatt, who pioneered the concept of cohesive crack model in 1959 (Barenblatt, 1959). At state 4, the nanoscale structure reaches the limit of stability, the bond force drops to zero, and the crack opens widely, which

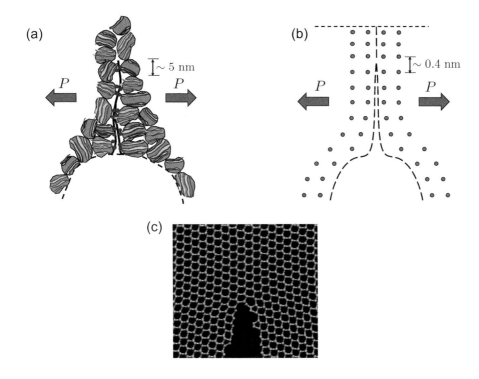

Fig. 6.4 Fracture of a nanoscale structure: (a) disordered system of C-S-H nano-particles in cement paste (Allen *et al.*, 2007), (b) regular atomic lattice block, and (c) deformation of a nano-crack simulated by molecular dynamics (Buehler and Keten, 2010).

represents the macrocrack front. The critical state of stability is characterized by the downward slope of the curve $\Pi_b(\delta)$ becoming equal in magnitude to the stiffness of the confinement within the surrounding solid. State 5, at which the separation of the opposite atoms or nano-particles greatly increases, must lie close to state 4 since the transition between these two post-critical states is unstable, and thus dynamic and fast.

Consider a continuum approximation of the fracture behavior of the nanoscale region surrounding one crack, subjected to remote load P. As mentioned earlier, the cohesive behavior of the bond can be described by a force-separation law. In a homogenized continuum approximation, we may convert this force-separation law to a cohesive law $\sigma_b = g(\delta)$, where $\sigma_b = F_b/\delta_a^2$ (δ_a = atomic spacing or spacing of nano-particle connection). Based on Sec. 3.8, the behavior of the cohesive zone can be described by the following two equations:

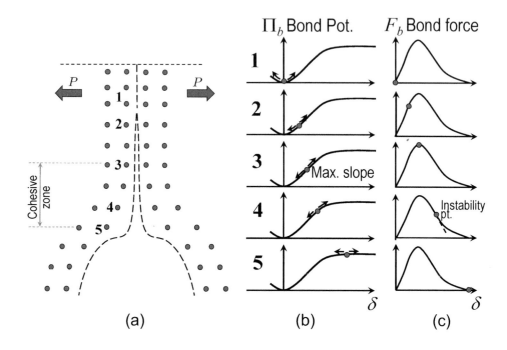

Fig. 6.5 Behavior of interatomic bond or inter-nanoparticle connection along a nanocrack under mechanical loading.

$$-\int_{a_0}^{a} C(x,x')\sigma_b(x')\mathrm{d}x' + C_\tau(x)\tau = g^{-1}[\sigma_b(x)] \tag{6.36}$$

$$-\int_{a_0}^{a} k(x)\sigma_b(x)\mathrm{d}x + k_\tau\tau = 0 \tag{6.37}$$

where $\tau = P/bl_a$ nominal applied stress (l_a = characteristic size of the nanoscale structure, b = width in the third dimension), a_0 = length of the traction-free crack, a = length of the cohesive crack, $C(x,x')$ = compliance function relating the crack opening at x produced by unit traction at x', C_τ = compliance function relating the crack opening at x produced by unit applied nominal stress, $k(x)$ = dimensionless stress intensity factor caused by a unit traction applied at x, and k_τ = dimensionless stress intensity factor caused by the applied nominal stress.

From Eqs. 6.36 and 6.37, one can, in principle, solve the cohesive stress profile $\sigma_b(x)$ and the size of the cohesive zone for a given far-field loading τ. If the remote applied stress τ is kept constant, this profile is stationary during propagation. The energy input to the cohesive zone can be calculated by

$$\Delta Q = \int_{a_0}^{a} \gamma x \int_{0}^{\delta(x)} \sigma_b(\delta')\mathrm{d}\delta'\mathrm{d}x \tag{6.38}$$

where γx = perimeter of the crack front at x if we consider axisymmetric crack growth. Note also that the equivalent linear elastic fracture mechanics can be used to calculate

$$\Delta Q = \gamma a_e \delta_a \frac{K_a^2}{E_1} \tag{6.39}$$

Here α_e = equivalent crack length defined in such a way that Eq. 6.38 gives the same energy dissipation as Eq. 6.39, K_a = stress intensity factor of the equivalent crack, γa_e = perimeter of the front of the equivalent crack (γ_a = constant, equal to 2π is the crack is circular), and E_1 = elastic modulus of the nanostructure. Furthermore, $K_a = \tau \sqrt{l_a} k_a(\alpha_e)$, where $\alpha_e = a_e/l_a$ = relative crack length, and $k_a(\alpha_e)$ = dimensionless stress intensity factor. Clearly α_e is governed by the cohesive law as well as the geometry.

Furthermore, stress τ applied on the nanoscale structure can be related to the macroscopic nominal stress σ by setting $\tau = c\sigma$ where c is a certain stress concentration factor. Substitution of the expression of K_a into Eq. 6.39 yields

$$\Delta Q = V_a(\alpha_e) \frac{c^2 \sigma^2}{E_1} \tag{6.40}$$

where $V_a(\alpha_e) = \delta_a(\gamma_a \alpha_e l_a^2) k_a^2(\alpha_e)$ = activation volume (note that, in the atomistic theories of phase transformations in crystals (Aziz *et al.*, 1991), one may generally define the activation volume as a tensor of the form $v_a = \partial\Delta Q/\partial\boldsymbol{\sigma}$, where $\boldsymbol{\sigma} =$ macroscopic stress tensor).

The foregoing analysis is based on a homogenized continuum. But, at the nanoscale, the crack propagation is discrete and represents a thermally activated process. So let us now focus on the transition between two metastable states of a nanoscale structure: State 1 before the equivalent nanocrack propagates, and state 2 after the equivalent nanocrack propagates by one spacing δ_a (which is either an atomic spacing in a crystal or the spacing between two nanoparticles of disordered structure).

It should be emphasized that the transition from state 1 to state 2 represents physically the propagation of the entire cohesive zone. The bond breakages are, of course, random, and so all the bonds in cohesive zone do not actually move forward simultaneously. But in the sense of statistical expectations they must, or else the length of the cohesive zone could not stay constant. Since this transition is a thermally activated process, it can be characterized by jumps over the activation energy barriers. As pictured in Fig. 6.6, the jumps occur in both forward and backward directions, in which the forward jump means crack propagation while the backward jump means crack or bond healing.

The energy input to the cohesive zone due to the applied load creates a directional bias on the potential energy of these two states. Therefore, the energies required for the forward and backward jumps differ and may be written as $Q_0 - \Delta Q/2$ and $Q_0 +$

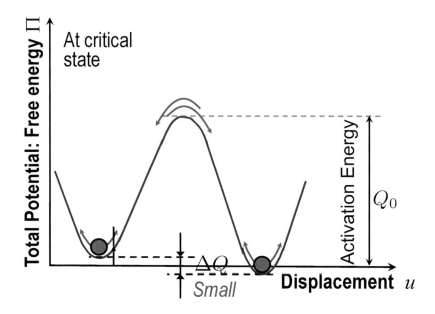

Fig. 6.6 Forward and backward jumps over a biased energy barrier.

$\Delta Q/2$, respectively (Fig. 6.6), where Q_0 = activation energy at no stress and ΔQ is proportional to σ^2 (Eq. 6.40). It is well known that the natural energy scale for chemical bonds, and thus also for the activation barriers to molecular rearrangements between long-lived well-defined molecular states, is the electron-volt (1 eV $= 1.60218 \times 10^{-19}$ J). This scale is more than an order-of-magnitude larger than the scale of thermal energy ($kT = 0.025$ eV) at room temperature (where $k = 1.381 \cdot 10^{-23}$ J/K $=$ Boltzmann constant and $T =$ absolute temperature). Therefore, we conclude that the energy barrier Q_0 of the entire cohesive zone must be much greater than the thermal energy, that is, $Q_0 \gg kT$.

Consequently, the transition between two adjacent states must be slow, relative to the thermal vibrations of atoms, and the process must be quasi-stationary. Due to the quasi-stationarity, the frequency of transition between two adjacent states must be proportional to the probability of the propagation or healing of the equivalent nanocrack. This is a crucial feature, already pointed out.

Meanwhile, we note that, for Eq. 6.40, macroscopic stress σ is in the order of MPa, elastic modulus E_1 is in the order of GPa, and atomic spacing δ_a is in the order of $10^{-9} - 10^{-10}$m. Therefore, ΔQ is about one order of magnitude smaller than the thermal energy kT, which indicates that $Q_0 \pm \Delta Q/2 \gg kT$. For the limiting case of a large free-energy barrier, the transitions between the subsequent local minima of free energy can typically be described by the asymptotic Kramers formula for the first passage time, which indicates an exponential dependence on the barrier energy relative to kT (which is also an Arrhenius dependence on temperature). The net frequency of the forward crack front jumps is given by (Kramers, 1940; Risken, 1989)

$$f_1 = \nu_T (e^{(-Q_0 + \Delta Q/2)/kT} - e^{(-Q_0 - \Delta Q/2)/kT}) \tag{6.41}$$

$$= 2\nu_T e^{-Q_0/kT} \sinh[\Delta Q/2kT] \doteq 2\nu_T e^{-Q_0/kT} \sinh[V_a(\alpha_e)/V_T] \tag{6.42}$$

where $V_T = 2E_1 kT/c^2 \sigma^2$; ν_T is the characteristic attempt frequency for the reversible transition, and $\nu_T = kT/h$ where $h = 6.626 \cdot 10^{-34}$ Js = Planck constant = (energy of a photon)/(frequency of its electromagnetic wave). Note that, generally, there are multiple activation energy barriers: $Q_0 = Q_1, Q_2, \ldots$ However, the lowest one always dominates the process. The reason is that the factor $e^{-Q_1/kT}$ is very small, typically 10^{-12}. Thus, if for example $Q_2/Q_1 = 1.2$ or 2, then $e^{-Q_2/kT} = 0.0043e^{-Q_1/kT}$ or $10^{-12}e^{-Q_1/kT}$, and so the higher barrier makes a negligible contribution. And if for example $Q_2/Q_1 = 1.02$, then Q_1 and Q_2 can be replaced by a single activation energy $Q_0 = 1.01Q_1$.

Earlier we estimated that $\Delta Q \ll kT$, which implies that the argument V_a/V_T in Eq. 6.41 is $\ll 1$. By considering $\sinh x \approx x$ for small x, we can rewrite Eq. 6.42 as follows(Bažant *et al.*, 2009):

$$f_1 \approx e^{-Q_0/kT}[\nu_T V_a(\alpha_e)/kT] \, c^2 \sigma^2/E_1 \tag{6.43}$$

It must be emphasized that the foregoing analysis considers the thermally activated jumps of the entire cohesive zone, in the sense of statistical expectation. The breaks of each atomic bond pair or each nano-particle bond pair are random and their rates momentarily differ but, because the advancing cohesive zone keeps constant length, Eq. 6.41 for frequency f_1 must give for all the bonds breaks the same statistical expectation. For our purpose, it is unimportant to determine the actual random distribution of cohesive stress. What matters is the statistical expectation (or mean), which characterizes the overall probabilistic behavior of the cohesive zone under external loading and is governed by the total energy input into the cohesive zone (Eq. 6.40).

In a homogenized continuum approximation, the load-deflection curve must be smooth as shown by the dashed curve in Fig. 6.7a. Since the equivalent LEFM crack in an atomic lattice or a disordered nanostructure advances by jumps, undulations are superimposed on the smooth load-deflection curve as shown in Fig. 6.7a. Since there are many bonds along the cohesive zone, many undulations must be superimposed along the descent of the curve of potential Π or the corresponding curve of load P. The large number of such undulations is another reason why the difference ΔQ between the subsequent valleys of the potential, shown in Fig. 6.7b and in Fig. 6.6, must be very small, which is the fact we used to justify Eq. 6.42 (Bažant *et al.*, 2009; Le *et al.*, 2011).

The maximum load P is characterized by the crack propagating at constant load or, at the nanoscale, at constant τ, which is what we assume from the outset. Having argued that Eq. 6.43 for jump frequency f_1 must apply, in the sense of statistical expectation, to the movement of the cohesive zone as a whole, we may conclude that the probability P_f of failure load or maximum load must be proportional to f_1, i.e.,

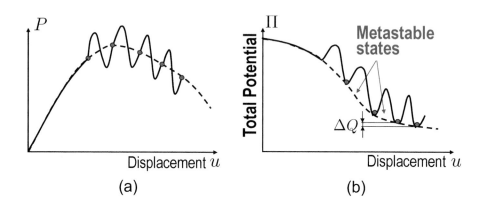

Fig. 6.7 Overall mechanical response of a nanoscale structure subjected to loading: (a) load-displacement curve, and (b) total potential-displacement curve.

$P_f(\sigma) \propto f_1$. Setting aside the dependence on temperature, we arrive at the conclusion that, at or near the atomic scale,

$$P_f(\sigma) \propto \sigma^2 \tag{6.44}$$

An essential point here is that the power law describing the strength distribution of a nanoscale structure has a zero threshold. The zero threshold is the direct consequence of the consideration of both forward and backward jumps over the energy barrier.[3] A non-zero threshold would be incompatible with the Kramers rule of transition rate theory and thus physically impossible (which is a point recalled in Sec. 6.4 with regard to the three-parameter Weibull distribution).

In Eq. 6.42 we considered the net frequency f_1 of a single jump, and thus ignored the possibility of continuous backward jumps towards crack healing, which would affect the failure probability. A rigorous analysis recognizing this fact (Le *et al.*, 2011; Bažant and Le, 2017; Xu and Le, 2019) may be based on the Fokker–Planck equation (Redner, 2001; Risken, 1989) which rests on the fact that the nanoscale bond jumps represent an independent quasi-stationary process (Krausz and Krausz, 1988). This equation combines drift, i.e., forward stress driven propagation as considered above, with diffusion, i.e., propagation with no forward-backward bias, like a random walk.

The solutions of Fokker–Planck equation are characterized by a dimensionless number $\text{Pe} = 2(l_a/\delta_a)(V_a/V_T)$ called Péclet number. When $\text{Pe} \ll 1$, diffusion or random

---

[3]Studying structure lifetimes, Zhurkov (Zhurkov, 1965; Zhurkov and Korsukov, 1974) considered jumps over activation energy barriers, but ignored the backward jumps and assumed there was no potential valley, no stable state, after the jump. His calculation led to an exponential instead of sine hyperbolic function, which overestimated lifetime at low stress by orders of magnitude (Bažant and Le, 2017).

walk dominates. It was shown (Le *et al.*, 2011; Xu and Le, 2019) that for nanoscale fracture this would occur for $P_f \approx 10^{-13}$, which would introduce into the distribution of P_f an exponential tail. But this probability is so small that it is of no practical interest. In fracture mechanics, the Péclet number is in the order of 10, which suffices for the diffusion aspect to be negligible for $P_f \geq 10^{-6}$, and thus of no interest. Thus the solution of Fokker–Planck equation shows that the tail distribution of nanoscale structure strength that is of practical interest (i.e., $P_f > 10^{-13}$) indeed follows a power law: $P_f \propto \sigma^2$ (Le *et al.*, 2011; Xu and Le, 2019).

6.2.2   Multiscale Transition of Failure Statistics

Modeling the transition of the failure statistics from the nanoscale structure to the macroscopic RVE is a difficult task. Though various multiscale computational models have been developed, they are not suitable for modeling the scale transition of failure statistics because the uncertainties of physical laws across the scales are hard to quantify and meanwhile direct numerical simulations are unable to predict the tail behavior of the failure statistics. The existing stochastic finite element method used at one scale can hardly predict more than the coefficient of variation on the next higher scale. In view of these limitations, we need to rely on analytical models to describe this transition.

For modeling the failure statistics of materials and structures, two classical statistical models for the entire distribution are tractable analytically. The first is the chain model, as known as the weakest-link model or series coupling model, in which the structure is represented by a chain of material elements and the strength of each element is intrinsically random. The mathematical formulation of the chain model follows Eq. 6.1. Based on Eq. 6.1, it is evident that the strength cdf of the chain has the following simple but important asymptotic properties:

1. If the strength cdfs of all the elements have a power-law tail of exponent p, then the strength cdf of the entire chain would also have a power-law tail and its exponent is also p; and

2. When the number of elements in the chain tends to infinity, we obtain the Weibull strength distribution as indicated by Eq. 6.21.

In modeling the failure of quasibrittle structures, the chain model physically represents the damage localization mechanism, for example, the transverse localization of failure into a single FPZ among many potential FPZs (Fig. 6.8a). Meanwhile, within the FPZ, the deformation of several damaged elements located side-by-side must be compatible with the overall deformation of the FPZ on the next higher scale (Fig. 6.8b). This compatibility condition can be statistically represented by a fiber-bundle model, which is the second basic statistical model that we will discuss next.

The second classical analytically tractable model of the entire distribution is the fiber bundle (or parallel coupling) model, shown in Fig. 6.9a. This model captures a load redistribution mechanism in which, after one element (usually called the "fiber") fails, the load will get redistributed among all the surviving fibers. The load capacity of the bundle is reduced to zero when all the fibers break. However, the maximum load capacity of the bundle is attained when only a fraction of the fibers breaks. This

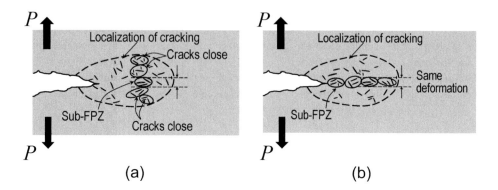

Fig. 6.8 Meso-scale behavior of a macroscopic FPZ: (a) localization in sub-FPZs, and (b) deformation compatibility of sub-FPZs.

fraction is an intrinsic random variable since the mechanical behavior of each fiber has a random nature.

Investigation of the strength distribution of the fiber-bundle has a long and rich history (Daniels, 1945; Coleman, 1958a,b; Harlow and Phoenix, 1978a,b; Phoenix, 1978b,a; Harlow *et al.*, 1983; Mahesh and Phoenix, 2004; Bažant and Pang, 2007; Le *et al.*, 2011; Salviato and Bažant, 2014; Bažant and Le, 2017). Previous studies have shown that the strength distribution of the fiber-bundle is influenced strongly by many factors, such as the strength cdf of fibers, the load-sharing rule, as well as the mechanical behavior of each fiber. In the literature, many intuitive hypotheses have been proposed for the load-sharing rule (Daniels, 1945; Phoenix, 1978b; Phoenix and Tierney, 1983; Mahesh and Phoenix, 2004). A more realistic approach, though, is to deduce the load-sharing rule from a mechanical model.

Here we consider initially elastic fibers connected by two parallel rigid plates. Since any fiber can be replaced by several fibers having different cross-sectional areas but the same combined elastic stiffness, it is not unduly restrictive to assume that all the fibers have equal elastic stiffness E_f and equal cross-sectional area A_f, and are subjected to the same displacement. The load-sharing rule is then fully determined by the mechanical behavior of the fibers. Three types of fiber behavior after reaching the strength limit are considered:

1. Brittle (Fig. 6.9b), in which case the stress drops suddenly to zero;
2. Plastic (Fig. 6.9c), in which case the fiber extends at constant stress; and
3. Softening (Fig. 6.9d), in which case the fiber exhibits a gradual strain softening.

The strength distribution of bundles with the aforementioned fiber behaviors has been studied extensively, for example, see Daniels (1945), Smith (1982), Bažant and Pang (2007) for brittle bundles; Bažant and Pang (2007) for plastic bundles; and Le

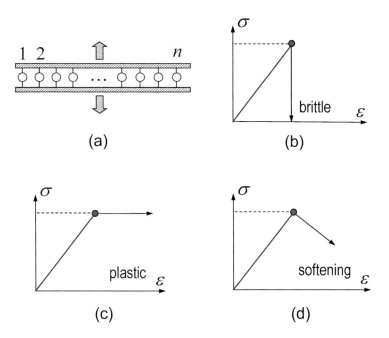

Fig. 6.9 Behavior of fiber-bundle model: (a) schematic of fiber-bundle model, (b) brittle fibers, (c) plastic fibers, and (d) softening fibers.

et al. (2011), Salviato and Bažant (2014), Bažant and Le (2017) for softening bundles. As shown in Sec. 6.2.1, the nanoscale structure has a strength cdf with a power-law tail. Therefore, the interest here is to study the strength distribution of a fiber-bundle, in which the strength cdf of each fiber has a power-law tail. It has been shown that, regardless of the fiber behavior, the strength distribution of this type of bundle has the following common properties:

1. If the power-law tail of the strength cdf of the i-th fiber has an exponent p_i, then the strength cdf of the bundle would also have a power-law tail and its exponent is equal to the sum of the exponents of the power-law tails of all the fibers. The preservation of the power-law tail and the additivity of the exponent were proven first for brittle bundles by using set theory (Harlow *et al.*, 1983; Phoenix *et al.*, 1997) and by the asymptotic expansion method (Bažant and Pang, 2007). This property was later shown to hold for plastic bundle (Bažant and Pang, 2007) as well as for general softening bundles (Le *et al.*, 2011; Salviato and Bažant, 2014; Bažant and Le, 2017).

2. As the number of fibers increases, the extent of the power-law tail is drastically shortened. The rate of shortening of the power-law tail depends on the mechanical behavior of the fibers. Let P_{t1} denote the probability at which the power-law tail of the strength cdf of a single fiber terminates, and P_{tn} denote the probability at which the power-law tail of the strength cdf of the bundle ends. It has been shown that the power-law tail of a brittle bundle shortens as fast as $P_{tn} \sim (P_{t1}/n)^n - (P_{t1}/3n)^n$, and the rate of shortening for the plastic bundles is about $P_{tn} \sim (P_{t1}/3n)^n$ (Bažant and Pang, 2007), where n is the number of fibers in the bundle. The behavior of the softening bundle lies in between the brittle and plastic bundles. The detailed analysis of the tail distribution of fiber bundles can be found in Bažant and Le (2017).

3. As the power-law tail drastically shortens with an increasing number of fibers, the core of the strength cdf converges to a Gaussian (or normal) distribution. For brittle bundles, the convergence to the Gaussian distribution was first proven by Daniels (Daniels, 1945). Smith (Smith, 1982) later determined the convergence rate to the asymptotic Gaussian distribution for large-size brittle bundles. For plastic bundles, the convergence to the Gaussian distribution is an obvious consequence of the central limit theorem. The most general proof (Le *et al.*, 2011) that, for any post-peak fiber properties, the limiting strength distributions is Gaussian, may be obtained by extracting from an infinite bundle two infinite sub-bundles such that, for example, every third fiber among fibers ordered by strength is placed into one bundle and the remaining ones into the other bundle. The force carried by the original bundle is the sum of the forces in the two extracted sub-bundles. Now note that all these bundles must have the same probability distribution of strength. This is known to be possible if and only if all the distributions are Gaussian (Le *et al.*, 2011).

The essential question is how to use the chain and bundle models to deduce the statistical properties of one macroscale RVE from the properties the nanoscale structures whose failure statistics has been derived in Sec. 6.2.1. It is evident that one microcrack (or micro-slip) in a RVE, or too few of them, would not cause the RVE to fail. Rather, a certain number of microcracks must form and coalesce to cause the RVE to fail. This implies that the behavior of one RVE must be close to the fiber bundle. But can a pure fiber bundle model be used to represent the behavior of one RVE?

To answer this question, we first note that, as implied by the weakest-link model, the Weibull modulus of the strength distribution of macroscale structures must be equal to the power-law tail exponent of the cdf of strength of one RVE. Meanwhile, in Sec. 6.2.1 we showed that the strength cdf of the nanoscale structure follows a power-law function with an exponent equal to 2 (Eq. 6.44). Now consider, as an example, concrete material, whose Weibull modulus is about 24 (Bažant and Novák, 2000*a*; Bažant, 2005). This means that the power-law exponent of the tail distribution of RVE strength is equal to 24. If we now modeled the RVE as a bundle of elements, each of which represents one nanoscale structure, then the bundle would have 12 elements (because $12 \times 2 = 24$ where 2 is the tail exponent of a nanostructure). According to the aforementioned properties of bundle strength, the power-law tail of the strength cdf of one RVE would reach only to the probability of about $P_f \approx 10^{-20}$. This implies that a chain that statistically represents the entire macroscale structure would have to consist of 10^{22} RVEs for the Weibull distribution to get manifested. What follows is that, on the scale of real structures, one would never observe Weibull distribution since the RVE size of concrete is in the order of 0.1 m, yet it is observed (Weibull, 1939).

As a matter of fact, based on experimental observations, the strength cdf of positive geometry structures is already much closer to the Weibull distribution than to the Gaussian distribution when the number of RVEs in the structure exceeds 10^3 or 10^4. Based on the weakest-link model (Sec. 6.1.1), this implies that the power-law tail distribution of RVE strength must reach up to $P_f \approx 10^{-4}$ or 10^{-3}. Therefore, one RVE cannot be statistically represented by a bundle model alone, or else the power-law tail would be too short by orders of magnitude. On the other hand, it is noted that the chain model extends the power-law tail roughly by one magnitude for each ten-fold increase in the number of elements in the chain (Bažant and Le, 2017). Therefore, the chain model must be active to increase the reach of the power-law tail.

Based on the foregoing discussion, we must introduce a statistical model consisting of both chains (i.e., series coupling) and bundles (i.e., parallel coupling). The model must satisfy two basic requirements:

1. the model must raise the tail exponent of the strength cdf from 2 at the nanoscale to about 10–50 at the macroscale, which is the range of Weibull modulus observed in real macro-scale structures, and

2. the model must yield a long power-law tail of correct length, reaching up to $P_f \approx 10^{-4}$ to 10^{-3}, so that a structure with more than about 10^4 RVEs would have a strength cdf close to the Weibull distribution, as revealed by analysis of experiments.

These requirements can only be achieved by a some sort of a hierarchical statistical model involving both parallel and series couplings. A simple and adequate model of this kind is shown in Fig. 6.10a. The first bundle (parallel coupling) must involve no more than two parallel elements, and each of them may then consist of a hierarchy of sub-chains of sub-bundles of sub-sub-chains of sub-sub-bundles, etc. Each element in the hierarchical model represents a nanoscale structure.

As seen in Fig. 6.10a, the elements, of identical power-law tails, are coupled in each sub-chain in order to extend the power-law tail to a realistic length (note that a long enough sub-chain would eventually produce a Weibull cdf, but this would be unrealistic). On the next higher scale, the parallel coupling of two or three of these sub-chains in a sub-bundle will raise their tail exponent but will shorten the power-law tail significantly. Then, on the next higher scale, a series coupling of many sub-bundles in a chain will again extend the power-law tail, and a parallel coupling of two of these chains will again raise the tail exponent and shorten the power-law tail significantly, until the macroscale RVE is reached. Thus, in each passage to a higher scale, the parallel coupling raises the tail exponent and the series coupling restores a realistic extent of the power-law tail.

Based on this hierarchical model, it was shown that the strength distribution of one RVE can be approximately described by a Gaussian distribution with a Weibull tail grafted on the left at the probability of about 10^{-4}–10^{-2}, with 10^{-3} as the mean

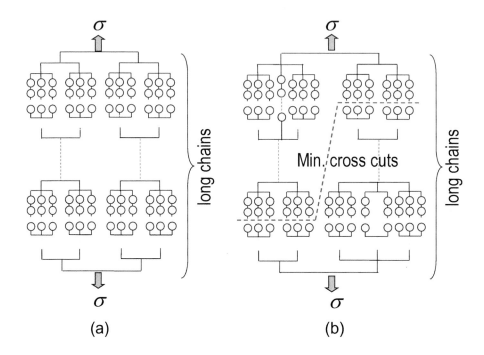

Fig. 6.10 Hierarchical model of failure statistics of one RVE.

estimate. The grafted cdf of strength of one RVE can be mathematically formulated as (Bažant and Pang, 2006, 2007; Le *et al.*, 2011; Bažant and Le, 2017)

$$P_1(\sigma) = 1 - e^{-\langle \sigma/s_0 \rangle^m} \qquad (\sigma \leq \sigma_{gr}) \qquad (6.45a)$$

$$P_1(\sigma) = P_{gr} + \frac{r_f}{\delta_G\sqrt{2\pi}}\int_{\sigma_{gr}}^{\sigma} e^{-(\sigma'-\mu_G)^2/2\delta_G^2} d\sigma' \qquad (\sigma > \sigma_{gr}) \qquad (6.45b)$$

where m (Weibull modulus) and s_0 are the shape and scale parameters of the Weibull tail, and μ_G and δ_G are the mean and standard deviation of the Gaussian core if considered extended to $-\infty$; r_f is a scaling parameter required to normalize the grafted cdf such that $P_1(\infty) = 1$, $P_{gr} = $ grafting probability, and $\langle x \rangle = \max(x, 0)$. Clearly,

$$P_{gr} = 1 - e^{-(\sigma_{gr}/s_0)^m} \qquad (6.46)$$

Finally, we require the continuity of the pdf at the grafting point: $[\mathrm{d}P_1/\mathrm{d}\sigma_N]_{\sigma_{gr}^+} = [\mathrm{d}P_1/\mathrm{d}\sigma_N]_{\sigma_{gr}^-}$

According to the aforementioned property of additivity of power-law exponent for the bundle model, the exponent of the power-law tail for one RVE is determined by the minimum of the sum of strength powers among all possible transverse cuts separating the hierarchical model into two halves.

Because random variations in couplings of the hierarchical model for the RVE must be expected, it would make hardly any sense to compute the structural failure probability directly from failure probability of the nanoscale structure. Nevertheless, the present statistical hierarchical model yields a physically based *functional form* of the strength distribution. The corresponding statistical parameters can be, and need to be, calibrated from macroscale experiments (Le and Bažant, 2012; Le *et al.*, 2013; Bažant and Le, 2017).

6.2.3 Macroscopic Strength Distribution and Experimental Validation

By knowing the entire distribution function of RVE strength, we can readily determine the probability distribution of the nominal strength of quasibrittle structures that fail under controlled loads at the macrocrack initiation from one RVE. Under the assumption of statistical independence of the random strengths of the RVEs (which requires the auto-correlation length of material strength not to exceed the FPZ size), the strength distribution of the structure can be calculated as

$$P_f(\sigma_N) = 1 - \prod_{i=1}^{N} \{1 - P_1[\langle s(\boldsymbol{x}_i) \rangle \sigma_N]\} \tag{6.47}$$

where $P_1(x)$ = strength cdf of one RVE given by Eqs. 6.45a and 6.45b, $P_f(\sigma_N)$ = failure probability of the entire structure, which is equivalent to the cdf of nominal structural strength, N = number of RVEs in the structure, and $s(\boldsymbol{x}_i)$ = dimensionless stress field such that $\sigma(\boldsymbol{x}_i) = \sigma_N s(\boldsymbol{x}_i)$ is equal to the maximum principal stress at the center of the i-th RVE with the coordinate $\boldsymbol{x}_i$.

Eq. 6.47 directly indicates that the probability distribution of structural strength would depend on the structure size, or equivalently, on the RVE number N. Fig. 6.11 presents the calculated cdfs of structural strength for different structure sizes in both the Weibull and Gaussian probability distribution papers. It is clear that the strength cdf of small-size structures is predominantly Gaussian. As the structure size increases but is not too large, the main portion of the strength cdf is primarily governed by the lower part of the Gaussian core of RVE strength distribution, and meanwhile the Weibull portion extends into the core of strength cdf. For sufficiently large structures (large N), what matters for P_f is the tail of the strength cdf of one RVE, that is, $P_1(\sigma) = (\sigma/s_0)^m$, and Eq. 6.47 naturally converges to the classical two-parameter Weibull distribution (Eq. 6.7).

Based on the foregoing analysis, it is evident that the finite weakest-link model (Eq. 6.47 and Eqs. 6.45a and 6.45b) is a generalization of the infinite weakest-link statistics. The dependence of the strength cdf on the structure size can be described by considering the following three asymptotic regimes (here the structure size is represented by the equivalent number of RVEs N_{eq} defined by Eq. 6.22, which contains the information on both the actual structure size and the applied stress field):

1. For small sizes [roughly $N_{eq} \in (1, 20)$], the cdf of strength can be approximated as Gaussian.

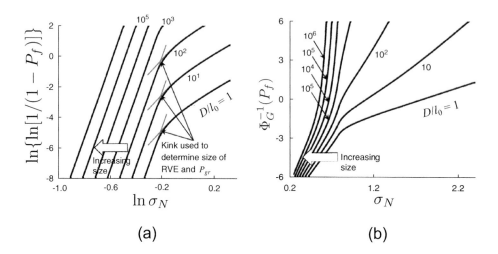

Fig. 6.11 Strength distributions of geometrically similar specimens of different sizes presented in (a) Weibull paper, and (b) Gaussian paper.

2. In the intermediate range of not too large sizes [roughly $N_{eq} \in (50, 500)$], the chance that the strength of the weakest RVE would fall into the power-law tail $P_f < 10^{-3}$ is still very small. Therefore, the strength of the weakest RVE, which is relevant to the overall failure statistics, is likely to be in the Gaussian core. This means, in theory, that the strength should approach the extreme value distribution of Gaussian variables, which is the Gumbel distribution (Fisher and Tippet, 1928; Gumbel, 1958; Ang and Tang, 1984; Kotz and Nadarajah, 2000), representing the so-called intermediate asymptotic regime in the sense of Barenblatt (Barenblatt, 1979, 2003).

3. For large sizes (roughly $N_{eq} > 5000$), there is a very high chance that the strength of the weakest element would be in the power-law tail $P_f < 10^{-3}$. This means that the strength distribution must converge to the Weibull distribution according to the theory of extreme value statistics (Fisher and Tippet, 1928; Ang and Tang, 1984; Kotz and Nadarajah, 2000).

Computations, however, show that the Gumbel distribution does not get manifested in most cases because the grafting probability of the RVE strength cdf is not small enough for the Gaussian core of the RVE strength to be relevant for a large value of N_{eq}.

It is worthwhile to point out that, in the literature, the Weibull distribution was sometimes used to describe the strength distribution of one RVE. However, it is simple to prove that this is fundamentally incorrect. Consider that the strength of a presumed

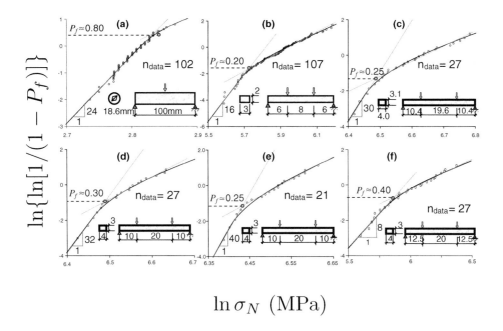

$$\ln\{\ln[1/(1-P_f)]\}$$

$$\ln \sigma_N \text{ (MPa)}$$

Fig. 6.12 Optimum fits of strength distribution of ceramics by the finite weakest-link model: (a) porcelain, (b) sintered α-SiC, (c) sintered Si_3N_4, (d) sintered Si_3N_4-Al_2O_3-Y_2O_3, (e) sintered Si_3N_4-Al_2O_3-CTR_2O_3, and (f) alumina/glass composite (Bažant and Le, 2017).

"RVE" follows the Weibull distribution. But this distribution can arise only from the weakest-link model, which implies that the damage will localize into only one of the links, which will thus represent the true RVE unless it also follows the Weibull distribution, in which case an even smaller material element would be the RVE. Therefore, the presumed "RVE" cannot be the true RVE. Rather it is the smallest failing link which must represent the true RVE, that is, the smallest material volume whose failure can trigger the failure of the entire structure.

Clearly the finite weakest-link model is limited to the case where the structure is larger than one RVE. Recent studies have shown that the strength distribution (or, equivalently, the failure probability) of quasibrittle structures can also be calculated as a first passage probability through the level excursion analysis (Xu and Le, 2017, 2018; Le, 2020). By explicitly taking into account the spatial correlation features of the material strength, the level excursion analysis calculates the failure statistics of a material point and, therefore, it can be applied to any structure size. The finite weakest-link model, which uses the RVE as the basic unit for calculating the overall failure statistics, can be considered as a discrete form of the level excursion model.

For most quasibrittle structures, where the structure size is not much larger than

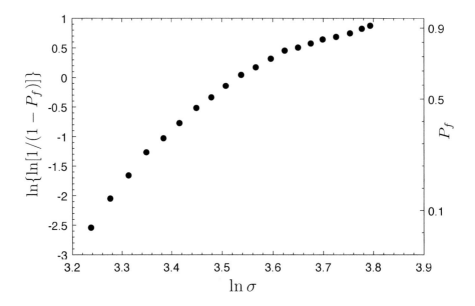

Fig. 6.13 Replot of Weibull's strength data on Portland cement (stored in water for 7 days).

the FPZ size, the strength distribution falls in the intermediate regime. This explains why the measured strength distributions of many quasibrittle structures consistently deviate from the Weibull distribution. We will now compare the finite weakest-link model with the experimentally measured strength statistics of quasibrittle structures. Figs. 6.12a–f present the plots of the strength histograms of various industrial ceramics with the optimum fits by the finite weakest-link model. The details of these histogram tests are as follows:

(a) Weibull investigated the strength distribution of porcelain by performing three point bend tests of 102 cylindrical unglazed porcelain rods with diameter 18.6 mm and length 100 mm (Weibull, 1939).[4]

(b) Silicon carbide (SiC) ceramics, which exhibit very high hardness and wear resistance and are widely used in automotive components, heat exchanger tubes, me-

---

[4]For 80 years, Weibull's experiments were thought to reach unprecedentedly low probability in the histogram. However, in 2020 Luo and Bažant (2020) it was noticed that upper ends of the histogram plotted in Weibull (1939) could be obtained only if the number of test repetitions was at least 10^5, which is hard to believe. Checking the data source it was then discovered that the natural (or Napier) logarithm was used in plotting the vertical scale and the base 10 (or common, or Brigg's) logarithm in plotting the horizontal scale. This caused a systematic shift of all the data. As a result it must be concluded that Weibull's tests did not actually reach below failure probability 0.01, or 1%. See the corrected plot of Weibull data in Fig. 6.13.

chanical seals, etc., were studied by Salem *et al.* (Salem *et al.*, 1996). They performed four-point bend tests on 108 prisms of sintered α-silicon carbide (Carborundum of Hexoloy, α-SiC), with dimensions $2 \times 3 \times 25$ mm; 36 of their prisms were produced with $0°$ grinding angle without annealing, the next 36 the same but with high temperature annealing, and 36 more with a $90°$ grinding angle and with high temperature annealing. For all the three groups, they found almost the same strength distribution (Salem *et al.*, 1996).

(c)–(e) Silicon nitride ceramics (Si_3N_4) (developed in the 1960s and 1970s in a search for dense, high strength, and high toughness materials, and are used, for example, for reciprocating engine components and turbochargers, bearings, and metal cutting or shaping tools) were investigated by several researchers (dos Santos *et al.*, 2003; Gross, 1996; Okabe and Hirata, 1995). Gross (Gross, 1996) conducted four point bend tests of 27 prismatic beams of sintered silicon nitride (SNW-1000), of dimensions $3.1 \times 4 \times 40.4$ mm. dos Santos *et al.* (dos Santos *et al.*, 2003) tested two types of silicon nitride ceramics with additives: one was a sintering additive (Si_3N_4-Al_2O_3-Y_2O_3), and another was an aluminum additive (Al_2O_3/Y_2O_3). dos Santos *et al.* (dos Santos *et al.*, 2003) also tested 21 four point bend beams, with dimensions $3 \times 4 \times 45$ mm, made of silicon nitride sintered with rare earth oxide additive (Si_3N_4-Al_2O_3-CTR_2O_3), which yields similar improvements but at lower costs.

(f) Lanthanum-glass-infiltrated alumina glass ceramics (which are attractive for restorative dentistry owing to their aesthetics and bio-compatibility, high strength, and fracture toughness (Ironside and Swain, 1998)) were studied by Lohbauer *et al.* (Lohbauer *et al.*, 2002). They tested alumina-glass composites dry-pressed and pre-sintered α-Al_2O_3 with a medium grain size. The specimens were CAD/CAM machined into prisms with dimension $3 \times 4 \times 45$ mm before infiltration with 25 wt% (by weight) of glass of the following weight percentages (Lohbauer *et al.*, 2002): 39–41% of La_2O_3, 16–17% of SiO_2, 15–18% of Al_2O_3, 15% of B_2O_3, 5% of TiO_2 and 4% of CeO_2. 27 four-point bend specimens were tested under dry condition.

Figs. 6.14 a–f present the optimum fitting of strength histogram of dental ceramics, tested by Tinschert *et al.* (Tinschert *et al.*, 2000). The specimens were made of six dental restorative ceramics commonly used in crown, veneer and inlay construction. The six ceramics materials are: (a) Dicor (tetrasilicic fluoromica glass-ceramic), (b) IPS Empress (IE & leucite-reinforced porcelain), (c) Vitadur Alpha Core (alumina-reinforced feldspathic porcelain), (d) Vitadur Alpha Dentin (feldspathic porcelain), (e) Zirconia-TZP (partially stabilized zirconia ceramic) and (f) Vita VMK 68 (feldspathic porcelain). For each material, 30 four-point bend beams with size $1.5 \times 3 \times 30$ mm were tested.

For the fitting of the aforementioned strength histograms, we first estimate the RVE size l_0 as approximately two to three times the size of material inhomogeneities. The Weibull modulus can easily be determined from the lower straight portion of the strength histogram in the Weibull scale. By knowing l_0, we can calculate the Weibull scaling parameter s_0. The statistical parameters of the Gaussian part of the RVE strength can be determined by the optimum fit of the upper curved portion of the strength histogram using the Levenberg–Marquardt nonlinear optimization algorithm.

Despite rather low numbers of the specimens tested, we observe from Figs.6.12a–f

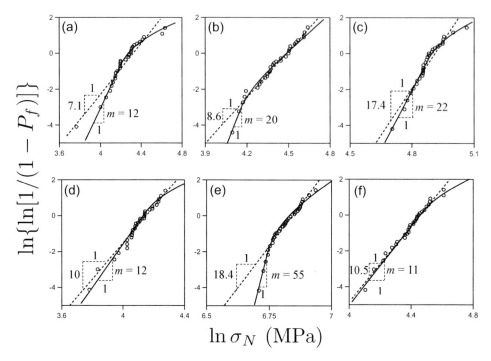

Fig. 6.14 Optimum fits of strength distribution of dental ceramics by the finite weakest-link model: (a) Dicor, (b) IPS Empress, (c) Vitadur Alpha Core, (d) Vitadur Alpha Dentin, (e) Zironia-TZP, and (f) Vita VMK 68 (Bažant and Le, 2017).

---

and 6.14a–f that the strength histograms plotted in Weibull scale are not straight lines. The histograms consist of two parts separated by a relatively abrupt kink. The left portion is a straight line signifying a Weibull distribution, and the right portion deviates from the straight line to the right. This clearly indicates that the two-parameter Weibull model cannot provide an optimum fit. Instead, it is seen that these histograms can be fitted, as closely as the scatter permits, by the finite weakest-link model, which is shown as solid curves in Figs.6.12a–f and 6.14a–f. Based on this model, the core part of histogram can be calculated by using a chain model of Gaussian elements. The kink is smooth but so abrupt that it may be approximated by a point transition at which both parts are grafted with a continuous cdf slope. The height of the grafting point characterizes the degree of brittleness of structure, which is governed by the equivalent number of RVEs, N_{eq}.

6.3 Mean Size Effect on Structural Strength

Based on the cdf of structural strength, we can calculate the mean structural strength as $\bar{\sigma}_N = \int_0^1 \sigma_N(P_f)\,\mathrm{d}P_f$. Since the nominal strength is positive by definition, we can

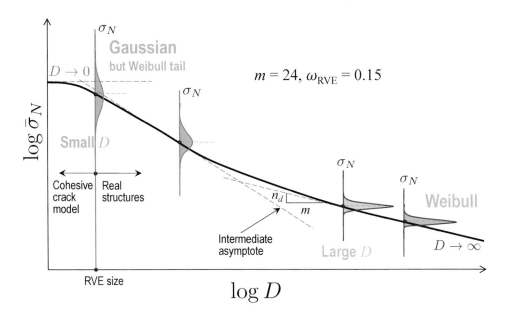

Fig. 6.15 Mean size effect curve predicted by the finite weakest-link model (Bažant and Le, 2017).

evaluate the mean strength alternatively as

$$\bar{\sigma}_N = \int_0^\infty \prod_{i=1}^N \{1 - P_1[\sigma_N \langle s(\boldsymbol{x}_i) \rangle)]\} \mathrm{d}\sigma_N \qquad (6.48)$$

where P_1 = cdf of RVE strength, which is given by Eqs. 6.45a and 6.45b. By considering geometrically similar structures of different sizes, we readily obtain the size effect curve of the mean structural strength.

Fig. 6.15 presents the mean size effect curve together with the corresponding strength pdf, varying from the Gauss–Weibull grafted distribution at the small size to the Weibull distribution at the large size. It is seen that, in the logarithmic plot, the large-size asymptote of the size effect curve follows a straight line, which agrees well with the classical power-law size effect of Weibull statistics. For small-size structures, the size effect curve deviates significantly from the straight line. This can be attributed to the fact that the RVE size (or equivalently the size of material inhomogeneities) is not negligible compared to the structure size and therefore the strength cdf is non-Weibullian.

It is interesting to note that the mean size effect curve predicted by the finite weakest-link model agrees well with the type 1 size effect curve shown in Fig. 5.7 (Eq. 5.67), even though in Chapter 5 (Sec. 5.5.2) the type 1 size effect was derived by using the equivalent linear elastic fracture mechanics (LEFM). This agreement is because both models capture the stress redistribution mechanism. In the equivalent LEFM, the nonlinear softening damage in the FPZ is represented by an equivalent LEFM crack, which introduces a characteristic length scale. In the finite weakest-link model, the failure mechanism is represented by the functional form of the strength distribution of one RVE, which is derived from a multiscale scale transition model (as discussed in Sec. 6.2.2). This transition model consists of statistical bundles and chains that statistically represent the damage localization and load redistribution mechanisms at different scales (only approximately, though, since only the elastic stresses are used).

It is evident that both models become inapplicable as the structure size becomes very small. As mentioned in Sec. 5.5.2, the type I size effect curve exhibits a vanishing size effect at the small-size limit. This asymptotic property can be derived mechanistically by using the cohesive crack model. Recent studies have shown that this small-size asymptote can alternatively be obtained statistically by using the level excursion analysis (Xu and Le, 2017; Le, 2020).

According to the finite weakest-link model, it is clear that, for a given specimen geometry, the mean size effect curve of structural strength is directly related to the strength cdf of RVE, which can be fully determined by four statistical parameters: the mean and deviation of the Gaussian core (μ_G and δ_G), Weibull modulus m, and Weibull scaling parameter s_0. However, it is impossible to obtain an analytical expression for the mean strength $\bar{\sigma}_N$ in terms of these parameters. We may use an approximate analytical formula similar to Eq. 5.63 to describe the mean size effect curve, that is,

$$\bar{\sigma}_N = \left[\frac{N_a}{D} + \left(\frac{N_b}{D} \right)^{n_d \psi / m} \right]^{1/\psi} \tag{6.49}$$

where N_a, N_b, ψ and m are constants to be determined by asymptotic properties of the size effect curve, and n_d = number of spatial dimensions in which the structure is scaled. Evidently, Eq. 6.49 converges to a power-law large-size asymptote $(N_b/D)^{n_d/m}$, which is manifested by a straight line of slope $-n_d/m$ in the size effect plot of $\log \bar{\sigma}_N$ versus $\log D$.

The Weibull asymptote of strength cdf at the large-size limit indicates that the exponent, m, must be equal to the Weibull modulus of strength distribution. Parameter N_b can then be directly related to the Weibull scaling parameter. The remaining parameters N_a and ψ can be determined by solving two simultaneous equations expressing the asymptotic conditions, $[\bar{\sigma}_N]_{D \to l_m}$ and $[d\bar{\sigma}_N/dD]_{D \to l_m}$, where l_m = the minimum structure size for which the finite weakest-link model is applicable. Solving these two equations directly yields the mean and standard deviation of the Gaussian core, and it has been shown that the solution is unique (Le *et al.*, 2013).

The foregoing discussion indicates that, with the finite weakest-link model, we can determine the strength distribution of quasibrittle structures directly from the mean size effect analysis. The size effect curve can be determined by testing the mean strengths of geometrically similar specimens of four to five sizes, which is far more efficient than the conventional histogram testing. This indirect method of determining the strength statistics was recently validated through a comprehensive set of size effect tests on the strength distribution of asphalt mixture specimens (Le *et al.*, 2013; Bažant and Le, 2017).

6.4 Problem with Applying Three-Parameter Weibull Distribution

As demonstrated in the previous section, the measured strength histograms of quasib-rittle structures (Figs. 6.12 and 6.14) cannot be fitted by the two-parameter Weibull distribution. As a remedy, the three-parameter Weibull distribution, in which a non-zero threshold is introduced, has been widely accepted as an empirical probability distribution. It can be written as

$$P_f(\sigma_N) = 1 - \exp\left(-\int_V \frac{\langle \sigma_N s(\boldsymbol{x}) - \sigma_u \rangle^{m_w}}{l_0^{n_d} s_w^{m_w}} \, dV \right) \qquad (6.50)$$

where σ_u = strength threshold, m_w = Weibull modulus of the three-parameter Weibull distribution, and s_w = scale parameter. Though the three-parameter Weibull distribution improves the fit of measured strength histograms as compared to the two-parameter Weibull distribution, we will argue that it is physically unsound and misleading in terms of safety.

The application of three-parameter Weibull distribution brings about a severe error in predicting the design strength for a low probability (e.g. $P_f = 10^{-6}$). As a demonstration, Fig. 6.16 shows a schematic plot of the strength cdfs of geometrically similar specimens predicted by both the finite weakest-link model and the three-parameter Weibull model. It is clear that, compared to the finite weakest-link model, the three-parameter Weibull model will grossly overestimate the design strength at $P_f = 10^{-6}$. Recent studies have shown that, for laboratory specimens, the three-parameter Weibull model could overestimate the design strength by 34%, and for large-size specimens, the overestimation could be as large as 96%. This would severely undermine the structural reliability (Bažant and Le, 2017).

6.4.1 Theoretical argument against three-parameter Weibull distribution

Based on Sec. 6.2.1 and Sec. 6.2.2, the zero strength threshold for the Weibull distribution is a consequence of (a) the transition rate theory, where the rates of forward and backward jumps over the activation energy barrier of the nanoscale structure must differ very little; and (b) the hypothesis that bridging up the ascending scales is equivalent to some combination of parallel and series couplings.

By reverse reasoning, the three-parameter Weibull distribution implies a finite stress threshold for the failure statistics of nano-scale structures. But this is inconsistent with the transition rate theory for stress-driven fracture.

6.4.2 Mean size effect analysis

Since it is difficult to use direct histogram testing of specimens of a single size to examine the validity of the distribution function unambiguously, we have to seek and analyze experimental evidence for other predictions of the theory. Among these, the size effect analysis is the most effective and efficient.

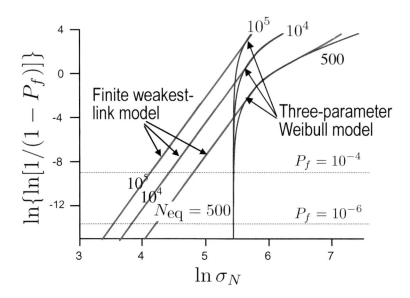

Fig. 6.16 Strength distributions of geometrically similar specimens predicted by the finite weakest-link model and the three-parameter Weibull model.

___

We first compare the size effect curves predicted by the finite weakest-link model and the three-parameter Weibull distribution (Fig. 6.17). It can be seen that a significant deviation occurs for the large specimens. This difference can be explained by considering a simple case where all the RVEs are subjected to the same stress. The three-parameter Weibull distribution yields the mean structural strength as

$$\bar{\sigma}_N = \int_0^\infty \left[\exp\left(- \left\langle \frac{\sigma_N - \sigma_u}{s_w} \right\rangle^{m_w} \right) \right]^{V/V_0} \mathrm{d}\sigma_N \tag{6.51}$$

$$= \sigma_u + s_w \Gamma\left(1 + 1/m_w\right) \left(V_0/V\right)^{1/m_w} \tag{6.52}$$

Clearly the three-parameter Weibull distribution predicts a strength threshold at the large-size limit. In contrast, the finite weakest-link model indicates that, at the large-size limit, the size effect curve must follow a power-law size effect, that is, $\sigma_N \propto D^{-n_d/m}$. For structures with a non-uniform stress distribution, the three-parameter Weibull distribution predicts a vanishing size effect at the large size limit, which leads to a constant value of the nominal strength. This large-size asymptote of size effect curve is inconsistent with the experimental observations. For example, the Weibullian asymptote at the large-size limit is clearly observed in the size effect curve of the modulus of rupture of concrete (Fig. 5.7b). Recent size effect tests on

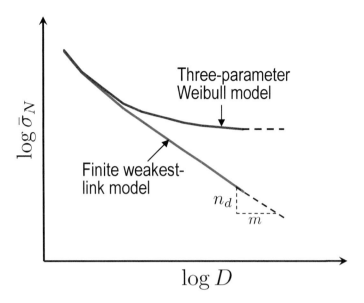

Fig. 6.17 Mean size effect curves predicted by the finite weakest-link model and the three–parameter Weibull model.

---

strength histogram of asphalt mixtures also shows that the three-parameter Weibull distribution would overestimate the mean strength of large-size specimens, due to its vanishing size effect at the large-size limit.

6.5 Aperçu of Fishnet Statistics for Biomimetic, Architectured, Octet-Lattice and Some Particulate Materials

As seen in the preceding exposition, until recently there existed only two analytically tractable models for material failure probability: (a) the weakest link chain model, either infinite (Weibull, 1939) or finite (Bažant and Pang, 2006, 2007; Bažant *et al.*, 2009; Le *et al.*, 2011), and (b) the fiber bundle model (Daniels, 1945). Recently a third analytically tractable model, called fishnet statistics (Luo and Bažant, 2017*a,b*, 2018), was found while studying the nacre of pearl oyster or abalone (Fig. 6.18), a material known for its amazing strength, an order of magnitude higher than the macro-scale strength of its main constituent, the aragonite (a form of $CaCO_3$). In deterministic terms, the mean strength was explained by fracture mechanics (Wang *et al.*, 2001; Gao *et al.*, 2003; Chen *et al.*, 2007; Shao *et al.*, 2012; Dutta *et al.*, 2013; Askarinejad and Rahbar, 2014; Dutta and Tekalur, 2014; Wei *et al.*, 2015). But, as revealed by fishnet statistics, the lamellar "brick-and-mortar" architecture of nacre brings about major advantages in the tail of very low failure probability. In this section we present an overview of recent studies trying to clarify these advantages (Luo and Bažant, 2017*a,b*, 2018, 2019, 2020)

Fig. 6.18 (a) Nacre inside a nautilus shell, and (b) nacre in abalone shell (image courtesy Wikimedia Commons/Doka54).

On the sub-micrometer scale, nacre's architecture consists of a lamellar system of overlapping imbricated (or staggered) nanoscale lamellea (or platelets), depicted in Fig. 6.19a and idealized in Fig. 6.19b. The tensile forces transmitted between two adjacent lamellae of the same row are negligible, and almost all of the longitudinal load gets transmitted by shear through thin biopolymer layers that bond parallel lamellae. The force transmission in the system is idealized as a system of bar links connecting the centroids of adjacent stiff lamellae Fig. 6.19b. The links form a system resembling a diagonally pulled fishnet (Fig. 6.19c and d) which can be simulated by a finite element program for a pin-jointed truss. Because what matters in computations is only the alternation of staggered node links, the fishnet is in computations allowed to collapse under longitudinal tension into one line, as depicted in Fig. 6.19c and d (alternatively, one could prevent this collapse by propping each fishnet eye by a transverse strut, but for statistics the result would be about the same).

6.5.1 Brittle Fishnet Statistics

First consider that the fishnet links are perfectly brittle, that is, fail suddenly when their strength limit is reached. The link failure causes stress redistribution into the adjacent links. To keep calculations simple, the redistribution is described deterministically. The equilibrium equations of fishnet nodes represent finite difference equations, which may be approximated by the Laplace equation (Luo and Bažant, 2017b), $\nabla^2_{(x,y)} u = 0$, in which u is the longitudinal displacement of the fishnet nodes, and (x,y) are nodal coordinates in the initial stress-free state (i.e., before the fishnet collapses into a line); x = longitudinal coordinate, in the load direction.

Consider now a rectangular fishnet with k rows and n columns (Fig. 6.19d). The load consists of a uniform uniaxial stress σ imposed at the row ends. Let $P_f(\sigma)$ denote the failure probability of the fishnet loaded by σ. The failure probability of each link, $P_1(\sigma)$ is considered to follow the Gauss–Weibull graft distribution (Eqs. 6.45a and

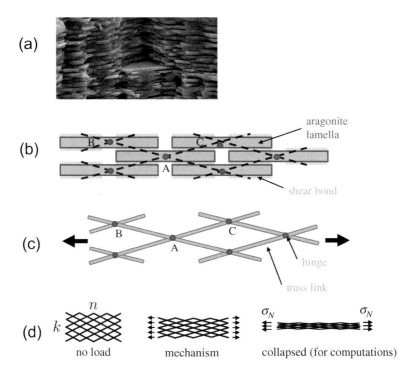

Fig. 6.19 Fishnet modeling of nacre: (a) electron microscopy image of a fractured surface of nacre, (b) idealized imbricated (staggered) lamellar architecture of nacre at sub-micrometer scale, (c) fishnet truss capturing the essential force transmission mode, and (d) deformation mechanism and lateral contraction of fishnet (Bažant, 2019).

b). Let $P_{S_n}(\sigma)$, with $n = 0, 1, 2, 3, ...$, denote the survival probabilities of the fishnet when exactly n links have already failed. Since $P_{S_n}(\sigma)$ characterize mutually exclusive events, the survival probability of the fishnet is

$$1 - P_f(\sigma) = P_{S_0}(\sigma) + P_{S_1}(\sigma) + P_{S_2}(\sigma) + \cdots \tag{6.53}$$

Except when the link strengths have an unusually low coefficient of variation, the terms in this sum form a rapidly decreasing sequence, and the terms with $n > 3$ are normally insignificant.

The first term, $P_{S_0}(\sigma)$, is simply the weakest-link model. To calculate the second term, we note that the probability that one link has already failed, which is $N P_1(\sigma)$, must occur jointly with the probability that all the remaining $N - 1$ links survive. Therefore, which gives

$$P_{S_1}(\sigma) = N P_1(\sigma) \cdot \prod_{i=1}^{N-1} [1 - P_1(\eta_i \sigma)] \tag{6.54}$$

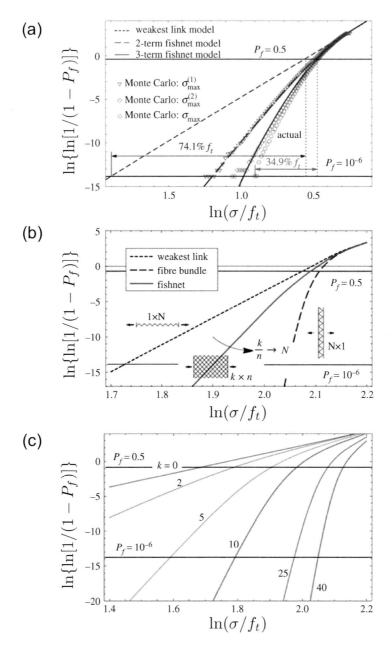

Fig. 6.20 (a) Weibull scale plots of two-term and three-term fishnet distributions compared to the weakest-link model (dashed line) and actual distribution via million Monte Carlo simulations (CoV = 7.8% for each link); (b) continuous transition from the Weibull line (dashed) to Gaussian distribution for pure parallel coupling obtained by changing aspect ratio of fishnet; and (c) failure probability distributions for a fishnet with softening links obtain from order statistics (or order 2, 5, 10, etc.) (Bažant, 2019).

where η_i is the stress redistribution factor of the i-th link due to prior link failure, which is calculated deterministically, by solving from the aforementioned Laplace equation the spatial decay of a disturbance; η_i equals 1 except in a small neighborhood of the previously failed link.

The third term, despite being less simple, can also be calculated analytically (Luo and Bažant, 2017b), by applying to link failures the theorems of probability of joint and disjoint events, while the stress redistribution within the neighborhood of previously failed two links is taken into account. An analytical expression for the fourth term, $P_{S_3}(\sigma)$, could be formulated but would be very complicated and would make a significant difference only for $P_f < 10^{-6}$ which is hardly of any interest.

The simulated failure probability distributions for the cases where 0, 1, or 2 links have already failed generated the three curves plotted in Weibull scale in Fig. 6.20a for $k = 16$ (rows) and $n = 32$ (columns). The uppermost curve is a straight line representing Weibull distribution although on top it deviates towards the Gaussian, approximately. The figure also shows, by circle points, the curve obtained by 1 million Monte Carlo random simulations of the fishnet failure. About 10^4 simulations are aggregated into each data point, and so the points are virtually exact for $P_f > 10^{-5}$ (the link strength distribution is assumed to be Gauss–Weibull, with a graft at $P_f = 0.015$ and CoV $= 7.8\%$). The third curve of computed points for the cases with up to two links failed is seen to be very close to the exact solution.

The second and higher term in Eq. 6.54, representing the deviations from the finite weakest-link chain, increase the survival probability at low P_f. Aside from the localized transition from the Gaussian curve to the Weibull straight line of slope m occurring at high P_f, there is, while moving to lower P_f, a transition to a doubled slope (Fig. 6.20a) when the second term of Eq. 6.53 kicks in. At even lower P_f, there is a transition to tripled slope, etc. Thus the size effect curve may be viewed as an envelope to a sequence of the so-called intermediate asymptotes (Barenblatt, 1979, 2003), each corresponding to one term in the sum of Eq. 6.53. The added term raises the slope of the intermediate asymptote by m (Fig. 6.20a). Obviously, the nacreous material architecture leads in the Weibull scale to a progressively steeper dipping curve deviating more and more from the Weibull straight line (Fig. 6.20a). With enough terms, the slope would become almost vertical. But this would occur when the corresponding probability is orders of magnitude below 10^{-6} which is practically irrelevant.

Changing the aspect ratio (or shape) of the fishnet, k/n causes a change in the distribution. For $k/n \to 1$, the cdf converges to the finite weakest-link chain (with Gauss–Weibull cdf), and for $k/n \to \infty$ to the fiber bundle (with Gaussian cdf). The fishnet thus provides a continuous transition between these two basic classical models (Fig. 6.20b).

An important point to note is that, aside from its well-known extraordinary mean strength, the nacreous material architecture brings about a major additional advantage at the low probability tail. In comparison to the Weibull straight line of slope m, applicable, for example, to fine-grained ceramics, the intersections of the horizontal line $P_f = 10^{-6}$ with the Weibull line and with the line of points computed for the fishnet (Fig. 6.20a), show the tail strength at $P_f = 10^{-6}$ to be 72% higher. This is an enormous safety gain.

Another noteworthy fishnet feature (Luo and Bažant, 2017a,b) is that the cdf of fishnet strength always lies, in Weibull plot, between the curves for the weakest-link chain and the fiber bundle, when the total number of links is fixed. So, these two curves represent bounds (Fig. 6.20b).

The higher the scatter of link strength, the higher the number of links that are expected to fail before the maximum load is reached. This clarifies another advantage of fishnet connectivity of the microstucture—when the scatter of link strength increases (which probably has a similar effect as an increase of scatter in the overlap length of adjacent lamellae of nacreous material), the lower left tail of the cumulative probability distribution in Weibull scale plot becomes steeper. Consequently, the ratio of the mean failure stress to the stress at the 10^{-6} tail gets decreased, in fact greatly so. Thus the fishnet scatter may increase structural safety, unless the mean strength gets decreased by scatter even more.

It is thus not surprising that, for geometric scaling, the mean strength of a fishnet exhibits a significant size effect. The size effect is similar, though not identical, to type 1 size effect (Bažant and Le, 2017) in quasibrittle random heterogeneous materials. In the log–log plot, the size effect curve descends at progressively decreasing slope. Compared to quasibrittle heterogeneous materials, it is steeper for medium sizes but less steep for large sizes. The slope is $1/3m$, rather than $1/m$, for the first three terms of Eq. 6.53, which mitigates the type 1 size effect at very large sizes. But for negligible scatter, the fishnet fails when the first link fails, in which case the finite weakest-link model applies.

6.5.2   Softening Fishnet, Order Statistics and Quantile Statistics

In the preceding section, the fishnet links were assumed to be brittle or almost brittle. However, the shear bond between the adjacent lamellae of nacreous material may exhibit progressive softening, and so may the fishnet links. In that case a different mathematical approach is needed (Luo and Bažant, 2018), resting on two ideas:

1. The post-peak softening curve of each link with random strength is described by a series of small sudden discrete drops, or partial damages, of stiffness and stress, and

2. the probability of failure due to these drops is treated according to the theory of order statistics (Leadbetter *et al.*, 2012), in which the failure probability is decided by failure of the link with the k-th smallest strength ($k = 1, 2, 3, ...$).

If many small stress drops are considered, the stress redistribution due to each drop of strength and stiffness is so feeble that it can be ignored, which is a great simplification. For various k, the analysis then leads to a series of progressively steepening cdf curves. The strength problem is thus converted to determining the k corresponding to the maximum load (see Luo and Bažant (2018)).

The distribution of the k-th smallest minimum of strength (i.e., the k-th order statistic) is (Luo and Bažant, 2018, 2020)

$$P_f(\sigma) = \Pr(\sigma_{\max} \leq \sigma) = \sum_{k=0}^{N} \Pr(N_c = k) W_k(\gamma_k \sigma) \qquad (6.55)$$

where N_c = number of link damages (partial stress drops) at the moment of reaching the peak load, γ_k = average stress concentration factor for various k values, and $W_k(x)$ = cumulative distribution function of the distribution of the k-th strength order statistics, which is

$$W_k(x) = 1 - [1 - P_1(x)]^N \sum_{s=0}^{k} \frac{\{-N \ln[1 - P_1(x)]\}^s}{s!} \tag{6.56}$$

Because the damages tend to cluster, $\Pr(N_c = k)$ is properly modeled by the geometric-Poisson distribution (Aeppli, 1924):

$$\Pr(N_c = k) = \sum_{s=1}^{k} e^{-\lambda} \frac{\lambda}{s!} \binom{k-1}{s-1} \theta^s (1-\theta)^{k-s}, \quad k = 1, 2, 3, \cdots \tag{6.57}$$

$$\Pr(N_c = 0) = e^{-\lambda} \tag{6.58}$$

where λ and θ are empirical exponents to be calibrated by tests.

Further it was shown (Luo and Bažant, 2020) that the fishnet model with order statistics can be written in the form of a power series expansion in terms of $P_1(\sigma)$, where $P_1(\sigma) \sim (\sigma/\sigma_0)^m$ at the lower tail. The analysis showed that the failure probability of fishnet can generally be described by the approximation

$$P_f(\sigma) = 1 - e^{-N\gamma^m x^m} \cdot f(N\gamma^m x^m) \tag{6.59}$$

where $x = \sigma/\sigma_0$, and γ is the average of stress concentration factors γ_k, close to 1, and f is a semi-empirical function (Luo and Bažant, 2020). Also, it was shown that the distribution of the k-th order statistics has, in the Weibull scale plot, the lower tail of slope km.

The scaling of fishnet strength is easier to understand by considering, approximately, the separate scalings in the transverse and longitudinal directions (Fig. 6.21a,b). The longitudinal scaling is seen to cause a vertical upward shift of the central Weibullian part of the strength distribution. This is easily proven by noting that survival probability of a chain of n links satisfies the equation $1 - P_f(\sigma) - [1 - P_1(\sigma)]^n$, and taking twice the logarithm yields $Y = X + \ln n$ where $Y = \ln\{-\ln[1 - P_f(\sigma)]\}$ and $X = \ln\{-\ln[1 - P_1(\sigma)]\}$ are Weibull scale ordinates and $\ln n$ is the vertical shift of the distribution. The downward deviation due to fishnet action is seen to decrease with increasing longitudinal enlargement. The transverse scaling in Fig. 6.21a causes rotation of the entire distribution in Weibull scale about a certain point Q. At infinite enlargement, the longitudinal scaling yields the Weibull distribution, like for a chain, and the transverse scaling yields the Gaussian distribution, like for a bundle. The superposition of both cases yields the approximate geometrical scaling in Fig. 6.21c, in which the distributions for various sizes are seen to intersect at a certain point with the probability a little less than 10^{-2}. This suggests the possibility of a reverse size effect at $P_f = 10^{-6}$, that is, a larger structure might have a higher tail strength that a small one.

So far we have focused on the type 1 failure. A different statistical concept is needed to model the randomness in type 2 failure, for example the shear failure of

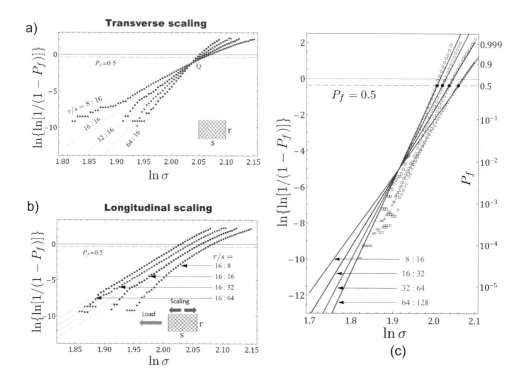

Fig. 6.21 Size effect on softening fishnet: (a) transverse scaling, (b) longitudinal scaling, and (c) 2D geometrical scaling (Luo and Bažant, 2019).

reinforced concrete (RC) beams and slabs. Monte Carlo finite element simulations with M7 microplane model (Luo *et al.*, 2021) showed that, after long stable crack growth, the size of the region of possible critical crack tip locations at maximum load is not constant in relative coordinates but decreases as the beam size increases until stabilizing at a relative size. This phenomenon requires replacing order statistics with quantile statistics. The basic idea advanced in Luo *et al.* (Luo *et al.*, 2021) is that, in a small critical region, the shear strength of RC beam follows the distribution of the sample p-quantile of strength, leading to normal distribution for the large size limit. The justification of this idea is that the number of critical material elements of random strength that can trigger the failure grows with the size of the near-tip critical region while this region is not constant but grows with the structures size, causing that what matters is not the number, but the percentage, or quantile, of the elements of random strength in the structure. This idea facilitated predicting the dependence of the CoV of the shear strength of RC beams. It turned out that the CoV is not constant, which would be the case for Weibull statistics. Rather, it first decreases with the structure size, like in ductile failures, yet at large sizes approaches a constant, like in brittle

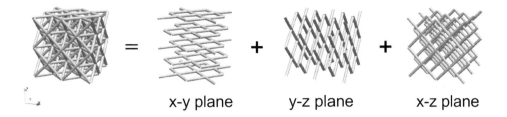

Fig. 6.22 Modeling of Octet-truss nanolattice as a union of three orthogonal fishnets (Luo and Bažant, 2020).

(weakest-link) failures.

6.5.3 Octet Lattice and Other Quasibrittle Materials

A further gain of safety at the 10^{-6} tail can be obtained with some architectured materials. Of particular interest has recently been Fuller's octet truss (Fuller, 1961), which minimizes material specific weight for a given elastic stiffness and can be effectively produced by three-dimensional (3D) printing. The spatial octet truss can be decomposed into three mutually orthogonal planar fishnets (Luo and Bažant, 2020) (Fig. 6.22), which helps simple analytical estimates. Fig.6.23 shows the strength distribution for a $8 \times 4 \times 4$ octet truss obtained by millions of Monte Carlo simulations (Luo and Bažant, 2020). It is remarkable that, at the tail of $P_f = 10^{-6}$, there is a significant additional safety gain compared to the planar fishnet, even if the distributions in Fig. 6.23 are shifted horizontally to have the same median strength.

A recent computational modeling with particle model (Nitka and Tejchman, 2015) showed that the force chain transmission in a tensile strength test of concrete includes many inclined force transfers, which are the hallmark of the fishnet. This suggests that, at the low probability tail, some degree of fishnet action might be present even in concrete and other quasibrittle materials (Luo and Bažant, 2020). This would increase the tail strength of these materials compared to extrapolation with the weakest-link chain model.

Finally, it should be pointed out that maximizing the material strength is not always optimal for structural safety. For quasibrittle or architectured materials it is possible that, for example (Fig. 6.24), some material that is in the mean 8% stronger than a fishnet material is actually 35% weaker in the 10^{-6} tail (Bažant, 2019), which is what usually matters for design.

6.6 Remark on Failure Probability of Concrete Specimens of Random Mean Strength in Large Database

It is a special feature of concrete engineering that structures are usually designed without knowing or specifying more than the required compression strength of standard

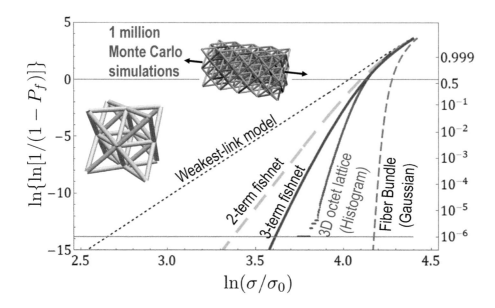

Fig. 6.23 Simulated strength distribution of Octet-truss nanolattice (Luo and Bažant, 2020).

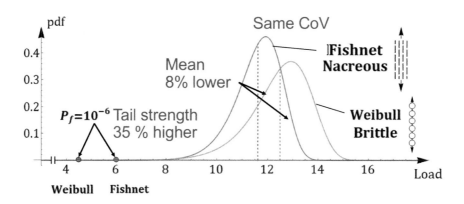

Fig. 6.24 Comparison between the strength distributions of brittle and fishnet material with different mean but same CoV.

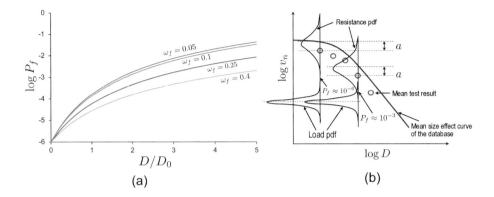

Fig. 6.25 Size effect in reliability analysis of diagonal shear failure of RC beams: (a) size dependence of failure probability, and (b) probability distribution functions of nominal shear strength for different specimen sizes in relation to the mean size effect measured from an individual test plotted in the logarithmic scale (Bažant and Yu 2009).

test specimens. The required strength is, of course, random, and it can be satisfied by many different concrete mixes for which the other concrete properties can vary widely. Safe design needs to take such uncertainty into account. This requires analyzing the failure probability of concrete specimens with the mean strength sampled from a large database of high variability.

A general framework to do that has recently been developed (Le and Bažant, 2020). It takes explicitly into account the level of uncertainty of the existing strength database, which consists of concretes of different mixes tested in different labs. This uncertainty has a strong effect on the probability distribution of the nominal strength of the designed structure. The energetic (deterministic) size effect on structural strength (Eq. 5.57) is incorporated in the calculations. An approximate solution procedure was developed to calculate the failure probability in the spirit of the first-order reliability method, which combines material randomness with the randomness of load. Such procedure was shown to yield and improved analytical solution of the safety factor to be used in reliability-based structural design.

The resulting framework was demonstrated by analyzing the reliability of reinforced concrete beams against diagonal shear failure. Four different levels of uncertainty of the mean strength database, with coefficients of variation $w_f = 5\%$, 10%, 25% and 40%, were considered. It was found (Le and Bažant, 2020) that the overall failure probability depends strongly on the specimen or structure size (Fig. 6.25a), and that the size effect further depends on the general uncertainty level of the database.

To calculate the failure probability from the mean nominal shear strength measured from an individual experiment, it has been shown that, on the logarithmic scale, the mean of the individual test is located at certain distance a below the mean of the database (Fig. 6.25b) and this shift a can, approximately, be considered independent of the specimen size (Bažant and Yu, 2009). Meanwhile, the randomness of a seems far smaller than the strength variability of the database. Therefore, for calculating the failure probability, the shift a can be approximately taken as a deterministic quantity. Recently, though, a more accurate analysis showed that, as the structure size increases, the CoV first decreases and only for large sizes stabilizes as a constant (Luo *et al.*, 2021).

Exercises

E6.1. State the stability postulate of extreme value statistics and name three probability distributions that satisfy it. Which one is limited to positive argument only?

E6.2. Why must the frequency of interatomic bond breaks be equal to their failure probability?

E6.3. Nanoscale fracture propagates by nanoscale jumps. What is the consequence for the decrease of the potential energy of a nanoscale region caused by fracture?

E6.4. What is the consequence of the fact that a nanoscale crack advance results form a small difference between forward and backward jumps over the activation energy barrier. Express it by an equation.

E6.5. How would the frequency of crack jumps change if there were no backward jumps, that is, if the potential difference between to subsequent valleys is large (this would lead to the classical 1960s theory of Zhurkov).

E6.6. Consider a system of elements whose strength tail distribution is Weibullian with exponent m. When these elements are coupled in series or in parallel, is the power-law tail always preserved? What happens to the exponent and to the extent (or reach) of the power-law tail?

E6.7. What is the role of post-peak softening slope of these elements in the series and parallel couplings?

E6.8. What happens to the grafting point of Gauss–Weibull distributions when the structure size is increased, assuming the structure geometry is such that the failure of one RVE causes a failure of the structure.

E6.9. Can the strength of one RVE have a Weibull distribution?

E6.10. What are the fundamental issues with the three-parameter Weibull distribution?

E6.11. How does the fishnet connectivity of the microstructure alter the Weibull distribution?

E6.12. Consider the chain composed of N links subjected to a tensile load X shown in Fig. 6.26. The cross-section area for each link is A_i. The failure probability of each link follows a power law as follows:

$$P_{f,1} = \langle \sigma_i/\sigma_0 \rangle^m$$

with σ_i = stress in the i-th link. Assume $P_{f,i}$ is small: (a) Derive the failure probability of the chain, $P_f(x)$; (b) Assume the number of links is increased to a very large number. Find the expression for $P_f(x)$; (c) For the case of a large number of elements, derive the average strength of the chain.

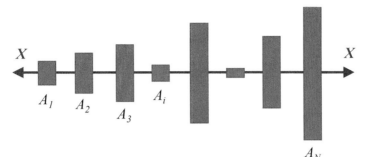

Fig. 6.26 Chain composed of N links investigated in problem E6.12.

E6.13. Tensile tests have been performed following ASTM A 370 on brittle steel specimens. The specimens were cylindrical of 12.7 mm and 76.2 mm effective length. The mean strength was 293 MPa and the standard deviation 7.3 MPa. (a) Make an estimate of Weibull modulus assuming zero threshold strength; (b) Find the characteristic strength σ_0 for the reference volume $V_0 = 10^{-3}$ m³; (c) Find the mean strength and the 99.9% probability strength of a tensioned bar of 30 mm diameter and 6 m length.

E6.14. Determine the mean strength for a three-point bend specimen with a span-to-depth ratio of 16, assuming an elastic-brittle material for which Weibull's analysis applies. Give it as a function of the beam volume, the mean strength of a tensile specimen of same volume, and the Weibull modulus. Particularize the results for (a) $m = 12$ and (b) $m = 25$. Use the beam bending theory (neglecting shear), and take the nominal strength defined as $\sigma_N = 6M/bD^2$ with M = maximum bending moment.

E6.15. Determine the mean strength for a cantilever beam with a length-to-depth ratio of 10, subjected to a uniformly distributed load. Assume an elastic-brittle material for which Weibull's analysis applies. Give the mean strength as a function of the beam volume, the mean strength of a tensile specimen of same volume, and the Weibull modulus. Consider (a) $m = 12$ and (b) $m = 25$. Use the beam bending theory (neglecting shear), and take the nominal strength defined as $\sigma_N = 6M/bD^2$ with M = maximum bending moment.

E6.16. Consider the parallel system shown in Fig. 6.9. Assume that each fiber has a brittle behavior. Daniels (Daniels, 1945) derived a recursive equation to describe the strength distribution of this parallel system:

$$P_n(\sigma) = \sum_{k=1}^{n} \frac{n!}{k!\,(n-k)!} F^k(\sigma)\, P_{n-k}\left(\frac{n\sigma}{n-k}\right)$$

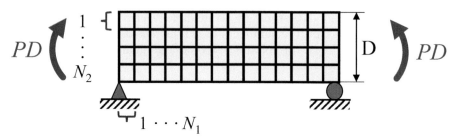

Fig. 6.27 2D weakest-link model of bending specimen.

where P_i = strength distribution of the parallel system with i number of fibers. $F(\sigma) =$ strength distribution of one fiber. Consider now $F(\sigma) = (\sigma/s_1)^p$ when $\sigma \to 0$. Show that $P_n(\sigma) \propto \sigma^{np}$ $(\sigma \to 0)$.

E6.17. Consider two-dimensional (2D) analysis of a rectangular beam subjected to moments $M = PD$ at its end as shown in Fig. 6.27. The beam is divided by a mesh of square elements of size l_0 with N_1 elements along the length and N_2 elements along the depth. D = beam depth. For each element, assume that the tail of failure probability follows $P_1 = \langle\sigma/s_0\rangle^2$ $(P_1 < 10^{-3})$, where σ = normal stress at the center of each element, Weibull modulus, $m = 24$, $\langle x \rangle = \max(x, 0)$, and scaling parameters $s_0 = 0.8f_t$. Calculate the loads corresponding to failure probability $P_f = 10^{-6}$ for:

- $N_1 = 30$, $N_2 = 6$;
- $N_1 = 300$, $N_2 = 60$;
- $N_1 = 3000$, $N_2 = 600$;
- $N_1 = 30000$, $N_2 = 6000$;

Plot the size effect diagram of the $\ln \sigma_N$ versus $\ln N_1$. Note that $\sigma_N = P/bD$, $b =$ width of the beam in the third dimension, which is kept constant).

7

Quasibrittle Size Effect Analysis in Practical Problems

A good theory is a source of happiness until one confronts the real world.

Previous chapters presented various essential concepts of mechanics of fracture and damage of quasibrittle structures. A particularly important result is the size effect in failure of quasibrittle structures, which has profound implications for the design of engineering structures. Over the last three decades, extensive efforts have been directed towards investigating the size effect in fracture of many quasibrittle materials, such as concretes (an archetypical case), composites, ceramics, rock, sea ice, snow, bone, cold asphalt mixtures, and polycrystalline silicon, and it has been found that the energetic and statistical size effect models presented in Chapters 5 and 6 are broadly applicable to all these materials.

This chapter provides an overview of various applications of the energetic and statistical size effect laws for many quasibrittle structures, including concrete and reinforced concrete (RC) structures, fiber-composite structures, nanocomposites, metal-composite hybrid structure, sea ice, bones, and microelectromechanical systems made of polycrystalline silicon (Poly-Si MEMS). The chapter also includes an analysis of failure of Malpasset dam, a real-world case study that elucidates the critical role of size effect laws in design of large-scale engineering structures.

Applications to concrete are the main focus of this chapter. Concrete is a material by far the most utilized in the world, not only by mass but also total expenditures. Due to its wide applications in building and construction industry, considerable attention has been paid to the size effect. The many developments of the size effect theory (Chapters 5 and 6) were motivated by numerous experimental observations of fracture and failure of concrete structures. This chapter discusses various important structural applications of the size effect laws in plain and reinforced concrete. Some have already been adopted for structural design code specifications (e.g. in ACI-318 design code, RILEM and ACI-446 recommendations) and are being used in daily design practice.

Quasibrittle Fracture Mechanics and Size Effect: A First Course. Zdeněk P. Bažant, Jia-Liang Le and Marco Salviato, Oxford University Press. © Zdeněk P. Bažant, Jia-Liang Le, Marco Salviato 2022. DOI: 10.1093/oso/9780192846242.003.0007

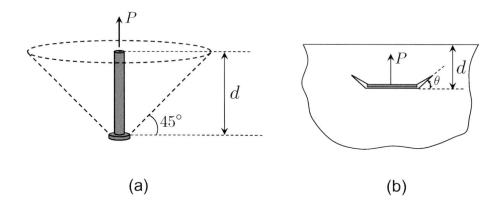

(a)   (b)

Fig. 7.1 Analysis of pullout failure of a single anchor embedded in concrete: (a) failure cone, and (b) progressive propagation of cracks emanating from the anchor head.

7.1 Tensile Fracture Problems

7.1.1 Anchor Failure in Concrete

The first fracture mechanics model that was adopted in the concrete design codes is a linear elastic fracture mechanics (LEFM) size effect for the load-carrying capacity of steel anchors embedded in concrete. The failure of anchor is caused by pullout of a concrete cone, as shown in Fig. 7.1a. Up to the mid-1980s, the load capacity was predicted on the basis of strength. It was assumed that the failure surface, shown in Fig. 7.1a, was subjected to a uniform tensile stress, and consequently it was predicted that the load capacity of the anchor scaled with the embedment depth quadratically, that is, $P_m \propto d^2$ (ACI Committee 349, 1989). In terms of the nominal strength of the anchor, defined as $\sigma_N = P_m/d^2$, the strength-based theory implies that σ_N would be independent of the embedded depth. However, experiments showed that the strength-based theory significantly overestimated the load capacity of the anchor for typical embedment depths (Klinger and Mendonca, 1982; Ballarini *et al.*, 1986; Eligehausen and Sawade, 1989; Ožbolt and Eligehausen, 1992; Ožbolt *et al.*, 1999). As it appeared, the anchor failure is accompanied by discrete cracks emanating from the anchor head Fig. 7.1b. It has thus become obvious that fracture mechanics is required.

The first to treat anchor pullout by fracture mechanics was Ballarini *et al.* (Ballarini *et al.*, 1986, 1987). They used LEFM to derive for the load capacity of the anchor the following relation, which includes the scaling:

$$P_m \approx K_c d^{3/2} \quad \text{or} \quad \sigma_N \approx K_c d^{-1/2} \tag{7.1}$$

where K_c = fracture toughness. Eq. 7.1 shows the classical "−1/2" LEFM scaling law. Interestingly, such scaling implies that the anchor failure is brittle rather than

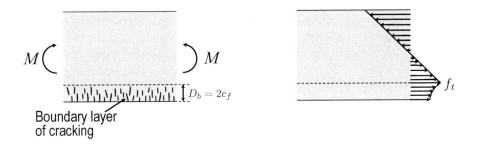

Fig. 7.2 Stress redistribution in the cracking layer of the bending specimen.

quasibrittle, and does not need the transitional scaling law. In fact, it appears to be the only type of failure of reinforced concrete that follows the LEFM.

Numerous experiments have subsequently confirmed that the LEFM-based Eq. 7.1 is sufficient for typical embedment depths, even those not much larger than the aggregate size. Clearly, it must be the special geometry of the embedded anchor that causes that the brittle behavior obeying LEFM. This example also confirms that the brittleness is governed not only by the structure size but also its geometry. Indeed, the brittleness number defined in type 2 size effect (Eq. 5.59) depends on both the structure size and geometry.

The fracture toughness in pullout has been empirically related to the compressive strength of concrete by introducing the so-called concrete capacity design (CCD) method, $K_c = k_c \sqrt{f'_c}$, where $f'_c =$ uniaxial compressive strength and $k_c =$ an experimentally determined factor introduced by Fuchs *et al.* (Fuchs *et al.*, 1995). By means of the CCD method, Eq. 7.1, has been incorporated into various design codes and provisions (Comité Euro-International du Beton, 1997; ACI Committee 318, 2008). Subsequent studies successfully extended the model to pullout failure of a single anchor in prestressed concrete (Piccinin *et al.*, 2010, 2012) and to an anchor group in reinforced concrete (RC) (Ballarini and Xie, 2017).

7.1.2 Modulus of Rupture or Flexural Strength

For RC structures, tensile cracks do not usually cause the ultimate structural failure, but they influence the serviceability and durability. In the case of unreinforced concrete structures, such as concrete pavements, foundation plinths, footings, or retaining walls, the failure is directly governed by the tensile strength of concrete. Different test methods are available for measuring the tensile strength of concrete: the direct tension test, the flexural strength test, and the splitting test. Among them, the test of flexural strength, for which the traditional name is the modulus of rupture (MOR) test, is the most popular in practice for its simplicity. The splitting test is also simple but is affected by crack-parallel compression.

The MOR test is essentially a bending test on a simply supported beam. Consider a three-point bend specimen ($L = $ span, $D = $ depth and $b = $ width). Let P_m denote the maximum of the concentrated load applied at midspan. The MOR is defined as $\sigma_N = (3L/2D)P_m/bD$, which is the maximum stress at the bottom face calculated by the engineering beam theory. However, due to stress redistribution in the evolving fracture process zone (FPZ) at the bottom face (Fig. 7.2), the maximum load is not decided by the condition that σ_N reaches the concrete's tensile strength f_t. Rather it is governed by the stress value $\bar{\sigma}$ roughly in the middle of the damage layer at the bottom face, representing the evolving FPZ (Bažant and Li, 1995). Approximately, $\bar{\sigma} = \sigma_N - \sigma'c_f$ where $\sigma' = 2\sigma_N/D = $ stress gradient and $c_f = $ half of the thickness of boundary layer of cracking, which is a material constant.

By setting $\bar{\sigma} = f_t = $ tensile strength of the material (Fig. 7.2), we have $\sigma_N(1 - 2c_f/D) = f_t$, which gives

$$\sigma_N = \frac{f_t}{1 - D_b/D} \tag{7.2}$$

in which $D_b = 2c_f$.

Eq. 7.2 for σ_N, however, is valid only when $D \gg D_b$ (that is, when the cracking zone is a small fraction of the beam depth). But since the foregoing derivation is valid only for small values of c_f/D (that is., up to the first-order term of the asymptotic expansion of σ_N in terms of $1/D$), one may replace Eq. 7.2 by the following asymptotically equivalent size effect formula:

$$\sigma_N = f_t \left(1 + \frac{rD_b}{D}\right)^{1/r} \tag{7.3}$$

which has (for any r) the same correct first two terms of the Taylor series expansion as Eq. 7.2, but has the advantage of being acceptable for the entire range of D (for positive r). Indeed, while Eq. 7.2 gives infinite σ_N for $D = D_b$ and negative σ_N for $D < D_b$, Eq. 7.3 gives a linear approach to a finite σ_N for $D \to 0$, and thus satisfies the first two terms of the small-size asymptotic expansion (Sec. 5.4, Eq. 5.53). Parameter r is empirical. It was taken between 1 and 2 for concrete (Bažant and Novák, 2000a), while $r = 1.47$ appeared optimum in Bažant and Novák's analysis of certain test data.

As discussed in Sec. 5.5.2, Eq. 7.3 can alternatively be derived by using the equivalent LEFM. Indeed Eq. 7.3 represents a classical example of type 1 size effect. The foregoing analysis ignores the contribution of material randomness. This is insignificant for small- and intermediate-size ranges (Bažant and Novák, 2000a; Bažant and Le, 2017; Le et al., 2018b), though not for the large-size range. For large beams, the Weibull statistical size effect must become dominant, which can be taken into account by modifying the large-size asymptote of Eq. 7.3 (see Eq. 5.68), that is,

$$\sigma_N = f_t \left[\frac{rD_b}{D} + \left(\frac{rD_a}{D}\right)^{2r/m}\right]^{1/r} \tag{7.4}$$

Fig. 7.3 shows that Eq. 7.4 agrees quite well with the existing test data on MOR of different concrete beams (note that D_b is different for each data set).

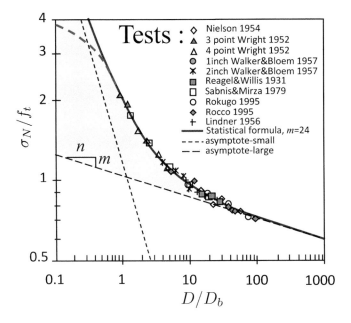

Fig. 7.3 Optimum fitting of MOR test results of concrete beams.

Eq. 7.3 is essential to the design practice of many concrete structures. For example, in the case of large-scale pavement construction, it is recommended for quality control to test the MOR of concrete (although the crack-parallel stress effect on the fracture energy, Sec. 4.1, which is high under wheel loads, may significantly reduce the MOR for pavements). Because a large number of test specimens is required, the depth (or size) of the flexural test beams is chosen to be approximately 150 mm. This is much less than the actual thickness of the road pavements, which is typically 300 mm, and for runways much more. In this case, the dependence of the flexural strength on the size becomes very important for design.

7.1.3 Plates under Biaxial Bending

The aforementioned MOR test essentially measures the flexural strength of concrete under uniaxial tensile stress. However, in most applications, the stress state is biaxial or triaxial. Recent studies (Zi et al., 2014; Kirane et al., 2014) investigated the size effect in fracture of concrete under biaxial stress states.

In a recent study (Zi et al., 2014), two types of biaxial experiments were performed. The first is the centrally loaded circular plate test (ASTM C1550). In ASTM C1550, a circular plate is supported at the rim by three equidistant round pivots, and is subjected to a concentrated load at the center (distributed over a small circle, Fig. 7.4a). At the center, under the load, the tensile stress is the maximum and is biaxial. The stress distribution in the plate is triply axisymmetric. The biaxial strength can

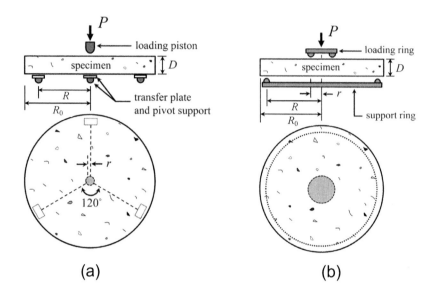

Fig. 7.4 Set-ups of biaxial test: (a) ASTM C1550 test, and (b) biaxial flexural experiment.

be expressed as (Ban *et al.*, 1992; Higgs *et al.*, 2001):

$$f_t = \frac{3P(1+\nu)}{4\pi D^2}\left\{1 + 2\ln(R/r_e) + \frac{1-\nu}{1+\nu}\left[\frac{2R^2 - r_e^2}{2R_0^2}\right]\right\} \tag{7.5}$$

where R = the distance from the center of the specimen to the support, D = plate thickness, R_0 = plate radius, $r_e = \sqrt{1.6r^2 + D^2} - 0.675D$ = equivalent radius (valid for $R < 0.5D$), r = radius of the loaded area, and ν = Poisson's ratio of the material.

The second set of experiments involved a recently developed biaxial flexural test set-up (Zi *et al.*, 2008). Similar to ASTM C1550, the test uses a circular plate specimen. The specimen is loaded through a ring (Fig. 7.4b). Because the support of the specimen at the rim is also an annular ring, the stress field is axisymmetric. Mechanical analysis showed that, within the area enclosed by the loading ring, the stress state caused by the applied load is uniform on every horizontal plane. Consequently, it is expected that the statistical randomness of local material strength would play a more pronounced role in the scaling behavior. The biaxial strength can be estimated by (Zi *et al.*, 2008)

$$f_t = \frac{3P}{4\pi D^2}\left[2(1+\nu)\ln(R/r) + \frac{(1-\nu)(R^2 - r^2)}{R_0^2}\right] \tag{7.6}$$

These experiments were performed on a set of geometrically similar specimens with the size ratio of $1:1.6:2.5$. The minimum dimension was chosen to be approximately only 4-times greater than the size of the coarse aggregate. During the biaxial flexural

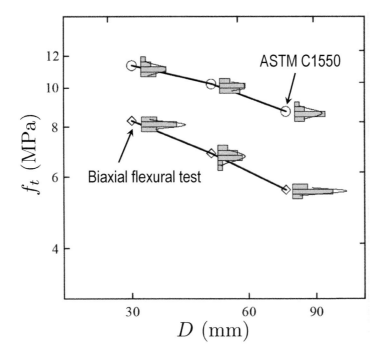

Fig. 7.5 Measured size effects on biaxial strength (Zi *et al.*, 2014).

test, the in-plane strain was measured to check whether the stress state was biaxial. Two strain gauges were attached to the bottom surface of the plates, perpendicular to each other. The strains in the two directions were found to be almost the same before the failure, for all the sizes. Fig. 7.5 shows the typical load-strain curves measured in the biaxial flexural test.

Fig. 7.5 further shows the size effect on the biaxial strength obtained from the ASTM C1550 test and the biaxial flexural test results with their scatter through five intervals. Despite the limited size range, the test results show a pronounced size effect. When the plate thickness increases from 30 to 48 mm and from 48 to 75 mm, the biaxial flexural strength decreases by approximately 17% and 19%, respectively. In the same study, the size effect on MOR was also measured for the same batch of concrete. The size effect on biaxial flexural strength is more pronounced than that on MOR.

For reasons of simplicity, it is in practice often preferred to use the uniaxial rather than biaxial tests, even if the actual stress condition is biaxial. Because the uniaxial strength is less than the biaxial strengths, the choice of the uniaxial test method would lead to an error on the side of safety. This might not be true for large sizes, however,

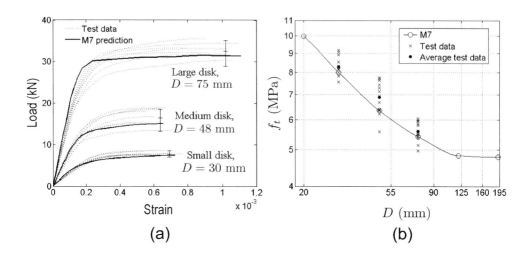

Fig. 7.6 Comparison between the simulated and measured behavior of biaxial flexural specimens: (a) load-strain curve, and (b) mean size effect curve (Kirane *et al.*, 2014).

because the size effect in biaxial loading is stronger. The safety margin would then be insufficient.

The remaining question is whether the biaxial flexural strength exhibits a type 1 or type 2 size effect. This question cannot be answered by the aforementioned experiments due to the limited size range. To explore a wider size range, the biaxial flexural tests were modeled numerically (Kirane *et al.*, 2014). In the finite element simulations, microplane constitutive model M7 was used for concrete, and was combined with the crack band model to prevent spurious mesh sensitivity. To avoid additional simulation error from post-peak adjustment, the element size was kept for all specimens sizes the same and equal to the crack band width w_c (it was 13 mm, which is double the maximum aggregate size of 6.5 mm).

The M7 free parameters were first calibrated by fitting the biaxial flexural test results. Fig. 7.6a shows the comparison between the measured and simulated load-strain curves of the biaxial flexural tests. The calibrated M7 was then used to simulate the behavior of specimens of other sizes, i.e. $D = 20, 120$, and 192 mm. The predicted size effect on the biaxial flexural strength is plotted in Fig. 7.6b.

The size effect curve resembles the type 1 energetic size effect. The basic feature of type 1 size effect is that a macroscopic continuous crack initiates from a smooth (unnotched) surface when post-peak softening begins. The simulation results confirm this feature for all the disk sizes. They show that, before reaching the peak load, there are no elements with zero stress and only few ones with very low stress (less than 0.5

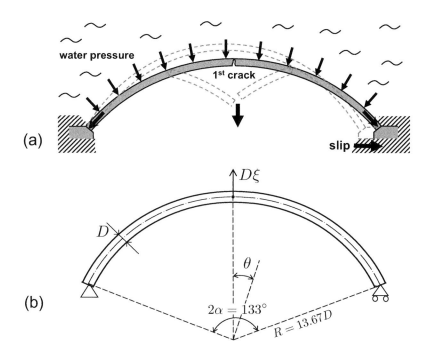

Fig. 7.7 Simplified 2D analysis of an arch dam: (a) failure mode, and (b) geometry of 2D model.

MPa).

7.1.4 Case Study of Type 1 Structural Failure—Malpasset Dam

The Malpasset arch dam was the tallest and slenderest arch dam on the world. It was built across the canyon of the Reyran Valley in France in 1954. In 1959, upon the first full filling of the reservoir after five days of heavy rain, it failed, unleashing a flood which killed all the inhabitants of the ancient town of Fréjus. Investigation attributed the failure to a vertical flexural crack caused by lateral displacement of the abutment (Bartle, 1985; Levy and Salvadori, 1992). A recent study examined several geological issues that could possibly cause the large displacement of the abutment (Duffaut, 2013).

The displacement of an abutment can never be exactly zero, and so the question is the maximum tolerable displacement that the dam could withstand, with the proper safety margin (Bažant *et al.*, 2007*a*; Le *et al.*, 2012; Le, 2015). This section summarizes a recent analysis of the failure statistics of the ill-fated Malpasset dam, with the aim of demonstrating the necessity of including the size effect in reliability-based analysis and design of quasibrittle structures in general.

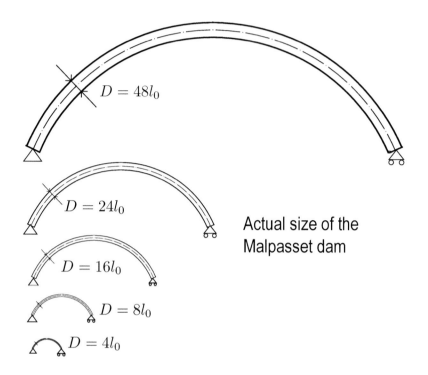

$D = 48l_0$

$D = 24l_0$

Actual size of the
Malpasset dam

$D = 16l_0$

$D = 8l_0$

$D = 4l_0$

Fig. 7.8 Geometrically similar arch dams of different sizes for the analysis (the arch dam with $D = 96l_0$ is not shown in this figure).

Here we consider the randomness of both the structural strength and the abutment movement, and evaluate the size effects on both the overall failure statistics of the dam and the design safety factors. For the sake of simplicity, the dam is modeled as a two-dimensional arch (Bažant *et al.*, 2007*a*), shown in Fig. 7.7. The dimensions of the arch correspond to the mid-height cross-section of the dam; they are $R = 92.68$ m, $D = 6.78$ m, and $\alpha = 66.5°$. The arch is loaded by a displacement at its right support. To investigate the size effect on the failure statistics of the dam, we consider a set of scaled geometrically similar arches with different sizes (Fig. 7.8), characterized by arch depth D and arch radius $R = 13.67D$. Based on the engineering beam theory, the elastic bending stress can be expressed by

$$\sigma(\theta, \xi) = \frac{2\xi(\cos\theta - \cos\alpha)}{1 - \cos\alpha}\sigma_N \tag{7.7}$$

where $\xi = x/D$, $x = $ coordinate across arch thickness measured from the center line, $\alpha = $ angle of half-arch (Fig. 7.7), and $\sigma_N = $ nominal structural strength of the arch dam, which is defined as the maximum elastic stress in the arch considered as a two-dimensional ring, caused by horizontal force in the abutment. Following Chapter 6, we can calculate the probability distribution of σ_N using the finite weakest-link model:

$$P_R(\sigma_N) = 1 - \exp\left(\frac{1}{l_0^2}\int_0^{1/2}\int_{-\alpha}^{\alpha}\ln\left\{1 - P_1\left[\sigma_N s(\theta,\xi)\right]\right\}RD\mathrm{d}\theta\mathrm{d}\xi\right) \quad (7.8)$$

where $P_1(x)$ = strength distribution of one RVE given by Eqs. 6.45a and 6.45b. The dimensionless stress field $s(\theta,\xi)$ can be written as

$$s(\theta,\xi) = \frac{2\xi(\cos\theta - \cos\alpha)}{1 - \cos\alpha} \quad (7.9)$$

Note that Eq. 7.8 is written by using the integral form of the original finite weakest-link equation (Eq. 6.47).

The nominal stress, σ_L, applied to the arch is defined as the maximum longitudinal elastic stress induced by the abutment movement Δ. Based on Castigliano's theorem, we have

$$\sigma_L = \frac{E(1 - \cos\alpha)D\Delta}{2R^2\int_{-\alpha}^{\alpha}(\cos\theta - \cos\alpha)^2\mathrm{d}\theta} \quad (7.10)$$

It is clear that the failure state of the arch can simply be written as $\sigma_N - \sigma_L < 0$. Since σ_N and σ_L are statistically independent, the overall failure probability of the structure can be calculated by Freundental's reliability integral (Freudenthal, 1956), that is,

$$P_{FL} = \int_0^{\infty} P_R(x)f_L(x)\mathrm{d}x \quad (7.11)$$

where $f_L(x)$ = probability density function (pdf) of σ_L.

Previous studies (Bažant *et al.*, 2007a; Le *et al.*, 2012) suggested the following parameters for the concrete strength distribution $P_1(x)$ of a RVE (Eqs. 6.45a and 6.45b): $m = 24, s_0 = 2.12$ MPa, $\mu_G = 2.91$ MPa, $\delta_G = 0.44$ MPa, $w = 0.15$, and a RVE size of 280 mm. However, the cdf of σ_L, which should be determined by the statistics of abutment movement, is not available. For the present size effect analysis, we estimate the cdf of σ_L by assuming that: (a) σ_L follows a Gaussian distribution with a coefficient of variation (CoV) of 40%, and (b) the failure probability of the dam must be 10^{-6}, i.e., the dam being designed on the basis of a scaled laboratory prototype with $D = 4l_0$ and standard tests of flexural strength. Based on these assumptions and on Eq. 7.11, we obtain the mean value of σ_L to be 0.76 MPa.

Based on the aforementioned probability distributions of σ_N and σ_L, we can use Eq. 7.11 to calculate the exact failure probability P_{FL} for geometrically similar dams of six sizes $D = 4l_0, 8l_0, 16l_0, 24l_0, 48l_0$ and $96l_0$ (Fig. 7.9), among which $D = 24l_0$ corresponds to the actual size D_a of the Malpasset dam. It is remarkable that the structural failure probability increases from 10^{-6} to 6×10^{-3} as dam size D increases from $4l_0$ to $96l_0$.

We may also use the Cornell reliability index to estimate the structural failure probability (Cornell, 1969; Ang and Tang, 1984; Haldar and Mahadevan, 2000). The

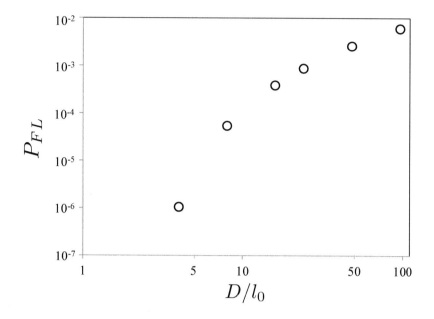

Fig. 7.9 Failure probabilities of geometrically similar dams of different sizes.

Cornell index, which is the most widely used approach for reliability-based analysis and design due to its simplicity, represents the original formulation of the mean-value first-order second-moment (MVFOSM) method. For a linear failure state, the Cornell index is calculated as

$$\beta = \frac{\mu_R - \mu_L}{\sqrt{\delta_R^2 + \delta_L^2}} \tag{7.12}$$

where μ_R, μ_L = mean values of the structural strength σ_N and of the applied nominal stress σ_L, respectively, and δ_R, δ_L = standard deviations of σ_N and σ_L, respectively. If both σ_N and σ_L follow the Gaussian distributions, then the Cornell index can be directly used to compute the theoretically exact failure probability of the structure, that is,

$$P_{FL} = \Phi(-\beta) \tag{7.13}$$

where $\Phi(x) = (\sqrt{2\pi})^{-1} \int_{-\infty}^{x} e^{-x'^2/2} dx'$ = standard Gaussian distribution function.

As discussed in Chapter 6, the finite weakest-link model of the nominal structural strength σ_N directly indicates that the mean and standard deviation (or variance) of σ_N vary with the structure size. Fig. 7.10 shows the mean and variance of σ_N for different size dams computed from Eq. 7.8. Based on this information, we can compute the Cornell reliability index β and consequent failure probability using Eqs. 7.12 and 7.13.

Table 7.1 presents the theoretically exact failure probabilities as well as those calculated from the Cornell index. It is found that the Cornell index gives a reasonable

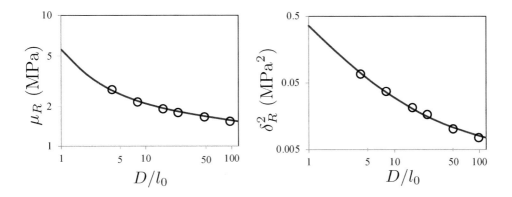

Fig. 7.10 Size effect on the mean and variance of nominal strength of the dam (Le, 2015).

Table 7.1 Comparison of calculated structural failure probabilities P_{FL}

Dam size	Exact P_{FL}	$\Phi(-\beta)$
$D = 4l_0$	1.05×10^{-6}	1.28×10^{-6}
$D = 8l_0$	5.37×10^{-5}	2.66×10^{-5}
$D = 16l_0$	3.84×10^{-4}	2.04×10^{-4}
$D = 24l_0$	8.72×10^{-4}	5.11×10^{-4}
$D = 48l_0$	2.57×10^{-3}	1.81×10^{-3}
$D = 96l_0$	6.12×10^{-3}	4.89×10^{-3}

estimation of the failure probability despite the fact that the probability distribution of σ_N is non-Gaussian. To explain this result, we plot the value of the integrand of Eq. 7.11 normalized by the failure probability, i.e. $P_R(x)f_L(x)/P_{FL}$ (Fig. 7.11). When the structure size is small, most of $P_R(x)$ follows a Gaussian cdf and the power-law tail makes a negligible contribution to the overall structural failure probability (For $D = 4l_0$, $\int_0^{\sigma_{gr}} f_L(x)P_R(x)\mathrm{d}x = 4.80 \times 10^{-8} = 0.045P_{FL}$). Therefore, the Cornell index yields a good estimation of the overall failure probability at the small-size limit.

At the large-size limit, Fig. 7.11 indicates that the entire $P_R(x)$ contributes to the structural failure probability due to the decreasing mean strength. Although $P_R(x)$ would follow a Weibull distribution, we note that, for the same mean and variance, Gaussian and Weibull distributions are close to each other except for the left and right tails. For this reason, the Cornell index can also predict the failure probability reasonably well. Based on Table 7.1, it is seen that, for the intermediate size range, the failure probabilities predicted by the Cornell index are within the same order of magnitude as the exact values.

It should be pointed out that the present analysis is based on the assumption that

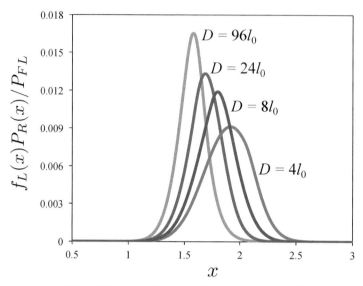

Fig. 7.11 Integrand of Eq. 7.11 normalized by the failure probability of the dam (Le, 2015).

the applied nominal stress σ_L follows a Gaussian distribution. For a non-Gaussian distribution of σ_L, a different reliability index (e.g. Hasofer–Lind index) would be needed to improve the estimation of P_{FL}.

The foregoing analysis documents that, if we design the dam based on the laboratory test results without considering the size effect, the failure probability of the actual dam would be intolerably high (on the order of 10^{-3}). To ensure an acceptable failure risk ($P_c \approx 10^{-6}$) for the actual design, the size effect on strength distribution must be taken into account. For the design purpose, the most straightforward approach is to use the concept of safety factors. The failure criterion is characterized by either the nominal values or the mean values of the structural strength and nominal applied stress. If the mean values are used, the failure domain can be written as

$$\mu_R < \bar{\zeta}\mu_L \tag{7.14}$$

where $\bar{\zeta}$ is referred to as the central safety factor. If the nominal values are used, the failure domain becomes

$$\mu_{RN} < \zeta\mu_{LN} \tag{7.15}$$

where μ_{RN} = nominal value of σ_N = $\mu_R - k_R\delta_R$, μ_{LN} = nominal value of σ_L = $\mu_L + k_L\delta_L$, k_R, k_L = constants, and ζ is the nominal safety factor. The safety factors $\zeta, \bar{\zeta}$ need to be determined to ensure that the overall failure probability of the structure be equal to a prescribed value P_c.

Fig. 7.9 clearly shows that, in order to keep $P_c = 10^{-6}$, the safety factors must vary with the structure size. Recent studies have proposed approximate equations for

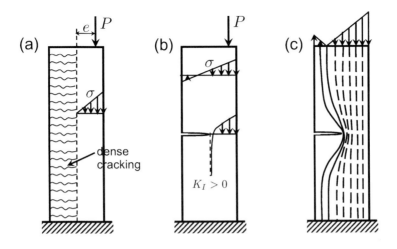

Fig. 7.12 (a) Stress distribution in a prismatic specimen without notch, (b) stress distribution in a notched prismatic specimen, and (c) approximate trajectories of minimum principal stresses.

size-dependent safety factors through asymptotic matching (Le, 2015; Bažant and Le, 2017). It was shown that, due to the size effects on the mean and standard deviation of structural strength, the safety factors must increase with the specimen size (Bažant and Le, 2017).

7.1.5 Is No-Tension Material a Safe Alternative to Fracture Mechanics?

Concrete dam and rock structures are often designed under the hypothesis that the material has no tensile strength. But is the result always on the safe side? An interesting point is that not always.

Consider the prismatic specimen of width D in Fig. 7.12a,b, subjected to a compressive load P with eccentricity $e = D/3$. If there is no notch and if the material is elastic with no resistance to tension, the solution is homogeneously cracked materials with zero stress to the left of centroidal axis, and a triangular elastic stress distribution to the right, as seen in Fig. 7.12a.

Consider now that the material is elastic even for tension, and that there is a notch spanning the depth, $a = D/2$, of the former zero-tension zone, as in Fig. 7.12b. At cross-sections farther from the notch, the same load P then produces the tensile stress of P/bD at the left face of prism (b = prism width) and the principal stress trajectories bend and crowd around the notch tip as shown in Fig. 7.12c. LEFM then shows the the stress intensity factor at the notch tip, K_I, is non-zero (Bažant, 1996). So, for $K_{Ic} = 0$, the crack tip must propagate, in fact dynamically. A stable equilibrium is

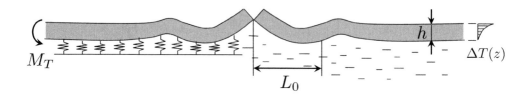

Fig. 7.13 Temperature-induced bending fracture of floating sea ice plate (Bažant, 2005).

---

found to exist $a = 0.549D$. However, if there is water pressure within the crack, as in a dam, no stable equilibrium exists and the prism fractures completely.

This example shows that, if the material is brittle rather than plastic, then the no-tension analysis is not always safe. Increasing the materials strength from zero to a finite value can cause the load capacity of structure to decrease (Bažant, 1990a, 1996). On the other hand, according to the theory of plasticity, increasing the material strength, or enlarging the material failure envelope, always causes the load capacity of structure to increase. In recognition of this fact, the US Army Corps of Engineers has required since the early 1990s that every dam design, usually made with the no-tension approach, must be checked by fracture mechanics.

7.2   Tensile Fracture of Sea Ice

Different types of size effect are exhibited by sea ice failures. The scaling of failure of floating sea ice plates in the Arctic presents some intricate challenges. One practical need is the load carrying capacity or vertcal penetration of sea ice sheet. Another is the force exerted by moving ice sheet against a fixed object. Still another is to understand the formation of very long fractures (of the order of 1 km to 1000 km) which cause the formation of open water leads or serve as precursors initiating the build-up of pressure ridges and rafting zones.

It must be pointed out that no adequate experimental data for these large scale fractures exist (except perhaps in private domain). Thus the analysis that follows is less experimentally grounded than that of concrete, composites and ceramics. Adequate experimental support exists only on the material scale—Dempsey *et al.*'s fracture tests on floating compact-tension specimens reaching dimensions up to $80 \times 80 \times 1.79$ m (Dempsey *et al.*, 1999).

7.2.1 Thermal bending fracture

A possible, and apparently the only known, mechanism that can produce sudden very long fractures on the Arctic sea ice cover is the thermal bending caused by rapid cooling of the surface of the ice plate (Fig. 7.13). Due to buoyancy, the floating plate behaves exactly as a plate on elastic Winkler foundation, in which the foundation modulus is equal to the unit weight of sea water. Under the assumptions that: (a) the ice plate is infinite and elastic, of constant thickness h, (b) the temperature profiles for various thicknesses h are similar, and (c) the thermal fracture is semi-infinite and propagates in a stationary manner, statically (i.e., with insignificant inertia forces), it was found that the critical temperature difference is (Bažant, 1992)

$$\Delta T_{cr} \propto h^{-3/8} \tag{7.16}$$

This equation was derived using LEFM. Despite the existence of a large FPZ, LEFM is justified because a steady-state propagation must develop. The FPZ does not change as it travels with the fracture front, and thus it dissipates energy at a constant rate, as in LEFM. It has been shown that Eq. 7.16 must apply to failures caused by any type of bending cracks, provided that they are full-through cracks propagating along the plate (created by any type of loading, e.g. by vertical load (Slepyan, 1990; Bažant, 1993)).

It may be surprising that the exponent of this large-size asymptotic scaling law is not $-1/2$. However, this apparent contradiction may be explained if one realizes that the plate thickness is merely a parameter but not actually a dimension in the plane of the boundary value problem, that is, the horizontal plane (x, y). In that plane, the problem has only one characteristic length— the flexural wavelength of a plate on elastic foundation, L_0. As it happens, L_0 is not proportional to h but to $h^{3/4}$. Since $\frac{3}{8} / \frac{3}{4} = \frac{1}{2}$, it thus follows that the scaling of thermal bending fracture with respect to L_0 does in fact obey the LEFM scaling law (Eq. 5.30):

$$\Delta T_{cr} \propto L_0^{-1/2} \tag{7.17}$$

Simplified calculations have shown that, in order to propagate such a long thermal bending fracture through a 1 m thick plate, the temperature difference across the plate must be about 25°C, while for a 6 m thick plate the temperature difference needs to be only 12°C (Bažant, 1992). This is a significant size effect. It may explain why very long fractures in the Arctic Ocean are often seen to run through the thickest floes rather than through the thinly refrozen water leads between and around the floes (as observed by Assur (1963)). Fracture due to in-plane tension would have to run through the thin leads, around the thick floes.

7.2.2 Vertical penetration failure

Another important practical problem, which exhibits a different scaling behavior, is the problem of load capacity, that is, the failure caused by vertical, downward or upward, penetration through the floating ice plate (Fig. 7.14a). In that case, the fractures are known to form a system of radial bending cracks (Fig. 7.14b) which propagate outward

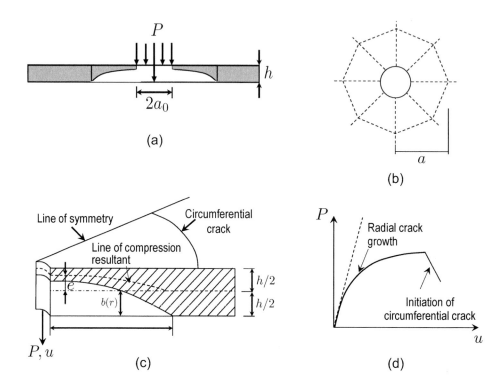

Fig. 7.14 Vertical penetration failure of ice plate: (a)–(b) cross-sectional and plan views of the radial and circumferential cracks caused by vertical penetration of an object through floating ice panel, (c) part-through radial crack and shift of compression resultant causing a dome effect, and (d) typical load-deflection diagram.

from the loaded area. The failure occurs when the circumferential bending cracks begin to form, as indicated by the load-deflection diagram in Fig. 7.14d.

This problem is fundamentally different from the aforementioned full-through bending cracks for which the asymptotic scaling law for large cracks is given by Eq. 7.16. Experiments as well as finite element analyses showed that the radial cracks formed before failure do not reach through the full thickness of the ice plate, as shown in Figs. 7.14a and c. This makes the analysis significantly more complicated.

This problem was first tackled through numerical simulation (Bažant and Kim, 1998a,b). In the model, the elasticity of one half of the sector of the floating plate limited by two adjacent radial cracks is characterized by a compliance matrix obtained numerically, and the radial cross-section with the crack is subdivided into narrow vertical strips. In each strip, the crack is assumed to initiate through a plastic stage of finite cohesive length (representing an approximation of the cohesive zone). This can be done by using a strength criterion with constant in-plane normal stress assumed to

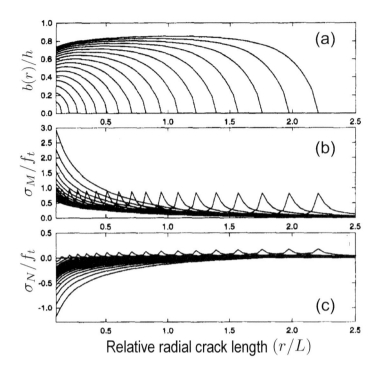

Fig. 7.15 Calculated profiles of (a) radial part-through crack, (b) normal stress due to bending, and (c) stress due to normal force (Bažant and Kim, 1998*b*).

develop within the cross section part where the strain corresponding to the strength limit is exceeded.

For the subsequent fracture stage, the relationship of the bending moment M and normal force N in each cracked strip to the additional rotation and in-plane displacement caused by the crack is described by the nonlinear line spring model of Rice and Levy (Rice and Levy, 1972). The transition from the plastic stage to the fracture stage is considered to occur as soon as the fracture values of M and N become less than their plastic values (to do this consistently, a plastic flow rule must be considered such that the ratio M/N would always be the same as for fracture).

This analysis produces the profiles of crack depth shown in Fig. 7.15a, where the last profile corresponds to the maximum load (the plate depth is greatly exaggerated in the figure). Figs. 7.15b and c show the radial distribution of the nominal stresses due to bending moment and to normal force. It is found that the normal forces transmitted across the radial cross-section containing the crack are quite significant. They cause a dome effect which helps to carry the vertical load. The analysis also indicates that the number of radial cracks emanating from the load point depends on the thickness

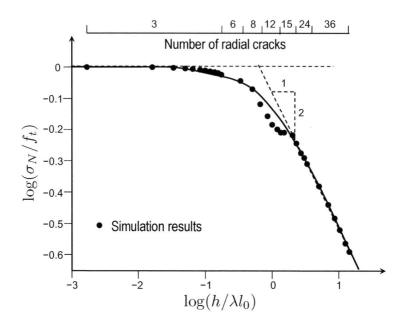

Fig. 7.16 The size effect curve obtained by analysis of growth of part-through cracks (Bažant and Kim, 1998*b*).

of the plate and has a significant effect on the scaling law.

Numerical solution of the integral equation along the radial cracked section, expressing the compatibility of the rotations and displacements due to crack with the elastic deformation of the plate sector limited by two radial cracks, yields the size effect plot shown in Fig. 7.16. As seen, the numerical results shown by data points can be relatively well described by the type 2 size effect (Eq. 5.57). That the size effect in this problem is of type 2 was also confirmed by an approximate analytical solution (Bažant, 2005).

The row on top of Fig. 7.16 shows the number of radial cracks for each range of crack thickness. The deviation of the numerical results from the smooth curve, seen in the middle of the range in the figure, is probably caused by insufficient density of nodal points near the fracture front. As confirmed by Fig. 7.16, the asymptotic size effect does not have the slope −3/8 but the slope −1/2. Obviously, the reason is that this is not a two-dimensional problem since, at the moment of failure, the cracks are not full-through bending cracks but grow vertically through the plate thickness.

7.2.3   Breakup of ice plate pushing against an obstacle

One important topic in offshore engineering is the prediction of a force applied by a moving ice plate on a fixed structure, such as an ocean oil platform. This is a

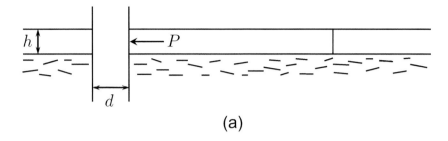

(a)

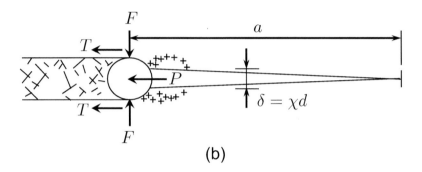

(b)

Fig. 7.17 Cleavage crack in ice plate pushing against a fixed structures: (a) cross-sectional view, and (b) top view (Bažant, 2005).

---

very complex problem since several different types of ice breakup mechanism can take place. One mechanism is the global buckling of the ice plate as a plate on elastic foundation, but it can govern only for rather thin plates. Another mechanism is the compression fracture of ice plate in contact with the structure, which leads to a size effect of ice thickness. The third possible mechanism is the overall fracturing of a finite-size ice floe impacting the legs of the structure. The fourth, and perhaps the most important, possible mechanism consists of a long cleavage crack in the ice plate, propagating against the direction of ice movement (Fig. 7.17). It leads to a size effect of the effective diameter d of the structure, which is different from the size effects presented in Chapter 5. Therefore, it will be discussed here in some detail.

The resistance of the crack to opening causes the ice to exert on the structure a pair of transverse force resultants F and a pair of tangential forces T in the direction of movement; $T = F \tan \varphi$ where φ may be regarded as an effective friction angle. Forces T have no effect on the stress intensity factor K_I at the crack tip (an update of what follows might be needed since the crack-parallel compression effect (Sec. 4.1), was unknown at the time of analysis). Considering the ice plate as infinite, we have

(Murakami, 1986; Tada *et al.*, 2000)

$$K_I = \frac{F}{h} \sqrt{\frac{2\pi}{a}} \tag{7.18}$$

To determine the crack length a (Fig. 7.17), we need to know the crack opening δ_x caused by F. To this end, we may calculate the load-point compliance function $C(a)$ of the pair of forces directly from the stress intensity factor (Eq. 2.111):

$$\frac{F^2}{2h} \frac{dC(a)}{da} = \frac{K_I^2}{E} \tag{7.19}$$

Substituting Eq. 7.18 into Eq. 7.19, we obtain

$$\frac{dC(a)}{da} = \frac{4\pi}{Eha} \tag{7.20}$$

This expression is now integrated from $a = d/2$ (surface of structure, considered as circular (Fig. 7.17)) to a (note that integration from $a = 0$, which would give infinite C, would be meaningless because a cannot be less than d). In this manner we obtain $C(a)$, and from it the opening deflection δ:

$$\delta = C(a)F = \frac{4\pi F}{Eh} \ln\left(\frac{2a}{d}\right) \tag{7.21}$$

If cleavage fracture were the only mode of ice breaking, we would have $\delta = d$. However, there is likely to be at least some amount of local crushing at, and ahead, of the structure. Consequently, the relative displacement between the two flanks of the crack is no doubt less that d. We denote it as χd ($\chi < 1$). Setting $\delta = \chi d$, we solve from Eq. 7.21:

$$a = \frac{d}{2} \exp\left(\frac{Eh\chi d}{4\pi F}\right) \tag{7.22}$$

Note that a/d is not constant but increases with d. Therefore, the fracture patterns are not geometrically similar, and so the LEFM scaling cannot be expected to apply. Substituting Eq. 7.22 into Eq. 7.18, setting $K_I = K_c = \sqrt{EG_f}$ (K_c = fracture toughness of ice), and solving for F, we obtain

$$\frac{2\sqrt{\pi}F}{h\sqrt{EG_f d}} = \exp\left(\frac{Eh\chi d}{8\pi F}\right) \tag{7.23}$$

The pair of forces F is related to ice load P on the structure (Fig. 7.17) by a friction law, which may be written as $P = 2F \tan\varphi$. Substituting $F = P/2\tan\varphi$ and $P = \sigma_N hd$ into Eq. 7.23, and solving the resulting equation for d, we have

$$\frac{d}{d_c} = \frac{1}{\tau^2} e^{1/\tau}, \quad \tau = \frac{\sigma_N}{\sigma_c} \tag{7.24}$$

in which τ is the dimensionless nominal strength, and d_c and σ_c are constants defined as

$$d_c = \frac{4\pi G_f}{\chi^2 E}, \quad \sigma_c = \frac{\chi \tan\varphi}{2\pi} E \tag{7.25}$$

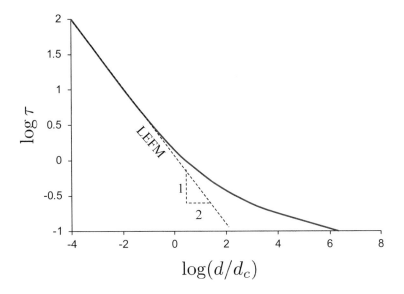

Fig. 7.18 Size effect in cleavage fracture (Bažant, 2005).

Eq. 7.24, plotted in Fig. 7.18, represents the law of cleavage size effect in an inverted form. The small-size asymptotic behavior is the LEFM scaling for similar structures with similar cracks:

$$\text{for } d \ll d_c: \quad \sigma_N \approx \sqrt{d_c/d}; \tag{7.26}$$

The plot of Eq. 7.24 in Fig. 7.18 shows that the size effect is getting progressively weaker with increasing structure diameter d (although, unlike type 1 size effect, no horizontal asymptote is approached by the curve). The reason for this is that the cracks of various lengths are dissimilar, that is, the ratio, a/d, of crack length to structure diameter is not the same for different sizes but increases with the structure size.

7.3 Compression Fracture with Shear and Size Effects

Compression fracture is a more complex phenomenon than tensile fracture and is understood less. It does not occur in materials such as steel, but is typical for heterogenous quasibrittle materials such as concrete. It is always combined with shear on planes inclined to the direction of compression. It is mitigated by lateral confinement, and never occurs when the compression is hydrostatic or when the lateral confinement is perfect, with zero strain enforced in all directions transverse to the direction of compression.

Compression fracture is of two types: (a) axial splitting in a field of homogeneous compressive stress, and (b) propagation of inclined bands of axial splitting cracks resulting in compression-shear failure (Fig. 7.19). The splitting is caused by lateral expansion of damage zone, engendered by shear on inclined planes within this zone.

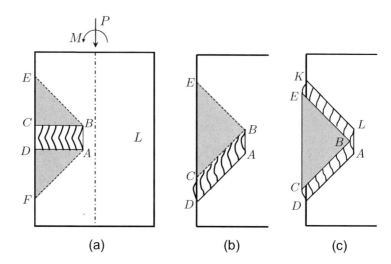

Fig. 7.19 Schematics of compressive fracture: (a) splitting cracks, (b) buckling of microslabs, and (c) stress relief zone.

The axial splitting alone causes no size effect because the energy is released from the undamaged material on the side of the axial splitting bands. The inclined bands do release stored energy from the undamaged material and thus cause size effect of type 2, provided that a long enough band can grow stably before $P_{\max}$. For details with calculations, see Bažant and Xiang (1997) and Sec. 5.1 in Bažant (2005). Extensive experimental evidence exists only for the role of compression fracture in shear failure of reinforced concrete (RC), which is discussed next.

7.3.1   Diagonal Shear Failure of Reinforced Concrete Beams

The diagonal shear failure of RC beams (Fig. 7.20a) has been a formidable problem. It has been studied experimentally since the 1890's (for example, by Mörsch (Mörsch, 1922) and Ritter (Ritter, 1899) who proposed the "truss analogy," recently renamed as "the strut-and-tie model"). For seven decades since 1899, plastic limit analysis was the only modeling approach, which made it impossible to capture the size effect. Analysis in terms of fracture mechanics began in the 1980s, but a satisfactory complete solution crystallized only by 2020, after the gap test clarified the role of crack-parallel compression.

When the load is increased up to about $0.5P_{max}$, parallel cracks of about $45°$ inclination and roughly regular spacing appear around the mid-depth of beam, sometimes accompanied at bottom by a delamination crack breaking the concrete cover away from the longitudinal bars. During further load increase, cracks grow upwards as well as downwards. On approach to P_{max}, they localize into a single dominant growing crack, which extends at P_{max} up to about 80% of cross-section depth (the localization

(a)

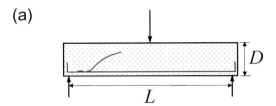

(b)

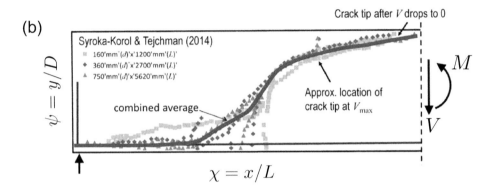

Fig. 7.20 Diagonal shear failure of RC beams: (a) diagonal shear crack in a beam with longitudinal but no shear reinforcement; and (b) observed cracking paths in geometrically similar beams of three different sizes, shown in normalized dimensionless coordinates (Bažant and Le, 2017).

of parallel cracks is a general phenomenon found in many situations, and is properly analyzed as a fracture stability problem (Bažant and Cedolin, 1991, ch. 12)). A complete break extending to the top surface appears only during post-peak softening.

In the initial stage up to about $0.75 P_{\max}$, the diagonal crack grows as a mode I opening crack and can be simulated by the cohesive crack model or, for large enough sizes, by LEFM. However, all the attempts to use a cohesive crack or LEFM up to $P_{\max}$ have failed. Why?—because line crack models are insufficient. On approach to $P_{\max}$, a band of high compression parallel to the crack, called "compression strut," develops on the upper side of the diagonal crack (Fig. 7.21). This has two effects:

1. The compressive stress is so high that a compression-shear failure, emanating from the front of the diagonal crack, runs across the strut (Fig. 7.21) releasing energy from the compression strut.
2. At the same time the effect of crack-parallel compression explained in Sec. 4.1

comes into play. The compression is high enough to reduce the mode I fracture energy to virtually zero, which means that, nearing P_{max}, there are no longer any cohesive stresses transmitted across the diagonal crack near its tip.

Both effects have been confirmed by FE simulations with microplane model M7 combined with the crack band model (Dönmez and Bažant, 2019, 2020).

The compression parallel to the crack produces a strong size effect of type 2. Its existence, inexplicable statistically, is what proved that plastic limit analysis does not apply (except for very small sizes) and that fracture mechanics must be used. And the fact that the size effect deviates from (size)$^{-1/2}$ is what proved that quasibrittle fracture analysis is necessary.

Most tests have been carried out on three-point loaded simply supported beams, longitudinally reinforced, with no stirrups. As the load in increased, many parallel cracks of inclination about 45° initiate, in the lower part of beam, at about $0.5P_{max}$. As P_{max} is approached, they localize into one dominant crack, whose shape and location is shown in Fig. 7.20b. The dots, plotted in relative coordinates, represent the paths of the cracks observed in the experiments of Syroka-Korol and Tejchman (Syroka-Korol and Tejchman, 2014) for geometrically similar beams of three different sizes, scaled in the ratio $1:2:4.69$, and prove that the cracks for different sizes, measured up to their tip locations, are geometrically similar. The cracks begin at the bottom of the beam at randomly scattered locations but, as they propagate towards the top, they localize into a single dominant crack. It is remarkable and important for modeling that the point of maximum load, marked in the figure, occurs for different sizes at approximately homologous locations, which show little random scatter and are dictated by mechanics. These features confirm the geometric similarity of the failures of beams of different sizes, which is crucial for the size effect analysis.

The stress redistribution during failure and the failure mode have been clarified by finite element analysis. For that purpose, it is essential to use a realistic triaxial damage constitutive law and counter the spurious mesh sensitivity by an effective localization limiter. Microplane models M4 and M7 (Bažant *et al.*, 2000; Caner and Bažant, 2013) along with the crack band model (Bažant and Oh, 1983; Cervenka *et al.*, 2005) have been used for this purpose (Bažant *et al.*, 2000; Bažant and Yu, 2005*a,b*). Recently the results and modeling of the gap test (Nguyen *et al.*, 2020*b,a*) helped to clarify the behavior under P_{max} at the crack front. Regarding the size effect and failure mechanism at P_{max}, the following points (stated in more detail in Yu *et al.* (2016)), should be noted:

1. While the beam shear crack initially propagates as a mode I crack, at P_{max} the effective K_I, as well as the critical K_{Ic}, becomes essentially zero (this explains why beam shear cracks with size effect have never been successfully predicted by LEFM, nor cohesive crack model).

2. According to crack band finite element simulations with M4 and M7 (Bažant and Yu, 2005*b*; Yu *et al.*, 2016; Dönmez and Bažant, 2019), the contribution of cross-crack cohesive (or aggregate interlock) stresses to the transmission of shear force V is small compared to crack-parallel compressive stresses, about $0.3V$ for beams 0.2 m deep, and about $0.05V$ for 3 m deep. Also, the contribution decreases with increasing beam size, as a manifestation of the size effect. This is documented by

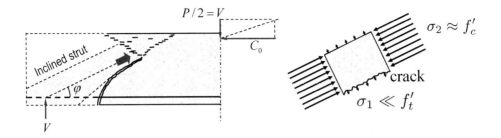

Fig. 7.21 Compression shear band formed at the crack tip.

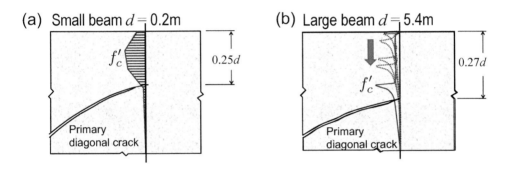

Fig. 7.22 Normal stress distribution across a vertical cross-section at the crack tip: (a) small beam, and (b) large beam (Yu *et al.*, 2016).

the calculated normal stress distribution across a vertical cross-section through the crack tip plotted in Figs. 7.22a and b. In the small beam (Fig. 7.22a), the strength of the ligament above the diagonal crack tip is almost uniformly mobilized at P_{max} while in the deep beam it is not, as seen in the localized stress spike traveling across the ligament during loading.

3. At P_{max}, the compressive principal stress near the crack tip is parallel to the crack and reaches the compression strength limit, f_c' (see Fig. 7.21). Consequently, and in view of the gap test (Sec. 4.1), the fracture energy G_f, or the effective K_{Ic}, is at P_{max} virtually zero. Hence, the crack can, after P_{max}, extend further only as a compression splitting crack, and cross-crack cohesive stresses play no role. As a result of compression crushing of the crack tip region, the compressed wedge above the crack tip slides along the crack path extension and is pushed upwards (Fig. 7.23).

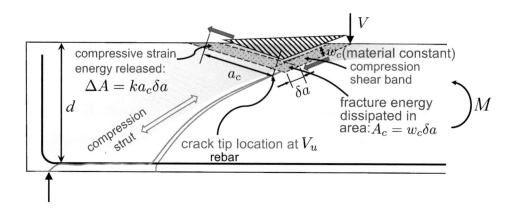

Fig. 7.23 Energy analysis of compressive damage mechanism at the crack tip in diagonal shear failure.

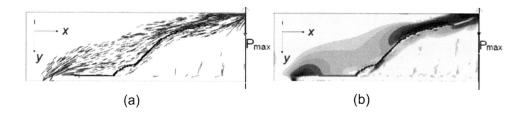

(a)   (b)

Fig. 7.24 Simulation results of diagonal shear failure of RC beams: (a) vectors of minimum principal compressive stresses (maximum in magnitude) at $P_{\max}$, and (b) energy release zones in relative coordinates (Dönmez and Bažant, 2019).

4. The $P_{\max}$, which represents beam failure under a load-controlled regime, is governed by compression-shear failure of the concrete above, and left of, the tip of the diagonal crack. The compression is transmitted to the beam support by a highly compressed strip of concrete visible in Fig. 7.24a. This strip corresponds to what is in the approximate strut-and tie model called the "compression strut"; Fig. 7.24a (in fact, the strut-and-tie model would be invalid if the cross-crack cohesive, or interlock, stresses on the diagonal crack were not negligible at $P_{\max}$; conversely, the strut-and-tie applicability implies that these stresses must be negligible).

5. Fig. 7.24b shows the contour plot incremental stored energy release as calculated in Dönmez and Bažant (2019). Note that most energy release driving the fracture is concentrated in a long strip (a sort of "compression strut") running above

and parallel to the diagonal shear crack. Since no energy is released from below this crack, no cohesive stresses participate. Clearly, the area of the energy release zone is proportional to the square of beam size, while the energy dissipated is proportional to the crack length. This mismatch must give rise to the size effect of type 2.

As complicated as the behavior is, a simplified analytical solution of the shear strength and size effect is nevertheless possible. Although we could use the general dimensional analysis presented in Sec. 5.5.3, let us present its variant based on a schematic picture of the diagonal shear failure in Fig. 7.23, for a beam without stirrups. At maximum (or ultimate) shear force V_u (equal to $P_{\max}$), the strain energy density in the area above the tip of the diagonal shear crack, highly compressed in the crack-parallel direction, may be expressed as $\bar{U} = Cv_u^2/2E$ where $v_u = \sigma_N = V_u/bd$, d is the beam depth from top face to reinforcement centroid, and C is a size-independent, though geometry dependent, constant. Balance of the energy released from the strip of length a_c and the energy dissipated during a small extension δa of the shear crack requires that

$$(A_c + \Delta A)\bar{U} = G_{cf}\delta a \tag{7.27}$$

where G_{cf} = effective fracture energy in compression. w_c = width of the damage band disintegrating in compression, $A_c = w_c\delta a$, and $\Delta A = ka_c\delta a$ where k = empirical size-independent factor enhancing the energy released from strip a_c by the energy released from the adjacent compressed zone. Substituting for $\bar{U}$ and solving the last equation for v_u, one gets, after rearrangements, the size effect law (SEL):

$$v_u = \frac{v_0}{\sqrt{1 + d/d_0}} \quad \text{where} \quad \alpha_c = \frac{a_c}{d}, \ v_0 = \sqrt{\frac{2EG_{cf}}{C}}, \ d_0 = \frac{w_c}{k\alpha_c} \tag{7.28}$$

This equation gives the size effect on the beam shear strength v_u, which was proposed first in 1984 (Bažant and Kim, 1984), immediately after the size effect factor was derived in general in Bažant (1984b). Subsequently, convincing verifications by a worldwide experimental database of about 800 tests, unbiased statistical interpretations of test data, laboratory testing, mesh-insensitive finite element simulations with a realistic material model, evaluations of many structural failures due to shear, and various design studies, were required to convince the code-making committees. Finally, the size effect factor, λ_s, based on Eq. 7.28 was in 2019 adopted for the design code of the American Concrete Institute (ACI), ACI Standard 318-2019:

$$\lambda_s = \min\left(1, \ \sqrt{\frac{2}{1 + d/d_0}}\right), \quad d_0 = 0.254 \text{ m} = 10 \text{ in.} \tag{7.29}$$

It was adopted for specifications governing not only beam shear but also the strut-and-tie model and punching shear of slabs (articles 22.5.5.1.3, 22.6.5.2, and 23.4.4.1 in ACI 318-2019), and optionally footings, too. The cutoff $\lambda_s \leq 1$ means that ACI allows ignoring the milder size effect on the mean strength for $d \leq d_0$. This might be justified not only by a desire for convenience but also by the fact that the scatter band of test data is widening toward smaller sizes such that, for $d < d_0$, its lower margin is roughly

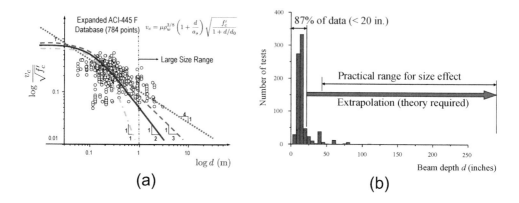

Fig. 7.25 Expanded ACI-445F database of beam shear strength: (a) 784 test results plotted along with different size effect asymptotes, and (b) number of beams tested versus beam depth, in inches (note that this plot does not represent the size effect because the means of secondary variables such as the steel ratio, shear span and aggregate size vary with d, which is often overlooked) (Yu *et al.*, 2016).

horizontal, independent of d (strictly speaking, this effect would better be represented by a size dependence of the safety factor).

The large-size asymptote of Eq. 7.29 is $v_u \propto d^{-1/2}$. However, *fib* Model Code 2010 has had a size effect formula with asymptote d^{-1}, which is the size effect proposed for ropes by Leonardo da Vinci and dismissed by Galileo (the Eurocode is proposed to have $d^{-1/3}$). Such asymptotes could be obtained with the foregoing calculation only if we started with stored energy density $\bar{U} \propto v_u$ (or v_u^3) instead of v_u^2. This would obviously be wrong. So asymptote d^{-1} (or $d^{-1/3}$), or any other than $d^{-1/2}$, is thermodynamically impossible. The same is concluded from the general dimensional analysis in Sec. 5.5.3.

Finite element simulations with a realistic damage constitutive model (M7) and the crack band model match closely both the SEL and the shear tests of concrete beams, including the size effect. This provides the most solid support of the SEL, Eq. 7.28, for beam shear.

Difficulties in Validating the Size Effect in Beam Shear: Doubling in size a previous database collected in Bažant and Kim (1984), ACI committee 445F assembled a database of 784 data on the beam shear strength from the laboratories around he world, shown in Fig. 7.25a. One problem is the enormous scatter, due to using different concretes, with different strengths, mix ratios, etc., which prevent distinguishing among various proposed size effect formulae directly. Another problem is that the data are crowded in a small-size range, as seen from the histogram in Fig. 7.25b. The third problem is that important secondary variables, including the reinforcement ratio, ρ,

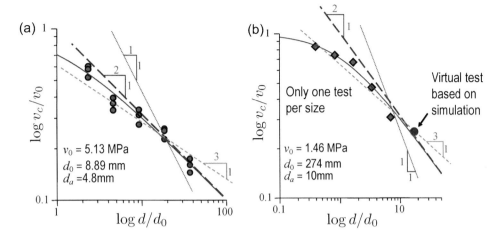

Fig. 7.26 Size effect tests of beam shear and fittings by different size effect models: (a) tests conducted at Northwestern University Bažant and Kazemi (1991), and (b) tests conducted at the University of Toronto (Podgorniak-Stanik, 1998).

the shear span a_s/d (a_s distance from the load to the end support) and aggregate size d_a vary not randomly but in a systematic trend with d. Consequently, optimization of a size effect formula requires a nonlinear weighted multivariate regression of the database, with weights assigned to counter the bias due crowding or sparsity of test data in various ranges. A formula obtained in this way in 2005 (Bažant and Yu, 2005a,b), shown on top of Fig. 7.25a, was then endorsed by ACI Committee 446 (Bažant *et al.*, 2007b), and its simplification led to what became in 2019 the current ACI Code.

   Another way to suppress the bias due to data crowding is to extract database subsets in which the secondary parameters are nearly constant when d is varied (Bažant and Yu, 2008; Yu *et al.*, 2016). To check correctness of the trend of the size effect effect formula, geometrically scaled tests for widely different sizes must be conducted on one and the same concrete. Two such tests are reported in Yu *et al.* (2016) and are shown in Fig. 7.26, where only Bažant and Kazemi's tests (Bažant and Kazemi, 1991) on the left adhered to geometrical scaling closely.

   As for vertical steel stirrups, the experimental evidence is scant. According to design codes, they are assumed to start yielding at $P_{\max}$, and so their force is simply added to V_u. However, FE simulations indicate that: (a) rather than eliminating the size effect, the stirrups push it to larger sizes, causing a tenfold increase of d_0 in Eq. 7.28 Yu and Bažant (2011); and (b) for beams deeper than about 2 m, the stirrups do not begin to yield before $P_{\max}$, which means that their full yield capacity cannot be added to V_u (Bažant and Yu, 2005b). Further research is necessary.

   The aforementioned type of size effect also exists in shear failure of prestressed

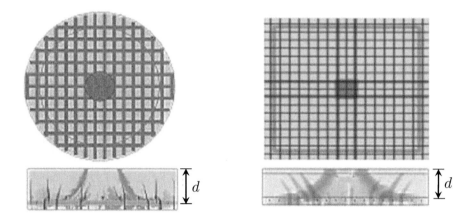

Fig. 7.27 Typical specimen geometries for punching shear tests.

concrete beams (Bažant and Cao, 1986), in torsion (Kirane *et al.*, 2016c) and various other types of concrete failure.

7.3.2   Punching Shear Failure in Reinforced Concrete Flat Slabs

Punching shear of RC slabs, for example, which occurs when a loaded column is breaking through the overlying slab, has caused many failures. It is a problem similar to the beam shear, but with one important difference—the failure occurs under biaxial in-plane confinement, and expansion of the FPZ increases the confinement.

The typical test specimens are shown in the inserts at the lower left of Fig. 7.27, in which the slab is punched from the top down. Based on dimensional analysis similar to that for beam shear, the following semi-empirical design formulas for the nominal (or average) strength of concrete in punching, $v_c = V_c/b_0 d$, were obtained by least-square optimization of the fit of database (Dönmez and Bažant, 2017):

$$v_c = v_0\lambda, \quad v_0 = (100\rho)^{0.3}\sqrt{2f_c f_1}\left(\frac{d}{b}\right)^{0.2}\left(\frac{c}{b}\right)^{0.4} \tag{7.30}$$

$$\lambda = \sqrt{\frac{2}{1 + d/d_0}} \tag{7.31}$$

where λ is the size effect factor, same as for beam shear (ACI code imposes the cutoff defined by λ_s in Eq. 7.29), d is the slab depth (from top to reinforcement centroid), d_0 is the transitional size (empirical, 60 mm according to database regression); $\rho = \sqrt{\rho_x \rho_y}$ (ρ_x, ρ_y are the flexural reinforcement ratios for two orthogonal directions), c = side length of square (or rectangular) column or diameter of circular column, b = perimeter

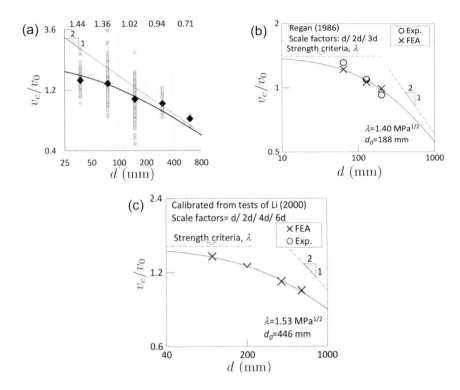

Fig. 7.28 Measured size effect on nominal shear strength v_c and its optimum fitting by Eq. 7.30: (a) entire database of punching shear strength, (b) test by Bažant and Cao (1987), and (c) test by Li (2000).

of loading area (or column), f_c = standard mean compression strength of concrete, and $f_1 = 1$ MPa.

Adding the load capacity of shear reinforcement, $V_s = \beta_s A_{sw} f_{yw}(d/s_w)(c/b)^{0.3}$ then gives to total shear capacity (ultimate punching capacity, or maximum load) $V_u = V_c + V_s$, where V_c, V_s = ultimate punching capacities due to concrete and to shear reinforcement; A_{sw} = cross-sectional area of one shear reinforcement layer around the column; s_w = radial spacing between shear reinforcement layers around the column.

Fig. 7.28 shows some of the fits of the database with this formula. In Fig. 7.28a, to eliminate bias due to data crowding or sparsity, the existing worldwide database of 440 tests was grouped into intervals of constant spacing in the $\log d$ scale. The interval means, shown by the bold data points, are seen to be fitted by the formula quite well. Fig. 7.28b,c show the fits of two sets of individual data obtained on one and the same concrete (Dönmez and Bažant, 2017).

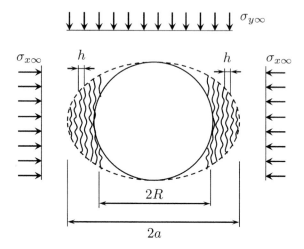

Fig. 7.29 Growth of an elliptical zone of axial splitting cracks during borehole breakout.

7.3.3 Size Effect in Borehole Breakout

The breakout of very deep boreholes in rock is a problem of compression fracture, which is known to exhibit an intricate size effect, as experimentally demonstrated by Nesetova and Lajtai (1973), Haimson and Herrick (1989), Carter (1992), and Carter *et al.* (1992). The studies of Kemeny and Cook (Kemeny and Cook, 1987, 1991) and others showed that the breakout of boreholes occurs due to the formation of splitting cracks parallel to the direction of the tectonic compressive principal stress of the largest magnitude, $\sigma_{y\infty}$ (Fig. 7.29). An approximate energy analysis of the breakout was carried out under the simplifying assumption that the splitting cracks occupy a growing elliptical zone (although in reality this zone is probably narrower and looks closer to a triangle) (Bažant *et al.*, 1993*b*). An important feature of compression fracturing is that this softening damage reduces the compressive stress to a finite critical residual stress σ_{cr} instead of zero stress.

The assumption of an elliptical boundary makes possible an analytical solution of the energy release from the surrounding solid, considered infinite. Eshelby's theory for uniform eigenstrain in an ellipsoidal inclusion (Bažant *et al.*, 1993*b*) may be used for this purpose. According to this theory, the change of potential energy when the stress in the two regions between the circumscribed ellipse and the original circle is reduced from $\sigma_{y\infty}$ to σ_{cr} is given by

$$\Delta\Pi_1 = -\frac{\pi}{2E'}[(a-R)R\sigma_{x\infty}^2+(2a+3R)(a-R)\sigma_{y\infty}^2-2R(R-a)\sigma_{x\infty}\sigma_{y\infty}-2a^2\sigma_{cr}^2] \quad (7.32)$$

in which $E' = E/(1-\nu^2)$, E = Young's modulus of the rock, and ν = Poisson's ratio; R = borehole radius, a = principal axis of the ellipse (Fig. 7.29); $\sigma_{x\infty}, \sigma_{y\infty}$ = remote

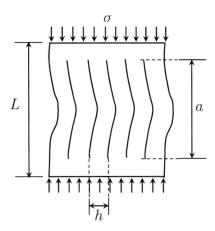

Fig. 7.30 Simultaneous buckling of rock slabs between parallel cracks.

principal stresses.

To calculate the critical residual stress σ_{cr}, we imagine the compression splitting cracks as a system of thin rock slabs between parallel cracks (Fig. 7.30). The elastic buckling analysis yields (Bažant *et al.*, 1993*b*)

$$\sigma_{cr} = -\frac{\pi^2 E' h^2}{12L^2} - \frac{h}{\lambda}G \tag{7.33}$$

where G = shear modulus of the rock, L = average (effective) half-length of the vertical cracks at the moment of failure, h = thickness of parallel slab-columns, and λ = empirical length (material property) representing the thickness of an intact rock layer in which the elastic shear relative displacement caused by unit shear stress is the same as that between the crack faces.

The residual strain energy, per unit thickness in the z-direction, stored between the ellipse and the initial circle, is equal to the bending energy of all the slab-columns. It may be approximately calculated as

$$\Pi_{cr} = (\pi a R - \pi R^2)\frac{\sigma_{cr}^2}{2E'} = \frac{\pi R(a - R)}{2E'}\left(\frac{\pi^2 E' h^2}{12k^2 R^2} + \frac{h}{\lambda}G\right)^2 \tag{7.34}$$

where $k = L/R$. The energy dissipation during the fracturing process is the sum of the energies dissipated by all the vertical splitting cracks (Fig. 7.29),

$$\Delta W_f = \left(\pi a R - \pi R^2\right)\frac{G_f}{h} \quad \text{per unit thickness in the } z\text{-direction} \tag{7.35}$$

where G_f is the fracture energy of the rock and h is the crack spacing. The net energy loss due to passing from a circular borehole in intact rock to an elliptical damage zone with vertical splitting cracks now is $\Delta\Pi = \Delta\Pi_1 + \Pi_{cr}$. The energy balance requires $\Delta\Pi + \Delta W_f = 0$. Assuming that the parallel cracks form progressively, we may write the incremental energy balance condition:

$$-\frac{\partial\Delta\Pi}{\partial a} = \frac{\partial W_f}{\partial a} \tag{7.36}$$

By substituting Eqs. 7.32, 7.34, and 7.35 into Eq. 7.36, we obtain

$$\sigma_{ef}^2 = F(R,h) \tag{7.37}$$

$$\text{where:} \quad \sigma_{ef} = \sigma_{y\infty}\left(1 - \frac{2\sigma_{x\infty}}{5\sigma_{y\infty}} + \frac{\sigma_{x\infty}^2}{5\sigma_{y\infty}^2}\right)^{1/2} \tag{7.38}$$

$$F(R,h) = \left(\frac{\pi^2 E'h^2}{12k^2R^2} + \frac{h}{\lambda}G\right)^2 + \frac{2E'G_f}{h} \tag{7.39}$$

To obtain the failure load, or strength, we need to determine the spacing of the vertical splitting cracks. The spacing h that will occur is that which minimizes the applied effective stress σ_{ef}. In other words, the splitting cracks will occur at the lowest compressive stress they can (this concept can be proven on the basis of Gibbs' statement of the second law of thermodynamics in the manner shown in Chapter 10 of Bažant and Cedolin (1991)). The necessary condition of minimum $\partial F(R,h)/\partial h = 0$ yields an algebraic equation of fifth degree for h, which cannot be solved analytically. However, it will suffice to examine the asymptotic cases.

For very small R, the solution of h reads

$$h = C_s R^{4/5} \quad C_s = \left(\frac{72k^4 G_f}{5\pi^4 E'}\right)^{1/5} \tag{7.40}$$

while for sufficiently large R, we have

$$h = \left(\frac{E'G_f\lambda^2}{5G^2}\right)^{1/3} = \text{constant} \tag{7.41}$$

Substituting Eqs. 7.40 and 7.41 into Eq. 7.37 gives the small- and large-size asymptotes of the effective strength. Although it is not possible to get a closed-form expression for the intermediate values of R, the following combination has the right asymptotic properties for both small and large R and is probably a good approximation that should suffice for practical purposes (Bažant *et al.*, 1993*b*):

$$\sigma_{ef} \approx C_1 R^{-2/5} + C_0 \tag{7.42}$$

A similar problem is the explosive rock burst experienced in the stopes of very deep mines, for which a similar analysis should be possible.

7.3.4   Kink Band Compression Failures in Fiber Composites

Kink band failure is a complex failure mode of unidirectional fiber-polymer composites under compression. Compression microbuckling in kink bands has been studied

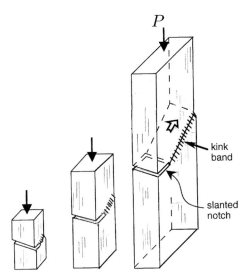

Fig. 7.31 Geometrically similar single-edge notched specimens tested by Bažant *et al.* (1999).

extensively for over several decades (Rosen, 1965; Argon, 1972; Budiansky, 1983; Budiansky and Fleck, 1993; Soutis *et al.*, 1993; Sutcliffe and Fleck, 1994; Fleck and Shu, 1995; Schultheisz and Waas, 1996; Fleck, 1997). In Bažant *et al.* (1999), a fracture mechanics-based analysis of kink band failures revealed an intricate size effect, which has significant implications for the design of composites.

Consider specimens with unidirectional (axial) fiber reinforcement shown in Fig. 7.31. The kink band has length a, which can be long or short compared to the specimen width D taken as the characteristic dimension. The width of the kink band, considered to be small, is denoted as w, and its inclination as β (Fig. 7.32a). The bending stiffness of fibers is neglected.

The loading is assumed to produce cohesive shear cracks parallel to the fibers. The cracks have a certain characteristics spacing s. The axial normal stress transmitted across the kink band (or band-bridging stress) is denoted as σ (Fig. 7.32b). Although Fig. 7.31 shows an in-plane fiber inclination, the behavior is similar for the out-of-plane fiber inclination in the test specimens because what matters for the analysis is the reduction of axial stress across the kink band, which is essentially the same for both cases.

The experimental observations of kink band propagation indicated that the relationship between the shear stress transmitted by the shear cracks and the slip displacement must exhibit softening (Fig. 7.33a). For the sake of simplicity, the stress-displacement diagram of the shear cracks is considered to be bilinear. The peak and residual strengths are denoted by τ_p and τ_r, respectively. Based on Palmer and Rice

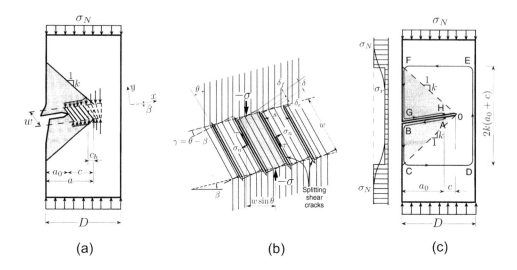

(a)   (b)   (c)

Fig. 7.32 Analysis of kind band fracture: (a) idealized kink band in a notched specimen, (b) microcracking of fibers in the kink band, and (c) path of J-integral with stress relief zones (in grey).

(1973), the area of the diagram above the residual strength represents the mode II fracture energy, G_f. The resulting relation between the axial normal stress versus the axial displacement across the kink band is shown in Fig. 7.33b. The fracture energy of the kink band, that is, the energy dissipated by fracture per unit length of the band is

$$G_b = G_f w/s \qquad (7.43)$$

Fig. 7.32a shows a schematic of an idealized kink band formed in a notched specimen. The kink band spans a length of c, in which the axial shear cracks are forming within the FPZ at the kink band front. In the context of the equivalent LEFM, we denote the equivalent FPZ length as c_b, which represents approximately the distance from the center of the FPZ of the kink band to the point where the stress is reduced to its residual value σ_r. The essential point is that c_b should be regarded as a material property, almost independent of the specimen size and geometry. For notched specimens, we may consider that, at the peak load, the FPZ at the maximum load is still attached to the notch, i.e., $c = c_b$.

The energy release due to propagation of the kink band in a notched specimen can be approximately estimated by using the J-integral. Consider the rectangular closed path $ABCDEFGH$ shown in Fig. 7.32c. The start and end points of the path lie at the FPZ boundary since the residual stress across the band does no work. Path segments CD, DE, and EF are far from the kink band, and therefore the initially uniform stress state remains undisturbed.

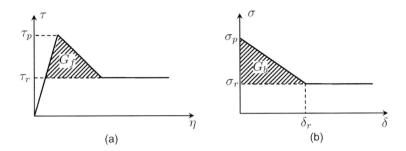

Fig. 7.33 Mechanical behavior of the kink band: (a) diagram of shear stress versus total shear displacement accumulated over distance s between cracks, and (b) diagram of axial normal stress versus axial displacement across the kink band.

For paths FG and BC, the axial stress distribution has a certain curved profile as sketched in Fig. 7.32c. The exact shape of this profile is unimportant. What matters is that, asymptotically for large sizes $D \gg w$, the profiles for various sizes must become geometrically similar. For simplicity, this stress profile may be approximated by a stepped piecewise constant profile shown in Fig. 7.32c, where the far-field applied stress σ_N suddenly drops to the residual stress σ_r. Points F and C can be considered to lie on inclined rays of a certain slope k. These rays may be imagined to emanate from the tip of the equivalent crack of length $a = a_0 + c$ (Fig. 7.32c) where c denotes the length of the FPZ of the kink band. Slope k depends on the structure geometry and on the orthotropic elastic constants. The resulting J-integral can the be written as

$$J = \oint \left(\overline{W} dy - \vec{\sigma} \cdot \frac{\partial \vec{u}}{\partial x} ds \right) = \frac{k}{E_y} (a_0 + c) \left[\sigma_N^2 - \sigma_r^2 - 2(\sigma_N - \sigma_r)\sigma_r \right]$$

$$= \frac{k}{E_y} (a_0 + c)(\sigma_N - \sigma_r)^2 \tag{7.44}$$

where E_y = effective elastic modulus of the orthotropic fiber composite in the fiber direction y.

Meanwhile, the energy dissipation can be calculated again using the J-integral whose path, based on Palmer and Rice (1973), runs along the equivalent crack surface and around the crack tip. This leads to

$$J_{cr} = G_b + \sigma_r \delta_r \tag{7.45}$$

Setting Eq. 7.44 equal to Eq. 7.45, and solving for the nominal strength σ_N of the specimen, we obtain:

$$\sigma_N = \sigma_R + \sqrt{\frac{E_y(G_b + \sigma_r \delta_r)/kc}{1 + D/D_0}} = \sigma_R + \frac{\sigma_0}{\sqrt{1 + D/D_0}} \tag{7.46}$$

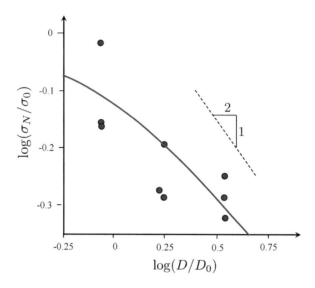

Fig. 7.34 Optimum fits of the measured size effect by Eq. 7.48.

in which

$$D_0 = \frac{c}{\alpha_0}, \qquad \sigma_0 = \sqrt{\frac{E_y(G_b + \sigma_r \delta_r)}{kc}}, \qquad \alpha_0 = \frac{a_0}{D}, \qquad \sigma_R = \sigma_r \qquad (7.47)$$

(note that for other geometries, σ_R need not be equal to σ_r). Eq. 7.47 is valid when a long enough kink band transmitting constant residual stress σ_r develops in a stable manner before the maximum load is reached. Due to the existence of residual stress σ_r, such stable propagation can happen even in specimens of positive geometry (that is, for increasing $g(\alpha)$). Stability of propagation is assisted by rotational restraint of specimen ends.

In the case of notched test specimens (of suitable geometry), the maximum load is achieved while the FPZ of the kink band is still attached to the notch. Except for the sign of the band-bridging stresses, the situation is analogous to tensile fracture of notched specimens. As discussed in Sec. 5.7, for laboratory size specimens only a short initial portion of the softening stress-displacement curve of the cohesive crack comes into play. It is only the initial downward slope of this curve which matters for the maximum load (while the tail of the post-peak load-deflection diagram, of course, depends on the entire stress-displacement curve of the cohesive crack).

A similar situation must be expected for kink bands in notched specimens. Since the shape of the entire softening stress-displacement curve of the cohesive crack model is irrelevant for the maximum load and only the initial downward slope of the curve

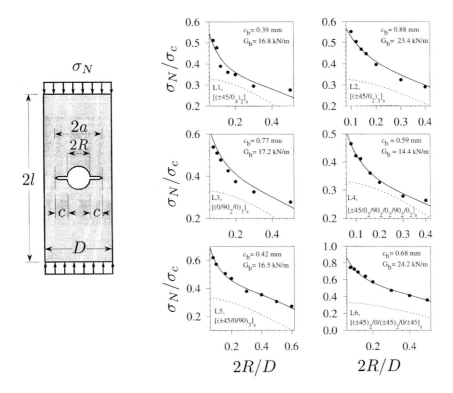

Fig. 7.35 Comparison between model predictions and test results of Soutis *et al.* (1993). Solid curves: optimum fits by the size effect equation. Dashed curves: predictions by the strength theory.

matters, the maximum load must be the same as that for a linear stress-displacement diagram, shown by the descending dashed straight line shown in Fig. 7.32c.

In this case, the residual stress σ_r can be ignored and the fracture energy G_B that mathematically governs the kink band growth at maximum load of a notched specimen corresponds to the entire area under the extended descending straight line in Fig. 7.32c. Obviously, $G_B > G_b$ if $\sigma_r > 0$. Consequently, Eq. 7.46 can be rewritten as

$$\sigma_N = \frac{\sigma_0}{\sqrt{1 + D/D_0}} \tag{7.48}$$

with

$$D_0 = \frac{c}{\alpha_0}, \qquad \sigma_0 = \sqrt{\frac{E_y G_B}{kc}} \tag{7.49}$$

Eq. 7.49 coincides with the type 2 SEL discussed in Sec. 5.5.1.

To demonstrate the existence of a size effect in kink band failure and validate the foregoing analysis, tests of relatively large carbon fiber-PEEK specimens of three different sizes (Fig. 7.31) were performed at Northwestern University (Bažant *et al.*, 1999). Slanting of the notches was found desirable to achieve that no axial shear-splitting would precede or accompany the kink band growth (Fig. 7.31). Rigid restraint against rotation at the ends made it possible for the kink band to grow stably for a considerable distance before attaining the maximum compression load.

The results of individual tests are shown in Fig. 7.34. Despite high scatter, which is probably inevitable in the case of fiber composites, one can see that the present theory does not disagree with the test results (Bažant *et al.*, 1999).

The foregoing size effect analysis may also be verified by comparison with the experiments of Soutis *et al.* (1993), which used rectangular prisms made from quasi-isotropic and orthotropic carbon-epoxy laminates of six layups, with circular holes in the middle. Although the specimen size was kept constant, the variation of the hole size could be considered as a combination of size effect and shape effect. The equivalent LEFM analysis captures both the size and shape effects, and therefore Eq. 7.46 could be used to fit the data. The comparison showed that the theory exhibiting size effect allows much better fits than a theory lacking the size effect (Fig. 7.35).

The results make it clear that the strength-based theories of kink band failure are adequate only for small structural parts. For large structures, the size effect must be taken into account.

7.4   Tensile Fracture and Size Effect in Fiber Composites

Fracture of composites is a complex problem, not yet settled and still evolving. Clearly, LEFM doe not apply and in most current practice (except large aircraft), the structural strength is still assessed by means of material failure criteria expressed in terms of stress and strain as if composites followed plastic limit analysis, with no size effects. The cohesive crack and crack band models have been widely distrusted. The main reason has been the difficulty in observing post-peak softening, which we need to explain first.

7.4.1   Measurement of Post-Peak Softening in Fiber Composites

Until 1963, the test specimens of concrete or rock failed dynamically as soon as the peak load was reached. That year, Rüsch and Hilsdorf in Munich for concrete, and later separately Fairhurst and Waversik in Minneapolis for rock, realized that the frames of the testing machines were much too soft (Rüsch and Hilsdorf, 1963; Waversik and Fairhurst, 1970). Upon greatly increasing the testing frame stiffness by an order of magnitude, gradual post-peak softening in concrete and rock suddenly became observable. It has been perplexing, though, that even in stiff testing machines the post-peak softening could not be observed in composites, although it would have to be if the cohesive or crack band model were valid. Consequently, fracture mechanics was widely distrusted and the use of plasticity-type strength criteria persisted (except in design of large airframes). It all changed in 2016 (Salviato *et al.*, 2016*b*), when it was found at Northwestern University that the tension grips used for the compact-tension specimens of composites were much too soft. After increasing their stiffness 250 times and weight 10 times, the post-peak softening became stable and easily observable. The pretext for distrusting fracture mechanics has been thus removed.

(a)

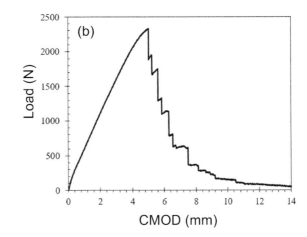

(b)

Fig. 7.36 (a) Fracture tests on woven carbon-epoxy laminate using new stiff and massive grips (US patent 10,416,053), and (b) measured load-deflection curve with post-peak softening.

To explain it, note that the compliance of the testing system is $1/K_{mg} = 1/K_m + 1/K_g$ where K_m, K_g = stiffnesses of the machine and the grips, respectively. The total stiffness resisting load P is $K_{mg} + K_s$. The displacement due to δP is $\delta v = \delta P/K$ and the work to produce δv is

$$\Delta W = -T\Delta S = \tfrac{1}{2}\delta P\delta v = \tfrac{1}{2}K_t(\delta v)^2 \tag{7.50}$$

where T = absolute temperature and S = entropy. The equilibrium of the system is stable if and only if $\Delta W > 0$ (or $\Delta S < 0$), that is, if

$$K_t = \frac{1}{1/K_m + 1/K_g} + K_s > 0 \tag{7.51}$$

where K_s = specimen stiffness, $K_s < 0$ in post-peak. It follows that the minimum stiffness of the grips is limited by

$$|K_g|_{\min} = \left(|K_s|_{\max}^{-1} - K_m^{-1}\right)^{-1} \tag{7.52}$$

where $|K_s|_{\max}$ = post-peak slope of maximum magnitude. Normally the test with post-peak is controlled by the crack mouth opening displacement (CMOD) or crack tip opening displacement (CTOD). The stability conditions for such test control have already been presented in Sec. 2.11. We omit discussing here controllability, a further important aspect of design and testing; see Salviato *et al.* (2016*b*).

Upon designing and fabricating the new stiff grips (Fig. 7.36a), the post-peak softening of compact tension specimens made of a twill 2×2 carbon fiber fabric was

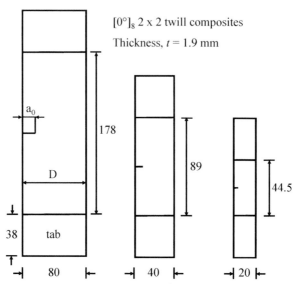

Fig. 7.37 Geometry of single-edge notch tension (SENT) specimens investigated. Dimension in mm.

---

easily observed; see the curve in Fig. 7.36b. The area under the curve gave the total fracture energy $G_F = 78$ N/mm. It may be noted that the size effect tests (Sec. 7.4.2) of this composite yielded $G_f = 73.7$ N/mm. The tail area to the right of the initial slope in Fig. 7.36b happens to be here rather small. Adding it to the G_f value, one gets an almost perfect match of G_f. These and further similar results for other composites confirmed that cohesive crack model (at negligible crack-parallel stress) and the crack band model apply to fiber composites perfectly.

7.4.2   Characterization of Fracture Energy and Size Effect in Textile Composites

An indirect evidence of the quasibrittleness of composite materials is the strong type 2 size effect on the structural strength. An example is provided by a recent size effect study on a $[0°]_8$ laminate made of T800 twill 2×2 layers (Salviato *et al.*, 2016c). In this investigation, radially-scaled single-edge notch tension (SENT) specimens of three different sizes (see Fig. 7.37) were tested to failure in order to measure their structural strength, $\sigma_{Nc} = P_c/tD$ where P_c = tensile load at failure, t = laminate thickness, and D = laminate width. The data was fitted by means of the transformation $X = D$ and $Y = 1/\sigma_{Nc}^2$ using the type 2 size effect as shown in Fig. 7.38a. Linear regression analysis enabled the estimation of the width, $D_0 = 7.93$ mm, marking the transition from quasiductile to quasibrittle behavior and the value of the horizontal small size asymptote characteristic of energetic size effect, $Bf_t = \sigma_0 = 0.643$ GPa. As can be noted from Fig. 7.38b, the double-logarithmic plot of textile composites displays a

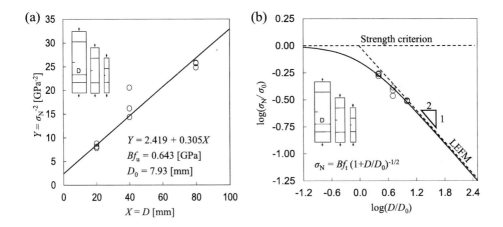

Fig. 7.38 Results of the size effect study: (a) linear regression analysis to characterize the size effect parameters, and (b) measured size effect for $[0°]_8$ twill 2×2 laminates (Salviato *et al.*, 2016*c*).

significant size effect with all the experimental data being close to the LEFM asymptote for the sizes investigated.

LEFM analysis with FEM furnished the dimensionless energy release rate, g_0, and its derivative, g_0' (Salviato *et al.*, 2016*c*). Real displacement boundary conditions as measured by digital image correlation (DIC) were used in the FE simulations. Thanks to accurate calculation of g_0 and g_0', Eq. 5.86 could be used to estimate the initial mode I fracture energy, G_f, and the equivalent FPZ size, c_f. Interestingly, the resulting fracture energy is 73.7 N/mm, which is in excellent agreement with the value calculated by means of the stable post-peak tests conducted on the same material and discussed in the previous section. Furthermore, the equivalent FPZ size was found to be 1.81 mm, which is comparable to the size of one tow of the twill 2×2 fabric.

The close agreement with the stable post-peak tests confirms the quasibrittleness of composites and supports the use of quasibrittle fracture mechanics for the description of their fracturing behavior. It is worth noting that the size effect method of measuring the fracture properties is easier to implement than other methods because only peak load measurements are necessary. The post-peak behavior, crack-tip displacement measurement, and optical measurement of crack-tip location are not needed, and even a soft testing machine without servo-control can be used.

The correct characterization of the fracture energy of composites is extremely important because it is one of the main parameters used in the progressive damage models utilized for the design of aerospace structures. The applicability of size effect law for

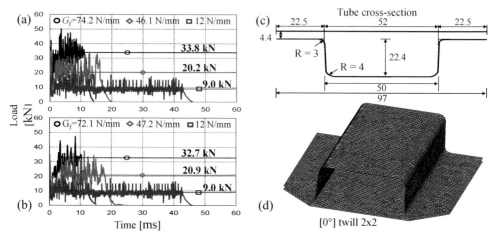

Fig. 7.39 Crashing of composite tubes (hat section: $[0°]_{11}$, plate: $[0°]_8$): (a) crashing load vs time predicted by the spectral stiffness microplane model (Salviato *et al.*, 2016a) for various values of intra-laminar fracture energy, G_f; (b) crashing load vs. time predicted by microplane triad model (Kirane *et al.*, 2016b,a) for various values of intra-laminar fracture energy, G_f; (c) geometric specifications of the crash-can cross-section (dimensions in mm); and (d) typical FE mesh used in the simulations.

---

measuring the fracture properties of the material was verified numerically by means of two recently proposed microplane models for textile composites (Kirane *et al.*, 2016a,b; Salviato *et al.*, 2016a). Both models matched the size effect data using the G_f-value estimated by the SEL. This method was thus shown to be accurate in the calibration of advanced computational models. The importance of using the correct value of G_f was discussed in Salviato *et al.* (2016c) by calculating its effects on the crashworthiness simulations of a textile composite crush-can for automobiles.

As shown in Fig. 7.39, both models were run with (a) the correct value of the fracture energy as estimated from the SEL, $G_f = 73.7$ N/mm, (b) with a value equal to 60% of the fracture energy estimated from size effect, and (c) with the value of $G_f = 12$ N/mm. The third case corresponds to the fracture energy obtained by assuming a vertical drop of the stress-strain response of the material after the strength is reached. This is equivalent to treating fracture by element deletion and erosion once the failure condition is met, which has unfortunately been a common practice in the composite community. Compared to the experimental results on the axial progressive crushing behavior of composite crush cans, the two models calibrated with the measured fracture energy provided an excellent prediction of the crushing load, 33.8 and 32.7 kN, respectively, compared to the experimental value of 35.6 kN. Furthermore, the parametric study showed that, for both models, a decrease of G_f to about 60% of the measured value can reduce the crushing load to almost a half. The assumption of a vertical drop of stress after the peak, typical of strength-based constitutive laws, led

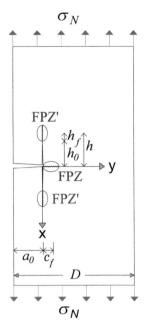

Fig. 7.40 Schematic of sideways crack propagation.

to an underestimation on the order of 70% for an element size of 2 mm. The error in this latter case was shown to be mesh dependent.

Similar size effects were found also in three-dimensional textile composites (Li *et al.*, 2017; Li, 2018; Li *et al.*, 2021) and other composites, including the discontinuous fiber composites (DFC) (Ko *et al.*, 2019*b,a*) and unidirectional composites (Bažant *et al.*, 1996).

7.4.3 Size Effect in Transverse and Sideways Fracture of Unidirectional Composites

The applicability of LEFM to composites was studied and repeatedly rejected for several decades. In 1996, though, Bažant, Daniel, and Li demonstrated experimentally the existence of a transitional energetic size effect, implying the applicability of cohesive crack or crack band model (Bažant *et al.*, 1996). These tests, already displayed in Fig. 2.17b and Fig. 7.37, measured the maximum loads of single- and double-edge notched tension (SENT and DENT) specimens of graphite/epoxy crossply or quasi-isotropic laminates subjected to longitudinal tensile load. Nevertheless, these tests did not suffice to convince most composites experts that the fracture mechanics was applicable. It was objected that the forward crack extension observed in these classical tests does not always take place in composites, and that in many situations on the SENT fracture specimens the crack propagates from the notch tip sideways, in the direction, x, of the

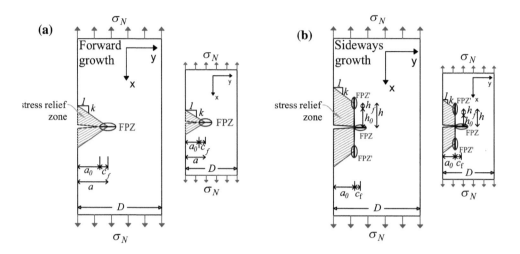

Fig. 7.41 Energy analysis of (a) forward crack propagation and b) sideways crack propagation.

tensile load as portrayed in Fig. 7.40. Eventually, though, it was shown by Dönmez and Bažant (Dönmez and Bažant, 2020) that the sideways cracks and their size effect are perfectly explicable and predictable by fracture mechanics.

In 1988, Nairn (Nairn, 1988) conducted fracture tests of unidirectional composites, probably the first ones in which sideways cracks and their size effect were observed and discussed (in terms of shear lag). In 2013, Tankasala, Deshpande, and Fleck (Tankasala *et al.*, 2018) showed that, if a potential sideways cohesive crack (called kinked crack by these authors) is placed at the tip of a transverse notch in a highly orthotropic composite, such a crack may open and propagate instead of the forward propagation (in the y direction). However, a potential cohesive crack had to be assumed to exist a priori and in the right place, and so the cohesive line crack was not predictive. As shown in Dönmez and Bažant (2020), prediction of sideways crack propagation is impossible without considering the tensorial nature and finite width of the FPZ and, of course, using a realistic damage constitutive model. The crack band model with an orthotropic damage law was used for this purpose.

In 2020, Dönmez and Bažant (Dönmez and Bažant, 2020) developed simple model for the sideways propagation. They used the idealized fracture estimates based on approximate energy release zones delimited by the so-called 'stress-diffusion' lines, of slope k, cross-hatched in Fig. 7.41. In these estimates, slope k, specific for the given geometry, needs to be calibrated by means of a FE calculation of the J-integral for one specimen size. Then these estimates give correct predictions of the size effect, with

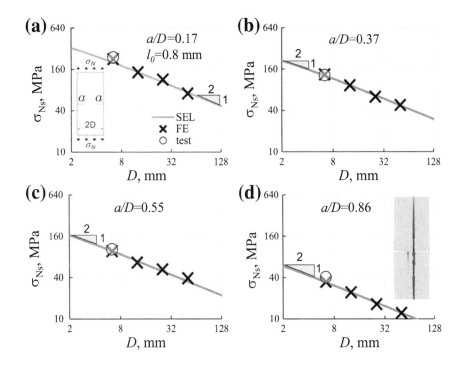

Fig. 7.42 Measured and predicted size effects on sideway strength for different notch depths (experimental data are shown in circles (Nairn, 1988)).

exact asymptotic scaling. Surprisingly simple rules were found in Dönmez and Bažant (2020) for the scaling of the nominal strength, σ_{Ns}, at sideways crack propagation and for the corresponding fracture energy, Γ_s:

$$\sigma_{Ns} = \sqrt{2\hat{E}\Gamma_s/a_0} \propto D^{-1/2} \quad \text{if} \quad a_0 \propto D \quad \text{or} \tag{7.53}$$

$$\sigma_{Ns} = \sqrt{C/D} \quad \text{where} \quad C = 2\hat{E}\Gamma_s(D/a_0) = \text{constant} \tag{7.54}$$

$$\frac{\Gamma_s}{G_f} = \frac{1}{2k(1+c_f/a_0)} \tag{7.55}$$

Here Γ_s = sideways fracture energy of the material; a_0 traverse initial crack or notch length at which the sideways propagation begins; and $\hat{E} = E_{xx} - E_{xy}^2/E_{yy}$ = effective elastic modulus of the orthotropic composite (estimated by means of Eqs. 2–3 from Dönmez and Bažant (2020)).

Eq. 7.55 means that sideways propagation occurs only if Γ_s is about 6 to 12 times smaller than G_f, which happens only in a strongly orthotropic materials and only if the initial transverse notch or crack is not too short. Eq. 7.53 means that the lateral propagation (orthogonal to the initial transverse crack) occurs at constant load, that is, at the limit of stability (or neutral equilibrium) if the load is controlled. For geometrically similar specimens, Eq. 7.54 predicts the size effect to be simply of the LEFM type, which is stronger than that for forward crack propagation.

As a limited check of the present model, optimal fits of size effect data measured by Nairn (Nairn, 1988) were obtained in Dönmez and Bažant (2020); see Fig. 7.42. The predicted LEFM slope of $-1/2$ is verified by these fits and the same model gives correct predictions for various a/D ratios (here $a = a_0$). For details and further verification, see Dönmez and Bažant (2020).

The same approach was also used in Dönmez and Bažant (2020) for studying sideways cracks that are inclined from the orthogonal x-direction. Such cracks occur when the fibers of the composite are inclined. Unlike the orthogonal sideways crack, the inclined ones are found to propagate at decreasing load, which means dynamically if the load is controlled.

7.4.4   Size Effect in Mode I and II Inter-laminar Fracture of Composites

Thanks to their fibrous reinforcement, laminated composites feature excellent in-plane moduli and strength. However, in the thickness direction of the plate, the reinforcement is either absent or present in small volume fractions making these materials rather prone to inter-laminar fracture. Interestingly, the study of delamination in composites represents one of the few cases in which LEFM was not only embraced by the composites community but also adopted by ASTM standard D5528 (ASTM, 2002).

To clarify the applicability of LEFM for interlaminar fracture of composites and to investigate the related size effects, a novel study was conducted in Salviato *et al.* (2019). The experiments covered both mode I and II delamination by testing radially-scaled double cantilever beam (DCB) and end notch flexure (ENF) specimens (Figs. 7.43 and 7.44).

The tests on geometrically scaled DCB and ENF specimens confirmed a remarkable size effect in both mode I and II interlaminar fracture. The analysis of the experimental data showed that, for the size range investigated in this work, the fracture scaling of the mode I interlaminar specimens was captured reasonably well by the LEFM. This is evidenced by Fig. 7.45 which shows that all the experimental data were close to the LEFM asymptote. Comparison of the mode I fracture energy, G_{If}, estimated via the SEL, to that ones estimated by LEFM, for each size, showed a difference of 18%, 4% and 2% for the small, medium and large sizes, respectively.

The case of mode II fracture, however, showed a more complicated scaling as shown in Fig. 7.46. As can be noted, the double logarithmic plot of the nominal strength as a function of the characteristic size of the specimens shows that the fracturing behavior evolves from quasi-ductile to brittle with increasing sizes. For sufficiently large specimens, the size effect data tends to the classical $-1/2$ asymptotic slope predicted by LEFM. However, for smaller sizes, a significant deviation from the LEFM scaling is found, with the data exhibiting a milder size effect, which is a behavior associated with a more pronounced quasi-ductility. This phenomenon was never found before in ENF tests and plays a key role in the correct characterization of mode II fracture. In fact, for the ENF specimens the LEFM predictions are about 55%, 43%, and 40%

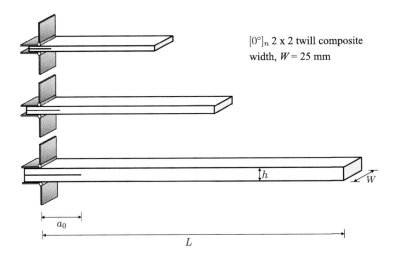

Fig. 7.43 Geometry of the double cantilever beam (DCB) specimens.

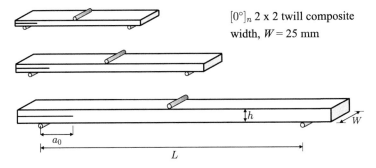

Fig. 7.44 Geometry of the end notch flexure (ENF) specimens.

lower compared to the SEL for the small, medium and large sizes, respectively. The difference decreases with increasing specimen sizes and tends to zero for sufficiently large specimens, as the FPZ becomes negligible compared to the specimen size.

The deviation from LEFM reported in the ENF experiments is related to the FPZ size. In mode I loading the damage-fracture zone at the crack tip, characterized by nonlinearity prior to reaching the peak load, is generally very small compared to the specimen sizes investigated. This is in agreement with the inherent LEFM assumption of negligible nonlinear effects during the fracturing process. However, the damage mechanisms such as matrix microcracking (Fig. 7.47a), crack path deflection and plastic yielding occurring under mode II loading, produce a significantly larger FPZ. For sufficiently small ENF specimens, the FPZ size is not negligible compared to the specimen characteristic size and thus highly affects the fracturing behavior. This leads to a significant deviation from the LEFM.

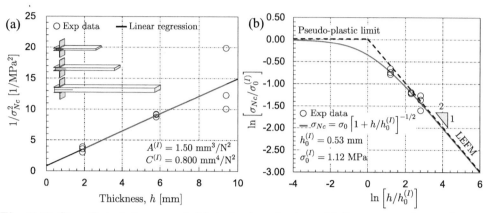

Fig. 7.45 Size effect study on mode I interlaminar fracture: (a) linear regression analysis to characterize the size effect parameters, and (b) size effect plot.

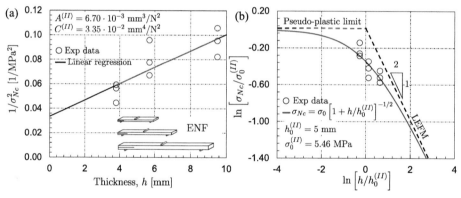

Fig. 7.46 Size effect study on mode II interlaminar fracture: (a) linear regression analysis to characterize the size effect parameters, and (b) size effect plot.

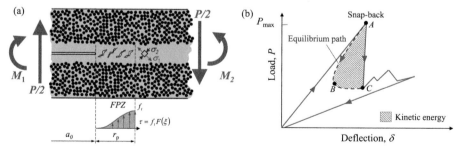

Fig. 7.47 (a) Schematic representation of the damage mechanisms in the FPZ of a mode II interlaminar crack leading to emergence of nonlinear cohesive shear stresses; (b) schematic illustration of the snap-back instability affecting the ENF tests.

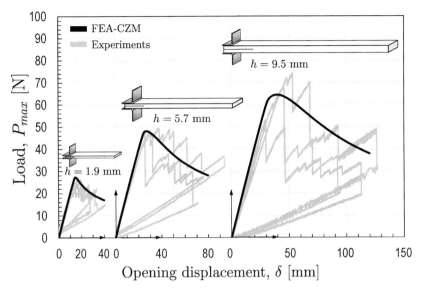

Fig. 7.48 Simulations by means of a cohesive zone model with a linear traction-separation law. The mode I fracture energy used as input is estimated by means of SEL.

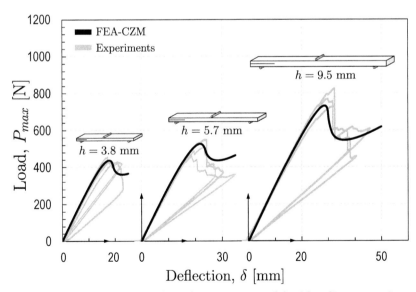

Fig. 7.49 Simulations by means of a cohesive zone model with a linear traction-separation law. The mode II fracture energy used as input is estimated by means of SEL.

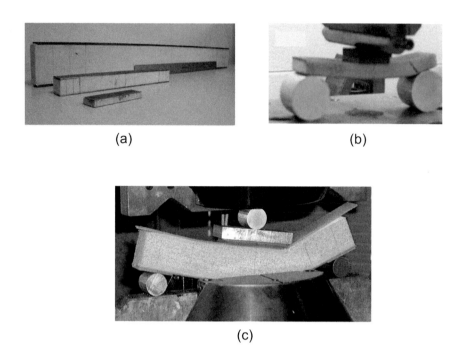

(a)   (b)

(c)

Fig. 7.50 Sandwich composite beams: (a) geometrically similar bottom notched beams, (b) a bottom notched beam under fracture test, and (c) a top notched beam under fracture test (Bažant *et al.*, 2006).

Finite element simulations of the DCB and ENF tests by means of a cohesive zone model featuring a linear traction-separation law support the use of SEL for the estimation of the mode I and II fracture energy. In fact, as shown in Figs. 7.48 and 7.49, by using the energy estimated by SEL as an input for the cohesive model, the agreement with the experimental load-displacement curves is excellent. This also confirms that the use of LEFM to calculate the fracture energy for cohesive zone models would lead to severe errors, especially in regards to the mode II cohesive law.

The foregoing evidence shows that particular care should be devoted to the understanding of the scaling of the fracture behavior of laminated composites. In particular, the fracture tests carried out to characterize, for example, the fracture energy, must guarantee objective results. The size effect testing on geometrically scaled specimens is a simple and effective approach to provide objective results. The size effect method of measuring the mode I and II interlaminar fracture properties is easier to implement than other methods because only the peak load measurements are necessary: the post-peak behavior, crack tip displacement measurement, and optical measurement of crack tip location, are not needed. This is particularly advantageous for interlaminar fracture tests which are often affected by snap-back instability, or by discontinuous crack

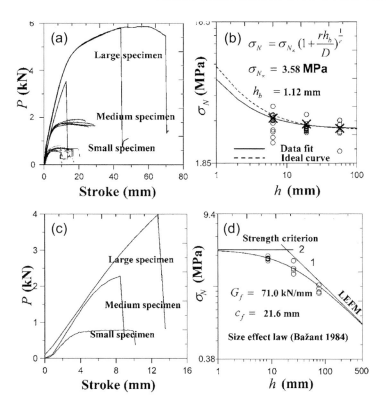

Fig. 7.51 (a) Measured load-deflection curves of unnotched sandwich beams, (b) size effect on nominal strength of unnotched beams, (c) measured load-deflection curves of bottom notched sandwich beams, and (d) size effect on nominal strength of notched beams (Bažant *et al.*, 2006).

propagation (Fig. 7.47b) which maked visual observations impractical and inaccurate.

7.4.5 Size Effect in Fracture of Sandwich Structures

Although sandwich plates and shells are traditionally designed by strength criteria, they show size effects that call for fracture mechanics. This was demonstrated, for example, in Bažant *et al.* (2006), in which sandwich beams consisting of laminate skins and vinyl foam, with two supports at bottom, were subjected to a concentrated load at center-span on top of beam (Fig. 7.50).

Sandwich beams without notches failed by a delamination crack propagating on top from an overhang at one end of the beam and terminating during post-peak softening by shear fracture across the foam core. Fig. 7.51a shows the experimentally observed load-deflection curves and the corresponding size effect on the nominal strength. Not surprisingly, the size effect is of type 1, as shown in Fig. 7.51b, since the maximum

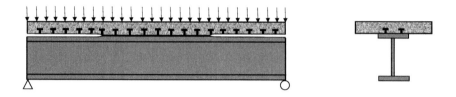

Fig. 7.52 Composite beam with connectors failed in symmetric crack-like regions.

load occurs at crack initiation from the FPZ still attached to the beam end. Fig. 7.51c shows the measured lead-deflection curves and the size effect plots of bottom-notched sandwich beams. These tests represented what could in practice be a preexisting stress-free fatigue crack. The size effect is strong and significant, and is, not surprisingly, of type 2 (Fig. 7.51d).

Finally it may be mentioned that a similar problem is the failure of composite steel–concrete, or concrete–concrete, beams (Fig. 7.52). Their fracture analysis has long been neglected and the possibility of size effect ignored. However, as argued in Bažant and Vítek (1999a,b), a significant size effect, due to the effect of connector size on its strength, is likely. The slab-beam connectors (or studs) fail by pullout and shear, and their post-peak response is at least mildly softening. This implies that the failure of connectors of various sizes must exhibit at least a mild size effect. This further implies that the failure of the connectors in a row cannot be simultaneous, as in plastic limit analysis, but must propagate from one connector to the next, along the interface, like a discrete crack propagating in jumps along the interface. This propagation, too, must introduce a size effect, of the beam size, superposed on the size effect of the connector size. The combined result may be a strong size effect (Bažant and Vítek, 1999a,b). This problem calls for large-scale experiments.

7.5   Bone Fracture and Size Effect

Fracture mechanics has also been applied to investigate the failure of many biomaterials. One subject of great interest is bone fracture, for which the basic question is the characterization of fracture properties.

Early attempts used the LEFM to model bone fracture (Norman *et al.*, 1995). In this case, the material fracture properties are characterized by a single parameter, the fracture energy G_f or, equivalently, by fracture toughness K_c. However, it has been shown that the observed FPZ size in human cortical femur bone is roughly 5 mm, which is far from negligible compared to the bone dimensions (Nalla *et al.*, 2004, 2005). Therefore, bone fracture must be modeled by nonlinear fracture mechanics.

Chapter 3 showed that the cohesive crack model is an effective model for quasibrittle fracture, of course, in absence of crack-parallel stresses. Chapter 5 (Sec. 5.7) demonstrated that the cohesive law for quasibrittle materials cannot be uniquely determined by optimum fitting of the load-deflection curve of a notched specimen of only

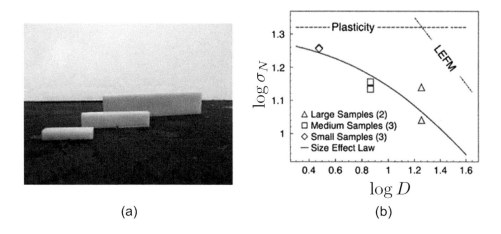

(a)   (b)

Fig. 7.53 Size effect tests on femur bone: (a) geometrically similar specimens, and (b) measured size effect on nominal strength (Kim *et al.*, 2013).

one size. This finding has recently been documented for bovine bones, and probably applies to human ones as well (Kim *et al.*, 2013).

In Kim *et al.* (2013), eight notched three-point bend beams of three sizes, geometrically scaled in two dimensions, were cut from a bovine femur bone—from one and the same bone, to minimize random scatter. All the specimens had a constant width $b = 4$ mm. The size ratio was $1 : \sqrt{6} : 6$, and the depth of the largest specimen was $D = 18$ mm (Fig. 7.53a). The tests were conducted in a standard closed-loop servo-controlled (MTS) testing machine. Because of viscoelastic behavior of the bone, the loading rates were selected so as to reach the peak load for all the sizes within about the same time, which was about 1,000 s.

Fig. 7.53b shows the measured size effect on the nominal strength of the specimens. It is seen that the overall mean slope of the data trend confirms that there is indeed a size effect. It is expected that the measured size effect should follow the type 2 SEL (Eq. 5.57), which gives, in the log-log plot, a curve of steepening downward slope (i.e., convex curvature). However, the data points indicate a decreasing slope (i.e. concave curvature). It was argued that this inconsistency is most likely a manifestation of random scatter of the large-size specimens, which is likely caused by the fact that different parts of the bovine bone do not have the same properties. Despite the scatter of the test results of large specimens, we may conclude from Fig. 7.53b that the measured size effect is transitional between the strength theory and the LEFM limit. This clearly confirms that bone follows quasibrittle failure mechanism.

In addition to the size effect analysis, a cohesive crack model was used to simulate the experiments. The simulation used two different cohesive laws as shown in Fig. 7.54b, to match the measured load-CMOD curve of intermediate size specimens (Fig.

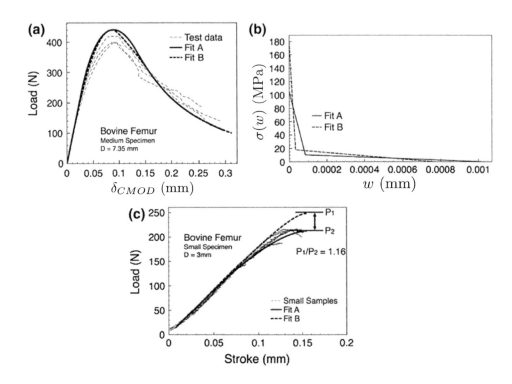

Fig. 7.54 Missing size effect data cause bilinear laws to be ill-conditioned; (a) post peak curves can be closely matched by two different bilinear softening laws, (b) two bilinear laws exhibit different f_t and G_f, and (c) difference in peak loads predicted by the two cohesive softening laws (Kim *et al.*, 2013).

7.54a). The difference between these two cohesive laws is 71% in terms of tensile strength f_t and 55% in terms of initial fracture energy G_f (relative to the lower value). Yet, the final load-CMOD curves predicted by these two models are very similar, and the difference between them is negligible and much smaller than the inevitable experimental scatter (Fig. 7.54a). These two cohesive laws were used to simulate the load-CMOD curves of the small size specimens, and a pronounced difference was seen (Fig. 7.54c). The difference between the predicted peak loads is about 16% (again of the smaller value).

This result reveals that the cohesive properties cannot be uniquely identified by relying on the complete load-deflection curve of specimens of one size. Instead, size effect tests are inevitable. It was shown that the fitting of size effect test data by type 2 SEL (Eq. 5.57) alone may be used to identify the initial fracture energy of the cohesive crack model of bovine bones, and probably human ones as well. Such fitting is what should normally suffice to compute the strength (or load capacity) of bones

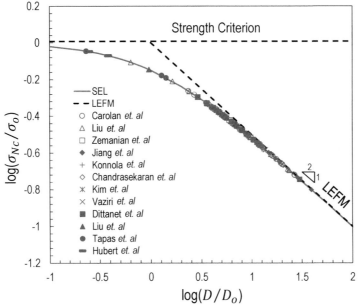

Fig. 7.55 Size effect in polymer nanocomposites: data taken from the literature (Kim *et al.*, 2008; Ashrafi *et al.*, 2011; Liu *et al.*, 2011; Vaziri *et al.*, 2011; Dittanet and Pearson, 2012; Jiang *et al.*, 2013; Zamanian *et al.*, 2013; Mirjalili *et al.*, 2014; Chandrasekaran *et al.*, 2014; Carolan *et al.*, 2016; Konnola *et al.*, 2016).

and the effect of implants. However, such fitting does not suffice to determine the total fracture energy and the tail of the cohesive law. As discussed in Sec. 5.7, for a unique determination of the cohesive fracture properties of bone, one needs both the tests of size effect on strength, with a sufficient size range, and measurement of the post-peak load-deflection curve of notched specimens of at least one size.

7.6 Size Effect in Polymer Nanocomposites

Polymer nanocomposites are gaining increasing interest from the scientific and industrial communities, owing to their outstanding combination of mechanical and functional properties. Over the last two decades, significant efforts have been devoted to improving the fracture toughness of thermoset polymers by the addition of carbon nanotubes (Ashrafi *et al.*, 2011; Mirjalili *et al.*, 2014; Konnola *et al.*, 2016), nanoparticles (Kim *et al.*, 2008; Zhang *et al.*, 2008; Vaziri *et al.*, 2011; Liu *et al.*, 2011; Dittanet and Pearson, 2012; Jiang *et al.*, 2013; Zamanian *et al.*, 2013; Carolan *et al.*, 2016), nanoplatelets (Quaresimin *et al.*, 2012; Zappalorto *et al.*, 2013a,b), and graphene layers (Chandrasekaran *et al.*, 2014; Kumar *et al.*, 2015; Mefford *et al.*, 2017; Chhetri *et al.*, 2017; Kumar and Roy, 2018). However, an aspect that has been often overlooked is that the addition of the nanoreinforcement leads to a less brittle response.

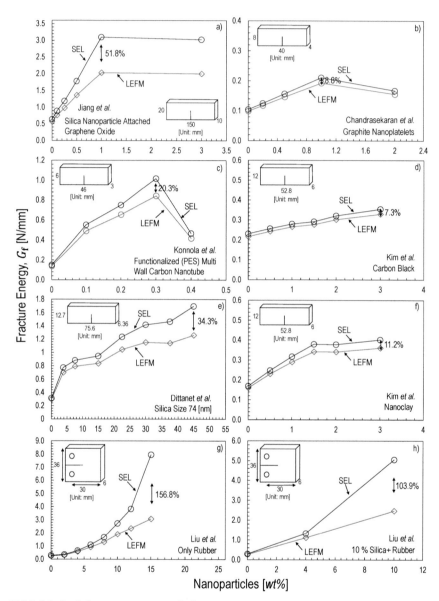

Fig. 7.56 Mode I fracture energy of thermoset nanocomposites estimated by LEFM and SEL. The latter formulation accounts for the finite size of the nonlinear FPZ. The data are re-analyzed from Kim *et al.* (2008); Liu *et al.* (2011); Dittanet and Pearson (2012); Jiang *et al.* (2013); Chandrasekaran *et al.* (2014); Konnola *et al.* (2016).

As was shown in, for example, Mefford *et al.* (2017), the addition of nanoparticles promotes the emergence of damage mechanisms at the nano- and micro-scales, including particle debonding, matrix yielding, shear banding, and crack deflection. This damage leads to the development of a FPZ several orders of magnitude larger than in the pristine polymer, leading to significant size effects. It is interesting to note that most of the investigations of the fracturing behavior of nanocomposites reported in the literature adopted LEFM, hence neglecting completely the effects of sizable FPZ.

As shown in Qiao and Salviato (2019*b*), neglecting the FPZ may be a source of significant errors in the evaluation of the fracture energy since most of the specimens investigated in the literature featured a FPZ of a size comparable to the specimen size. This can be clearly appreciated in Fig. 7.55 which reports a comprehensive study of most of the fracture tests available for polymer nanocomposites (Qiao and Salviato, 2019*b*). The double-logarithmic plot reports the structural strength normalized by the asymptotic strength σ_{Nc}/σ_0 as a function of the specimen width normalized by the transition size, D/D_0. It is noteworthy that the vast majority of the data fall exactly in the transition region between brittle and quasi-ductile behavior.

Overlooking the nonlinear effects induced by the FPZ has led to unobjective estimates of the increase of fracture energy induced by nanoparticles. This is because the use of LEFM, which considers the nonlinear FPZ as a mathematical point, *in-lieu* of the SEL, has led to underestimations that depend on the ratio between the FPZ and the specimen size. This important aspect has been discussed in Qiao and Salviato (2019*b*) where, assuming a linear traction-separation law, the following formula was proposed to correct the experimental results:

$$G_{f,SEL} = G_{f,LEFM} \left[1 - \frac{0.44E^* g'(\alpha_0) G_{f,LEFM}}{D f_t^2 g(\alpha_0)} \right]^{-1} \tag{7.56}$$

where $G_{f,SEL}$ is the fracture energy corrected via the type 2 size effect, $G_{f,LEFM}$ is the fracture energy as estimated from LEFM, f_t is the tensile strength, and $g(\alpha_0)$ and $g'(\alpha_0)$ are the dimensionless energy release rate functions. The re-analysis of the literature data showed that, for the majority of tests, the effects of the FPZ were not negligible, leading to values of the fracture energy deviating from LEFM up to more than 150% in some cases. Some of the re-analyses are shown in Fig. 7.56 revealing the severity of the errors induced by the use of LEFM for the characterization of the fracturing behavior of nanocomposites.

7.7 Interfacial Fracture of Metal-Composite Hybrid Joints

Many modern engineering designs involve combinations of different materials to achieve a better structural performance (Barsoum, 2003). The use of dissimilar materials naturally leads to bimaterial hybrid structures, which involve lap joints, butt joints, scarf joints, etc. In these structures, the strength of the corresponding stress singularities is generally weaker than the "−1/2" crack-like stress singularity. Over several decades, extensive research has been directed to analyzing the elastic near-tip field of the bimaterial corner (Bogy, 1971; Rice, 1988; Hutchinson and Suo, 1992; Desmorat and Leckie, 1998; Qian and Akisanya, 1998; Liu and Fleck, 1999; Reedy Jr., 2000; Labossiere *et al.*, 2002). In many bimaterial hybrid joints, both the adhesives and the base materials

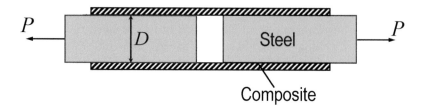

Fig. 7.57 Schematic of metal-composite double-lap joint.

exhibit a strain-softening behavior, which leads to the formation of a localized damage zone. Based on Chapter 5, one must naturally expect that the nominal strength of such joints would exhibit a strong energetic size effect.

Here we present recent results on scaling of fracture of metal-composite hybrid joints (Fig. 7.57) (Yu *et al.*, 2010a; Le *et al.*, 2010; Le, 2011; Le and Xue, 2013; Yu *et al.*, 2013; Bažant and Le, 2017), which are found in large ship hulls, aircraft wings and fuselages. In these joints, the metal is joined with the composite material using adhesives. The stress singularities at the bimaterial corner can be calculated by either the Mellin transform for isotropic materials (Bogy, 1971) or the complex potential method for anisotropic materials (Stroh, 1958; Lekhnitskii, 1963; Desmorat and Leckie, 1998). The asymptotic field at the bimaterial corner can be generally described by a separated power-law singular field. Depending on the corner geometry and the mismatch of elastic properties, the exponents of stress singularity λ_k $(k = 1, 2)$ could be either a pair of complex conjugates or two distinct real numbers. As explained in Sec. 2.12, a complex exponent implies displacement oscillations and overlaps of opposite crack faces, but it was concluded that these oscillations are well within the range of the FPZ in which the LEFM is not valid.

In recent studies (Le *et al.*, 2010; Le, 2011), an approximate size effect equation was derived for the nominal strength of hybrid joints through asymptotic analysis. At the small-size limit, the failure is quasi-plastic and so the stress in the fracture process zone (FPZ) must be almost uniform (Yu *et al.*, 2013). Therefore, the size effect must vanish in this limiting case. At the large-size limit, the joint experiences a brittle failure, for which LEFM is applicable. As shown in Sec 2.3, the stress singularity of the bimaterial joint is generally weaker than "$-1/2$", which implies that the energy release rate at the corner is zero (Bažant and Yu, 2009).

The physical interpretation of this result is that the bimaterial corner would not propagate as a corner. Instead, a FPZ must form at the corner, and at the peak load a macrocrack starts to propagate from the corner. Assume the bimaterial interface to be weaker than either of the base materials. One may approximate the FPZ at the large-size limit by an equivalent bimaterial interfacial crack of length l_c, which is entirely

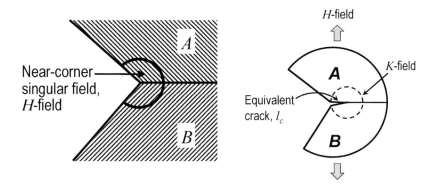

Fig. 7.58 Near-tip analysis of bimaterial corner.

embedded in the singular stress field of the bimaterial corner (Fig. 7.58). In this way, the energy release rate at the crack tip can be related to the stress intensity factor of the bimaterial corner (Akisanya and Fleck, 1997; Liu and Fleck, 1999; Labossiere *et al.*, 2002). If one considers that the peak load is attained once the energy release rate reaches the fracture energy, then, for structures with two real stress singularities, the following large-size asymptote of the size effect can be obtained:

$$\sigma_N = \frac{\sqrt{EG_f/l_c}}{[B_1(D/l_c)^{-2\lambda_1} + B_2(D/l_c)^{-2\lambda_2} + B_3(D/l_c)^{-\lambda_1-\lambda_2}]^{1/2}} \tag{7.57}$$

where $B_1, B_2, B_3 = $ constants depending on the joint geometry as well as the mismatch of the elastic properties of the base materials, $E = $ elastic modulus of one of the base material, and $G_f = $ fracture energy of the interface, which depends on the mode mixity angle characterized by Eq. 2.133.

Since we use the equivalent crack to approximate the FPZ, which actually consists of the two dissimilar materials as well as the adhesive, G_f and l_c depend on the properties of the adhesive as well as the base materials.

Motivated by the type 2 SEL (Eq. 5.57), the following approximate size effect equation, which bridges the small-size and large-size asymptotes, was proposed:

$$\sigma_N = \sigma_s \left\{ 1 + \left[(D/D_1)^{-2\lambda_1} + (D/D_2)^{-2\lambda_2} + (D/D_3)^{-\lambda_1-\lambda_2} \right]^\gamma \right\}^{-1/2\gamma} \tag{7.58}$$

where $\gamma = $ positive constant, which is close to 1; D_1, D_2, and D_3 are constants determined by the fracture properties as well as the structure geometry; and σ_s represents the nominal strength at the small-size limit.

If only one real stress singularity λ dominates the fracture process, then Eq. 7.58 reduces to

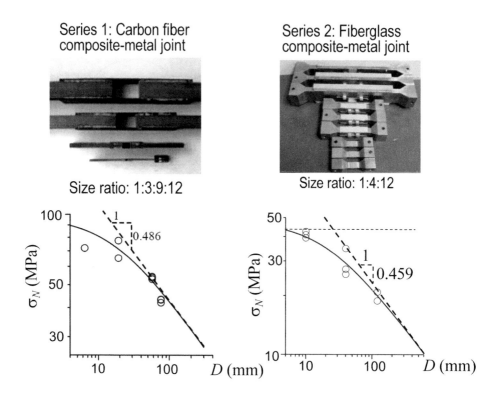

Fig. 7.59 Measured size effect on strength of metal-composite hybrid joints (Bažant and Le, 2017).

$$\sigma_N = \sigma_s \left[1 + (D/D_1)^{-2\gamma\lambda} \right]^{-1/2\gamma} \tag{7.59}$$

where γ is an empirical constant close to 1. Note that Eq. 7.59 can also be applied to structures with complex stress singularity, where λ now represents the real part of the exponents of stress singularities (Le *et al.*, 2010).

A series of size effect tests was performed on geometrically similar double-lap joints made of fiber composites and steel (Yu *et al.*, 2010a). Two series of different types of fiber composite were considered: The series 1 specimens were made of Newport NCT301 carbon laminates, and the series 2 specimens were made of fiberglass G-10/FR4 (Fig. 7.59). Based on the elastic properties of these two types of laminates, the near-corner stress field of series 1 specimens exhibits two real stress singularities: -0.486 and -0.264, and for series 2 specimens only one singularity which is complex, with exponent $-0.459 \pm 0.06i$ ($i = \sqrt{-1}$). For series 1 specimens, the size effect is primarily governed by the dominant stress singularity. Fig. 7.59 presents, for these geometrically similar joints, the measured size effect on σ_N and its optimum fit by Eq.

7.59 (with $r = 1$). Eq. 7.59 is seen to agree well with the test data.

7.8 Reliability of Polycrystalline Silicon MEMS Devices

Microelectronic mechanical systems (MEMS) devices made of polycrystalline silicon (poly-Si) are widely used in numerous technologies including transportation systems, energy conversion, biochemical threat detection, and medical devices (Ballarini, 1998; Espinosa *et al.*, 2005; Fitzgerald *et al.*, 2009; Jiang and Spearing, 2012; Le *et al.*, 2015; Xu and Le, 2017; Xu *et al.*, 2019). Preventing their failure has thus become essential. The strength of MEMS devices has been shown to exhibit considerable variability. Therefore, understanding and modeling the probability distribution of failure strength of poly-Si MEMS devices has become an important subject, especially for use under high stress and at large deformation.

Previous experiments showed that the surface of the sidewall of poly-Si structures is quite random and contains sharp corners which can initiate failure (Reedy, Jr. *et al.*, 2011). This randomness introduces uncertainty in the degree of stress concentration. Meanwhile, the local material strength also exhibits some spatial variability. The result is uncertainty in the location of fracture initiation as well as the overall strength of MEMS structures.

A renewal weakest-link model was recently developed for the strength statistics of poly-Si MEMS structures (Xu *et al.*, 2019). The model combines the finite weakest-link model presented in Chapter 6 and the renewal theory. Consider the uniaxial tension specimen, which is the commonly used configuration for testing the poly-Si material. The fabrication process creates sidewall grooves and, for the purpose of stress analysis, we model them as sharp V-notches. Based on experimental observations (Reedy, Jr. *et al.*, 2011), it is reasonable to consider that the specimen attains its peak load capacity once a crack initiates at any one of these notches. Since the sidewall V-notches have random geometries and the local material resistance is also assumed to vary spatially, the location of fracture initiation along the sidewall of the MEMS device is intrinsically random.

Consider a specimen of length L subjected to remote stress σ_N (Fig. 7.60a). Each sidewall of the specimen contains a number of segments whose length is a random variable (Fig. 7.60b). All the segments except the last one contain a V-notch at their mid-points. Based on the foregoing discussion, we may consider that only segments with V-notches would contribute to the failure of the specimen. For a given specimen length L, the number of V-notches along each sidewall is random. The failure probability of the specimen under stress σ_N, or equivalently the probability distribution of structural strength, can be calculated as

$$P_f(\sigma_N, L) = 1 - \left[\sum_{k=1}^{\infty} R(\sigma_N, L, k) \right]^2 \tag{7.60}$$

where k = number of V-notches along one sidewall, $R(\sigma_N, L, k)$ = survival probability of a sidewall of length L that contains k number of V-notches.

For a sidewall of length L to contain k number of V-notches, one must have

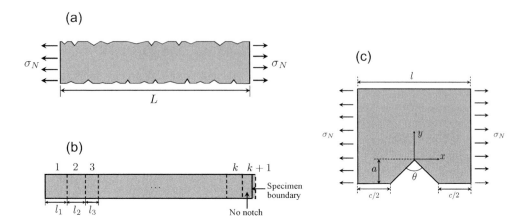

Fig. 7.60 Weakest-link model of failure of poly-Si MEMS specimen: (a) uniaxial tensile MEMS specimen with sidewall defects idealized by V-notches, (b) schematic of renewal weakest-link model, and (c) stress analysis of a single segment.

$$\sum_{i=1}^{k} l_i \le L \quad \text{and} \quad \sum_{i=1}^{k+1} l_i > L \tag{7.61}$$

where l_i = length of ith segment. Here the total length of $k+1$ segments could exceed the actual sidewall length L (Fig. 7.60b), but this is inconsequential since what matters for the failure statistics of the specimen is the first k number of segments with V-notches, whose total length is no greater than L.

Let $f(x)$ be the probability density function (pdf) of the segment length. The overall reliability of the first segment is given by $\int_0^L f(l_1)R_e(\sigma_N, l_1)dl_1$, where $R_e(\sigma_N, l) =$ reliability of the segment of length l. The length of the second segment must satisfy $0 \le l_2 \le L - l_1$, and therefore the overall reliability of the second one is given by $\int_0^{L-l_1} f(l_2)R_e(\sigma_N, l_2)dl_2$. Following the same analysis, the overall reliability of the kth segment is given by $\int_0^{L-l_1\cdots-l_{k-1}} f(l_k)R_e(\sigma_N, l_k)dl_k$. The foregoing consideration enforces the condition $\sum_{i=1}^{k} l_i \le L$. Furthermore, to satisfy that $\sum_{i=1}^{k+1} l_i > L$, we also require $l_{k+1} \ge L - l_1 \cdots - l_k$.

As already mentioned, the survival of each sidewall requires all the segments to survive. Assuming the segment failures to be statistically independent from each other, one has

$$R(\sigma_N, L, k) = \int_0^L \int_0^{L-l_1} \cdots \int_0^{L-l_1-l_2\cdots-l_{k-1}} \left[\prod_{i=1}^{k} f(l_i)R_e(\sigma_N, l_i)\right]$$
$$F_c(L - l_1 - l_2 \cdots - l_k)dl_k dl_{k-1} \cdots dl_1 \tag{7.62}$$

where $F_c(x)$ = complementary cumulative distribution function (cdf) of the segment length.

The reliability function $R_e(\sigma_N, l)$ of each segment must be calculated from the failure probability of the V-notch. Consider a segment of MEMS containing a V-notch under tensile stress σ_N (Fig. 7.60c). For uniaxial tension specimens, the primary failure mode is mode I fracture. It is noted that the crack begins to propagate from the V-notch tip once the nonlocal stress σ evaluated at the notch tip reaches the material tensile strength, that is,

$$\sigma = r_c^{-1} \int_0^{r_c} \sigma_{xx}(y)\mathrm{d}y = f_t \tag{7.63}$$

where r_c is an averaging length scale, σ_{xx} = tensile stress in $x-$direction, and f_t = tensile strength of the material. The choice of nonlocal stress in the present failure criterion takes into account a finite FPZ formed ahead of the notch tip, whose size is proportional to the length scale r_c.

Since the nonlocal stress is calculated from elastic analysis, the failure criterion in Eq. 7.63 can be rewritten by

$$\sigma_N z(a, \theta, l) = f_t \tag{7.64}$$

where z is a dimensionless stress, which depends on the notch depth a, notch angle θ, and segment length l. Clearly z is a random variable, which is independent of the random material strength f_t. Therefore, the reliability of a segment of length l can be written as

$$R_e(\sigma_N, l) = \mathrm{Pr}(f_t/z \geq \sigma_N | l) = 1 - \int_0^\infty F_{f_t}(x\sigma_N) f_z(x|l)\mathrm{d}x \tag{7.65}$$

where $F_{f_t}(x)$ is the cdf of material strength, $f_z(x|l)$ is the conditional pdf of the dimensionless stress for given segment length l. This conditional pdf can be calculated as

$$f_z(x|l) = f_{zl}(x, l)/f(l) \tag{7.66}$$

where $f_{zl}(x, l)$ is the joint pdf of dimensionless stress and segment length. To obtain $f_{zl}(x, l)$, one can first determine the dimensionless stress z as a deterministic function of segment length l, notch depth a, and notch angle θ through a series of finite element simulations. Meanwhile, one can generate a large number of groups of a, θ, and l, which are sampled from their known distribution functions. For each group of a, θ, and l, the corresponding value of z can be calculated by interpolating the simulated function $z(a, \theta, l)$. Based on these groups of z and l values, the pdf $f_z(z, l)$ can be determined.

The local tensile strength distribution, $F_{f_t}(x)$, is considered to follow a grafted Gauss–Weibull distribution (Eq. 6.45a and 6.45b). The grafted distribution contains six parameters, among which any four are independent and may be chosen. The most convenient choices are the Weibull modulus m, the Weibull scale parameter s_0, the mean value μ_G and the standard deviation δ_G of the Gaussian core.

Fig. 7.61 shows the optimum fitting of two sets of measured strength distributions of uniaxial tensile poly-Si MEMS specimens. The first data set consists of strength distributions of specimens of two gauge lengths ($L = 20$ and 70 μm) (Hazra *et al.*, 2009; Reedy, Jr. *et al.*, 2011). The nominal width of specimens of both lengths is 2 μm. The

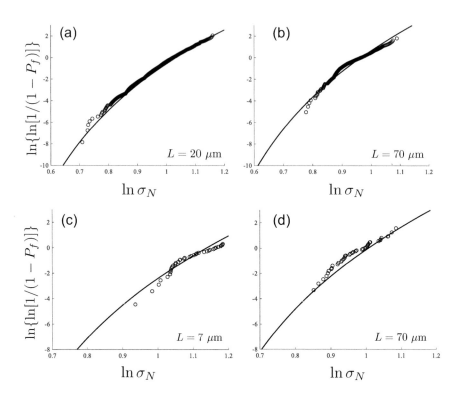

Fig. 7.61 Optimum fitting of measured strength distributions of poly-Si MEMS specimens: (a)–(b) dataset 1, and (c)–(d) dataset 2.

70 μm-long specimens were tested using an on-chip tester, which involved a Chevron thermal actuator using a prehensile grip mechanism. The 20 μm-long specimens were tested using a slack-chain tester, in which a number of specimens were placed in a chain and loaded by a custom-built probe station. The second data set consists of strength distributions of specimens of gauge lengths of 7 and 70 μm, which were tested using an on-chip tester (Saleh *et al.*, 2014). For both data sets, the specimens were produced by Sandia's SUMMiT V poly-Si microfabrication process (Sniegowski and de Boer, 2000).

As seen from Fig. 7.61, the model can match well the measured strength distributions of specimens of two different gauge lengths. Similar to many quasibrittle materials, the measured distributions cannot be fitted by the two-parameter Weibull distribution, which is represented by a straight line in the Weibull scale. To improve the fits of experimental data, the three-parameter Weibull distribution has been suggested. However, the functional form of the three-parameter Weibull distribution conflicts with physics on the nanoscale and rests solely on the extreme value statistics.

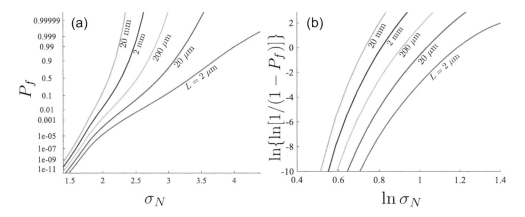

Fig. 7.62 Predicted strength distributions of MEMS specimens of different sizes plotted in (a) Gaussian paper, and (b) Weibull paper.

For uniaxial tensile specimens, the extreme value statistics would be valid if there were a large number (at least 10^5) of V-notches along the sidewall. However, this is by far not the case for specimens of a gauge length on the order of 10–100 μm, which contain only about several hundred V-notches. The agreement between the present model and the experimental data indicates that the strength threshold is unnecessary. The underlying reason for the deviation of the measured strength distribution from the two-parameter Weibull model is the finite random number of V-notches along the sidewall.

The model naturally predicts an intricate size effect on the strength distribution, as indicated by Eq. 7.62. Fig. 7.62 plots in both the Gaussian and Weibull distribution papers the predicted strength cdfs of specimens of different lengths by using the model parameters calibrated based on data set 1. It is seen that, for small-size specimens, the strength cdf deviates significantly from the two-parameter Weibull distribution, and in fact it can be better approximated by a Gaussian distribution except for its left tail. As the specimen size increases, the strength cdf starts to approach the Weibull distribution. Therefore, the specimen size influences not only the mean and standard deviation of the strength distribution but, more fundamentally, also the functional form of the distribution. This is the key feature of the strength distribution of quasibrittle structures as discussed in Chapter 6.

7.9 Analogy with Scaling of Small-Scale Yielding Fracture of Metals

It is interesting that the present mathematical approach to scaling has found analogy in metals (Nguyen *et al.*, 2021). The analogy was suggested by scaled notched tests of aluminum, performed in 1987 (Bažant *et al.*, 1987) before the quasibrittle scaling

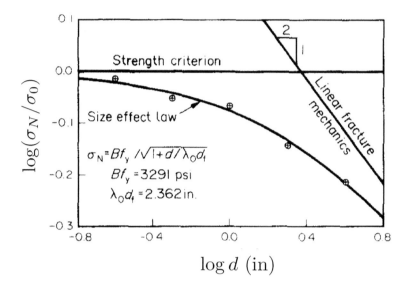

Fig. 7.63 Measured size effect on strength of aluminum specimens (Bažant *et al.*, 1987).

theory was fully developed. These tests indicated a size effect similar the quasibrittle one; see Fig. 7.63.

While in quasibrittle fracture the softening damage zone of the FPZ borders directly on the elastic zone of the structure, fracture of metals is complicated by the fact that a large plastic hardening zone is inserted in between. The analysis, which extends the classical HRR theory (Hutchinson, 1968; Rice and Rosengren, 1968), is based on the classical triaxial generalization of the uniaxial Ramberg–Osgood law $\epsilon/\epsilon_y = \sigma/\sigma_y + \alpha_p(\sigma/\sigma_y)^n$ where $n = $ plastic hardening exponent (≈ 10), $\sigma_y = $ yield strength of material, $\epsilon_y = $ yield strain limit; $\alpha_p = $ empirical parameter.

As in the HRR theory, the elastic part of strain is neglected. Then it is found that the nominal structure strength $\sigma_N \propto D^{-\frac{1}{n+1}}$ when the whole cross-section of structure is yielding and $D \gg$ FPZ size. In the usual case of small-scale yielding, $D \gg r_p$, the effective yielding zone size r_p is calculated by virtual work matching of the plastic yielding field and the remote elastic field with singularity of $r^{-1/2}$ type. Thus the overall scaling law represents a transition through three asymptotic regimes (Fig. 7.64):

$$\sigma_N \propto D^0, \quad D^{-\frac{1}{n+1}}, \quad D^{-\frac{1}{2}} \tag{7.67}$$

where the first regime, $\sigma_N = $ constant $\propto D^0$, is the same as on approach to the FPZ size in quasibrittle fracture, considered in this book, and the third regime corresponds

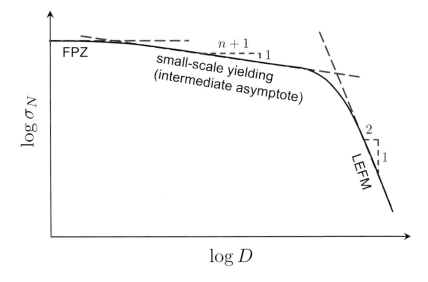

Fig. 7.64 Three asymptotes of the size effect curve.

to the LEFM. The scaling in the middle corresponds to what Barenblatt (Barenblatt, 1979) defined as the intermediate asymptote. The applicability of this asymptote is here broad since the widths of the yielding zone and the FPZ are separated by three orders of magnitude.

The SEL for small-scale yielding has been obtained by an energy approach similar to the derivation of the SEL in Sec. 5.5. But the energy balance equation had to be generalized as

$$\mathcal{G}_s + \mathcal{G}_b = G_f + G_p \tag{7.68}$$

where $\mathcal{G}_s$ is the energy release rate from the elastic structure, G_f is the fracture energy dissipated in the FPZ, $\mathcal{G}_b$ is the rate of release of elastic strain energy originally contained within the band of material traveled by the yielding zone, and G_p is the energy dissipated by unloading of plasticized materials that occurs behind the advancing yielding zone (G_p is absent from quasibrittle fracture). Note that no energy is dissipated within the yielding zone itself since, by definition, it contains no unloading region (this is also clear from the path independence of the J-integral within the yielding zone treated as nonlinearly elastic (Rice, 1968b)).

Fitting the scaling law derived from Eq. 7.68 to test results on notched specimens of different sizes D, one has a simple method to identify G_f and r_p, although this is yet to be verified experimentally. This method is useful for studying the effect of crack-parallel stress on the small-scale yielding fracture.

The foregoing discussion applies to type 2 scaling. The type 1 scaling for failures at the initiation of plastic fracture from a smooth structure boundary is different.

Exercises

E7.1. How does the nominal strength of an anchor, $\sigma_N = P_m/d^2$, scale with the embedment depth? Is it LEFM scaling?

E7.2. What is the cause of deterministic size effect on flexural strength?

E7.3. What are the deterministic (non-statistical) small-size and large-size asymptotic conditions for flexural strength? How do they change if we include the size effect of material randomness?

E7.4. Overcoming the biaxial flexural strength under aircraft wheel is one mode of fracture of airport pavement. Estimate the size effect in a pavement 0.50 m thick compared to the standard ASTM beam test of flexural strength.

E7.5. What was the mode of failure of Malpasset Dam? Was the size effect in the actual dam predominantly statistical or deterministic? What would be the effect of increasing the maximum aggregate size?

E7.6. Is it always safe to design plain concrete or rock structures under the classical no-tension hypothesis? If not, explain why.

E7.7. Why is the thickness exponent of critical temperature drop in thermal bending fracture of floating sea ice plate different from $-1/2$?

E7.8. What is approximately the FPZ length for in-plane propagation of fracture in floating arctic sea ice plates? Can the size effect in arctic sea ice be determined by normal laboratory fracture tests? What is roughly the FPZ length ratio compared to fine grained ceramics?

E7.9. Is the vertical load capacity of floating arctic sea ice subjected to size effect? Of what kind?

E7.10. Explain the energy mechanism of size effect due propagating compression-shear bands in compression fracture.

E7.11. Does mode I fracture play a role in RC beam shear failure at crack initiation? At maximum load? What kind of fracture controls the maximum load and the size effect? What is the size effect factor, λ_s, prescribed since 2019 by ACI design code?

E7.12. The large worldwide database on the RC beam shear strength has a bias. What kind of bias? How can it be filtered out?

E7.13. Give a simple explanation of the size effect on the breakout of very deep boreholes in rock.

E7.14. Do the compression kink bands in fiber composites propagate like cracks or form simultaneously along their length? Why?

E7.15. Explain the cause of the size effect in kind band propagation.

E7.16. What is the size effect of hole diameter in the failure of fiber composites?

E7.17. Given the testing machine stiffness K_m and the steepest (negative) tangential stiffness K_s for post-peak softening of a compact tension specimen of fiber composite, derive the minimum necessary stiffness K_g of the grips, necessary to ensure test stability in the post-peak.

E7.18. The size effect method for determining the fracture energy G_f requires knowledge of the dimensionless energy release rate $g(\alpha)$. Can this rate be obtained from standard tables of stress intensity factors? If not, why?

E7.19. Can the sideways cracks in highly orthotropic fiber composites be detected by the cohesive crack model? Do these cracks suffer from size effect?

8
Overview of History

History of big ideas is a struggle.

Previous chapters discussed the essential aspects of quasibrittle fracture mechanics and its associated scaling theory. In this chapter, we present a brief historical overview of the development of the classical theories of fracture mechanics and scaling, with an emphasis on quasibrittle fracture. The overview covers linear elastic fracture mechanics, cohesive crack model, crack band model, nonlocal continuum models, the classical theory of scaling and computational approaches.

8.1   Classical Theories of Fracture Mechanics and Scaling

The scaling of structural strength, which is a quintessential property of fracture mechanics, was discussed long before the emergence of fracture mechanics itself. In the 1500s, before the concept of stress existed, Leonardo da Vinci speculated that the strength of ropes was proportional to the inverse of their length (da Vinci, ca 1500). In 1638, Galileo dismissed this speculation but introduced the idea of size effect on structural strength, as the stress at failure, to explain why small animals have slender bones and large ones bulky bones (Galilei, 1638; Williams, 1951). He called the latter the "weakness of the giants."

Mariotte was the first to identify, correctly, the physical source of one type of size effect, namely the random distribution of material strength (Mariotte, 1686). Many other savants advanced various conjectures about the scaling (Young, 1807, for example) but the first consistent mathematical theory of the statistical size effect and of the corresponding probability distribution of structural strength had to wait until Weibull's epoch-making advance in 1939 (Weibull, 1939) (expanded by him in 1951 (Weibull, 1951). However, 11 years earlier, in a marvelously succinct paper (Fisher and Tippet, 1928) with no reference to structural strength, Fisher and Tippett derived Weibull distribution as one of three (and only three) possible extreme value distributions (in the writers' opinion, one should in fairness write the "Fisher–Tippett–Weibull" distribution).

Quasibrittle Fracture Mechanics and Size Effect: A First Course. Zdeněk P. Bažant, Jia-Liang Le and Marco Salviato, Oxford University Press. © Zdeněk P. Bažant, Jia-Liang Le, Marco Salviato 2022. DOI: 10.1093/oso/9780192846242.003.0008

Fig. 8.1 Brittle fracture of Liberty Ships (image courtesy: Wikipedia)

A harbinger of fracture mechanics was Inglis' conclusion that the stress at the apex of an elliptical hole in a perfectly elastic body approaches infinity as the ellipse is shrunken into a line crack (Inglis, 1913). The foundation of fracture mechanics was laid in 1921 by Griffith (Griffith, 1921). He was the first to formulate the energy criterion of crack propagation and explain the enormous diameter effect in glass fibers. Willams (Williams, 1952) is credited with obtaining in 1952 the infinite series expansion of the near-tip elastic stress field, but it is interesting that this expansion was obtained already in 1927 by Knein (Knein, 1927) who, unfortunately ignorant of the work of Griffith, deleted the first singular term, arguing that the stress cannot be infinite, and then proceeded to obtain an incorrect stress field in the structure.

Two major disasters after the Second World War ignited interest in fatigue fracture, and generally in nonlinear fracture mechanics of metals with an inelastic zone at the crack front. One was the failures of Liberty Ships, the first fully welded ships, built during the Second World War by the Navy in Seattle. Due to hull and deck cracks, 12 ships broke in half suddenly with no early warning (Fig. 8.1). In addition, 694 of Liberty Ships suffered major fractures, which required immediate repair (Biggs, 1960). Another disaster, in the early 1950s, was the mid-flight failures (Cotterell, 2002), each after about 1000 flights, of seven Comet airliners, the first commercial pressurized jet aircraft built by de Havilland Co. in England. A part of the fuselage of the seventh airliner was retrieved from the Tyrrhenian sea and revealed a long fatigue crack growing from a sharp window corner. These disasters provided a major impetus to the development of nonlinear fracture mechanics of metals (Orowan, 1945; Irwin, 1948; Orowan, 1949; Irwin, 1958; Cottrell, 1961; Wells, 1961, 1963) and culminated in 1968 with the Hutchinson–Rice–Rosengren (HRR) theory of small-scale yielding (Hutchinson, 1968; Rice and Rosengren, 1968), still remaining as the mainstay of fracture mechanics of plastic hardening metals.

A major theoretical milestone, the second after Griffith, was Irwin's 1957 discovery of the relation between the energy release rate and the stress intensity factor, and his estimate of the fracture process zone size as a material characteristic length. Invoking the principles of fracture mechanics, Freudenthal proposed a physical explanation of Weibull's theory in terms of the statistical distribution of the mesoscale flaws in the material (Freudenthal, 1968). The crucial fact that the material strength distribution must have a power-law tail was first shown during 2006–2009 (Bažant and Pang, 2006, 2007; Bažant *et al.*, 2009; Le *et al.*, 2011) on the basis of the transition rate theory of nanoscale crack propagation, particularly the Kramers rule of the frequency of breakage of interatomic bonds, coupled with the proof that scale transitions of the failure probability distribution tails from the nanoscale to the material macroscale maintain the power law tail and increase its exponent. Recent studies used a level excursion analysis to show the essential role of the power-law tail distribution of material strength in determining the tail behavior of the overall failure statistics of the structure (Xu and Le, 2018, 2019).

8.2 Development of Cohesive Crack Model

The third milestone was Barenblatt's concept of the cohesive crack model (Barenblatt, 1959). He proposed the novel idea of a softening stress-separation law as a material property, governing the gradual decrease of cross-crack cohesive stresses during increasing separation of the opposite crack faces. Building on Zheltov and Khristianovich's 1955 observation that the opposite faces of the crack must have a smooth closing at the crack tip (Zheltov and Kristianovich, 1955), Barenblatt (Barenblatt, 1959) used Sneddon's 1951 solution for the distribution of opening displacements along the radius of a circular crack in an infinite isotropic elastic solid, to calculate the cohesive stresses required to counteract the load and achieve the smooth closing, which is obviously equivalent to vanishing of the total stress intensity factor at cohesive crack tip.

Barenblatt's 1959 paper (Barenblatt, 1959) was published in Russian. Despite its prompt translation into English in the United States, it was only in 1962 that his follow-up, more detailed, paper in English (Barenblatt, 1962) gained a world-wide attention. Meanwhile, Dugdale in 1960 (Dugdale, 1960) published an apparently similar paper dealing with cracks (or, in his words, slits) that have a large plastic zone in front. The uniformly distributed stresses of plastic yielding ahead of the crack tip have a similar effect on crack closing as the non-uniform decreasing cohesive stresses. They also eliminate the singularity of stress and strain at the crack tip, and cause the total stress intensity factor due to load to vanish, which ensures smooth crack-tip closing.

Most subsequent authors, however, did not cite Barenblatt's original 1959 paper in Russian or translation and, in their superficial comments, cited only Dugdale's 1960 and Barenblatt's second (1962) paper, in English. This gave the erroneous impression that Barenblatt was not the first to make this discovery. However, even if Barenblatt's 1959 paper in Russian did not exist, his 1962 paper alone sufficed to make him the creator of the cohesive crack model (Bažant, 2020). Perfunctory readers did not realize that Dugdale's 1960 paper (Dugdale, 1960) dealt with a different problem—the plastic cohesive zone near the tip of a sharp slit, with no softening and no fracture criterion. Dugdale's plastic cohesive zone could extend indefinitely without any actual crack (or slit) growth (Tada *et al.*, 2000; Bažant and Planas, 1998) and, indeed, he called his model the "strip yield model".

It is noteworthy, though, that if a cutoff is introduced into Dugdale's model so that the material would suddenly break at a certain critical relative displacement, then the model would become equivalent to a limiting case of the cohesive crack model in which the stress-displacement function is rectangular, ending with a sudden stress drop. While such a function is not realistic, it has the advantage of allowing instructive analytical solutions of some interesting cases, as exploited in Bažant and Planas (1998).

It must nevertheless be recognized that Dugdale's model was a milestone in the development of ductile fracture mechanics, for which it was no less consequential than Barenblatt's model was for the development of quasibrittle fracture mechanics. Dugdale's model subsequently served as a foundation for the later development of the criterion of critical crack-tip opening displacement (CTOD) for fracture of tough ductile steels. This criterion has been related to the energy flux given by the *J*-integral. It was eventually introduced into the British standard (British Standards Institute, 1979) and many others. The capstone of ductile fracture mechanics was later laid by the HRR theory, developed in 1968 by Hutchinson, Rice and Rosengren (Hutchinson, 1968; Rice and Rosengren, 1968).

Although the effect of crack-parallel stresses recently revealed by the gap test (Sec. 4.1) severely restricts the applicability of cohesive crack model and all line crack models, the cohesive crack model has nevertheless been necessary as a stepping stone toward the crack band or nonlocal models. It remains to be also indispensable for teaching and understanding fracture mechanics, and for providing accurate benchmark solutions needed for checking special cases of other models.

8.3   Rice's *J*-Integral

The fourth milestone was Rice's 1966 discovery of the path-independent *J*-integral (Rice, 1966, 1968*b*) giving the energy flux into the fracture front. Rice and Rosengren (Rice and Rosengren, 1968), as well as Hutchinson (Hutchinson, 1968), demonstrated the *J*-integral application to nonlinear hardening metals. Rice proved, via his *J*-integral, that the flux of energy into an advancing fracture front is equal to the area under the curve of the softening cohesive law (Rice, 1968*b*). This fact has often been overlooked (e.g., int the formulation of the fictitious crack model for concrete, discussed in the next section).

In another pioneering contribution, Palmer and Rice (Palmer and Rice, 1973) studied landslides in overconsolidated dilatant granular soil. They analyzed them as a mode II shear fracture, in which the sliding does not occur simultaneously over the whole failure surface but has a propagating front. They showed that the fracture propagation necessarily predicts a strong non-statistical size effect. They noted that the shear fracture, rather than reducing the cohesive shear stress to zero, may terminate with a finite frictional shear stress plateau. Their essential result was to prove, via *J*-integral, that, if a residual plastic plateau exists, then the energy flux into the propagating fracture front equals only the area of the stress-slip curve lying above the plateau.

8.4 Quasibrittlenes, Scaling, and Fictitious Crack Model

Misunderstanding of the consequences of quasibrittleness was, for a long time, the source of various unnecessary controversies—for example, a famous clash, at the ASCE National Meeting in Milwaukee in 1925, about the stress concentration factor at a hole, between S. P. Timoshenko and G. F. Swain (a Harvard professor whose condescending letter to Timoshenko was circulated after the meeting). Timoshenko had in mind elasticity of a homogenous material, and Swain knew that the measured strength of concrete beams with a hole was correctly predicted by ignoring the stress concentration factor, as if concrete were plastic. Today, of course, we know that both were right, albeit in different ways. Timoshenko was right for holes that are much larger than the inhomogeneity size, while Swain (who has been unfairly derided) was right for concrete in which the hole is not sufficiently larger than the aggregate size.

Quasibrittle fracture mechanics was inspired by laboratory tests of concrete structures in which the fracture process zone (FPZ), roughly 0.5 m in length, was large enough to make quasibrittleness conspicuous. The first evidence of serious deviations from LEFM despite lack of plasticity transpired from experiments revealing severe deviations from LEFM, especially in terms of size effect in geometrically scaled notched specimens (Kaplan, 1961; Leicester, 1969; Kesler *et al.*, 1972; Walsh, 1972, 1976).

For concrete fracture, a major milestone was Hillerborg's formulation of the fictitious crack model (Hillerborg *et al.*, 1976). Although developed independently of the cohesive crack model (at that time there was virtually no interaction between the concrete researchers and those in metals and theoretical mechanics), the fictitious crack model was in fact identical (and so the term "fictitious" is now usually replaced by "cohesive"), except for a conceptual difference–the cohesive crack could form without any pre-existing crack or notch, and unlike metals with no need for nucleation, at any point where the strength limit was attained. This is justified by the fact that the concept of crack nucleation is inapplicable because, from the time of solidification, concrete is full of densely spaced cracks at all scales from 10 nm to macroscale. A pioneering contribution of Hillerborg and his student Petersson (Petersson, 1981) was that they formulated an effective algorithm for propagation of inter-element line cracks for FE programs and, most importantly, demonstrated experimentally and computationally that fracture mechanics was in fact applicable to concrete structures. However, their mathematical formulation of the fictitious crack model was incomplete. It omitted the fact that the fracture energy defined as the work of cohesive stress on complete crack opening is equal to the energy flux into the propagating crack front obtained by the *J*-integral (Rice, 1968*b*). In view of the new gap test (Sec. 4.1), the applicability of the fictitious (or cohesive) crack model is now severely limited to situations with negligible crack-parallel stresses.

An important improvement of the cohesive crack model was Planas and Elices' demonstration (Planas and Elices, 1992, 1993) that the cohesive softening curve is properly described as bilinear, beginning with a steep slope defining the initial fracture energy as the third parameter, and followed by a long tail, with the height of slope change point being the fourth parameter.

Recently, the need for a bilinear cohesive softening law with four parameters was verified by extensive tests of size effect and notch length effect performed by Hoover *et al.* at Northwestern University (Hoover *et al.*, 2013; Hoover and Bažant, 2013, 2014*b*). Various consequences of softening bilinearity were clarified in Bažant *et al.* (2002) for concrete and in Bažant and Yu (2011) and Kim *et al.* (2013) for bone. One is that in the absence of size effect data, the softening cohesive curve is non-unique because the one-size data can be fitted equally well with fracture energy and tensile strength values differing by as much as 70%. Consequently, Hillerborg's test of fracture energy, called the work-of-fracture test since it determines the fracture energy solely from the area under the load-displacement curve with complete post-peak softening, yields unambiguous results only if conducted at sufficiently different sizes or complemented by size effect tests of maximum nominal strength. The same test was developed for ceramics earlier in Nakayama (1965), and doubtless suffers from the same limitations.

8.5   Crack Band Model

The first attempt to model concrete fracture as an interelement crack was made, in 1962, by R.W. Clough (Clough, 1962), the inventor of finite element method. He published at the University of California at Berkeley a complete stress analysis of Norfolk Dam containing cracks of observed geometry. In 1968, Rashid (Rashid, 1968), also at Berkeley, analyzed a prestressed concrete pressure vessel for a gas-cooled nuclear reactor, modeling cracks by deletion of finite elements. In the 1960s, of course, it was not yet known that spurious mesh sensitivity must be countered, and that the element size and softening law must be matched to the fracture energy of the material. Nevertheless, Rashid's results were reasonable, doubtless because the element size and strength properties he used implied a realistic fracture energy value. The spurious mesh sensitivity, along with the necessity of introducing a finite material characteristic length and a localization limiter, was not appreciated until 1976, when it was simply demonstrated (Bažant, 1976) by stability analysis of a strain softening zone in a bar.

The cohesive (as well as fictitious) crack model represents a scalar (or uniaxial) stress approximation of the FPZ behavior. In reality, the FPZ is in a complex triaxial stress state, whose continuum approximation should properly be described by a triaxial (tensorial) softening damage constitutive law (the importance of this point is now accentuated by the results of the gap test, Sec. 4.1). In some situations, such as the standard fracture specimens, the scalar stress limitation does not matter, but in many others it does. In view of the gap test (Sec. 4.1), it does matter for quasibrittle (heterogeneous) material when there is a large compressive or tensile stress parallel to the crack plane. It also does matter for the progressive localization of distributed cracking into distinct parallel cracks, which is a problem important for reinforced concrete, rock, and hydraulic fracturing of shale.

The limitation of the cohesive (or fictitious) crack model to a scalar softening law was in 1982 overcome by the crack band model (Bažant, 1982; Bažant and Oh, 1983) (whose precursors were based on energy release per element-long crack advance and softening with a sudden stress drop (Bažant and Cedolin, 1979, 1980)). Aside from avoiding the scalar softening law, the second objective of crack band model was to achieve effective finite element modeling of distributed cracking, with localization to distinct cracks limited by a material characteristic length. For shear fracture of soils, a similar model was proposed by Pietruszczak and Mroz (Pietruszczak and Mróz, 1981).

The concept of strain localization was first introduced, by Rudnicki and Rice (Rudnicki and Rice, 1975), to explain shear banding in plastic-hardening metals. But the physics of this localization was different from concrete or other quasibrittle materials. The localization was caused by nonlinear finite strain effects in plastic-hardening material, rather than softening damage.

The essential ideas of the crack band model were four:

1. to characterize the FPZ (or cohesive zone) by a triaxial tensorial strain-softening constitutive law calibrated by test specimens not significantly larger than the length of the FPZ, which are in common use for concrete;

2. to consider the width of the crack band as a second independent material characteristic length (additional to FPZ length l_0, which is best defined by the minimum possible spacing of parallel cracks stable against localization);

3. to interpret the fracture energy G_f as the area under the complete stress-strain curve under transverse tension multiplied by the crack band width; and, crucially,

4. to make it possible to change the element size and the crack band width by adjusting the post-peak softening slope so as to ensure preserving the same fracture energy, G_f (in situations where damage is known to localize).

The crack band model is now virtually the only one used in concrete and geotechnical engineering industry and for fracture analysis of large composite airframes. Although it is embedded in some commercial programs such as ATENA or DIANA dedicated to concrete, it is usually implemented as a user's subroutine VUMAT of UMAT in commercial software such as ABAQUS.

8.6 Nonlocal Continuum Modeling of Softening Damage

Another way to overcome the limitations of the cohesive crack model is to use the nonlocal model (for its detailed review, see Bažant and Jirásek (2002)). The nonlocal concepts have been introduced into solid mechanics by Kröner, Mindlin, Eringen and others already in the 1960s (Mindlin and Tiersten, 1962; Mindlin, 1964, 1965; Eringen, 1966; Kröner, 1966, 1967; Eringen, 1972; Eringen and Edelen, 1972). But it was for a different purpose—to approximate a discrete elastic system by a continuum. Yet the nonlocal averaging of softening damage (due to microcracks and microslips) can alternatively be used to prevent the strain localization instability, to characterize highly scattered damage in the average sense and, most importantly, to impose a characteristic length on the simulated material, serving as a *localization limiter*. This concept, called strongly nonlocal, was introduced in 1984 (Bažant *et al.*, 1984).

For almost 80 years, the concept of strain-softening material was by continuum mechanicians regarded as insane. Hadamard in 1903 (Hadamard, 1903) showed that a material with negative tangential stiffness or, more precisely, a non-positive-definite tangential moduli tensor, is unstable, cannot propagate waves, leads to an ill-conditioned initial-boundary value problem, and thus cannot exist. This argument was reinforced later by Thomas (Thomas, 1961) and many others (Sandler, 1984; Read and Hegemier, 1984). However these arguments were based on several premises later shown to be incorrect:

1. Heterogeneity stabilizes the softening damage in a small enough zone.

2. The incremental material stiffness for unloading is always positive, which means that unloading waves do exist even in softening material.

3. Due to activation energy of atomic bond breakage, damage is always accompanied by rate effect which causes the tangential stiffness of material in a strain-softening state to be always positive under fast enough loading (Bažant and Gettu, 1992; Bažant *et al.*, 1993*a*, 1995) (this explains why even loading waves of sufficiently high frequency propagate through a softening medium, as confirmed by acoustic wave measurements).

4. Finally the opposition to the concept of strain softening was put to rest by showing that the ill-conditioned problem of waves in a strain-softening continuum does have a unique and rigorous mathematical solution, albeit with certain pathological unrealistic features if no localization limiter is introduced (Bažant and Belytschko, 1985).

Another way to impose a material characteristic length as a localization limiter is the gradient model, which is the case of weak nonlocality. The use of the first gradient, leading to a theory with couple stresses, appeared not to be realistic, but introduction of a second gradient (the Laplacian) of strain tensor into the constitutive law was. It led to a weakly nonlocal softening damage model (Aifantis, 1984; Bažant, 1984*a*).

The original strongly nonlocal model, though, suffered from spurious zero-energy periodic softening modes of deformation, which had to be suppressed by an overlaid (imbricated) local continuum. This problem was soon remedied by formulating, in 1987, the nonlocal damage model (Pijaudier-Cabot and Bažant, 1987; Bažant and Pijaudier-Cabot, 1988), in which the nonlocal averaging is applied only to the inelastic damage part of the strain tensor rather the the total strain tensor. This model turned out to be quite simple, has found wide usage, and was shown to predict quasibrittle fracture quite well. Nevertheless, the programming of the nonlocal averaging is cumbersome.

To improve the physical basis of nonlocal averaging, a nonlocal model based on microcrack interactions has been formulated (Bažant, 1994). Later it was implemented in a nonlocal finite element code (Ožbolt and Bažant, 1996) but, unfortunately, appeared to be relatively complex.

In 1996, Peerlings et al. (Peerlings *et al.*, 1996) greatly improved the gradient model by solving the nonlocal strain from a simultaneous Helmholtz differential equation in which the nonlocal, rather than local, strain is subjected to the Laplacian operator.

A weak point of all nonlocal models, including the gradient models, is the lack of physical basis of the boundary conditions. To overcome it, a boundary layer treated similarly to a crack band has been introduced (Bažant *et al.*, 2010) (in the context of statistical modeling). Alternatively, Pijaudier-Cabot and coworkers tackled the problem by modification of nonlocal averaging of damage near the crack faces (Krayani *et al.*, 2009).

The recently developed gap test injected a serious obstacle to the nonlocal and gradient models—the material characteristic length should probably be different in the directions normal and parallel to the direction of damage band propagation, i.e., to the principal directions of the damage tensor. This is a problem still to be tackled.

Recently, some researchers in fracture of heterogenous solids adopted the phase-field model, which has been coopted (including the name) from earlier computational models for scalar and vector fields in physics. The phase-field model rests on a similar idea as the crack band model but has a computational advantage—the phase-field crack band can run in any direction through the mesh (Borden *et al.*, 2012, 2014; Lo *et al.*, 2019). However, the phase field as it exists is restricted to softening damage constitutive laws controlled by only one damage parameter. Yet for most materials, such as concrete, several damage parameters are needed (e.g., the microplane model has many). The phase-field model contains a length parameter l_0 governing the width of the zone, over which the damage parameter changes from 1 to 0 to 1. The model predicts that the length parameter l_0 is proportional to the Irwin characteristic length (Wu and Nguyen, 2018; Lo *et al.*, 2019), and the proportionality constant is governed by the damage evolution law. The physical justification of this dependence is lacking. Besides, the damage parameter changes from 1 to 0 to 1 cannot capture the transition from uniformly distributed damage to a localized damage band.

A recent fad of nonlocal type has been the dynamic fracture simulation by the so-called peridynamics (Silling, 2000; Silling and Askari, 2005; Silling *et al.*, 2007; Silling and Lehoucq, 2010). In peridynamics, inspired by the highly efficient molecular dynamics program LAMMPS, one assumes that any particle in a heterogeneous material interacts by a potential with all the particles within a so-called "horizon" whose radius is a material property. In one of two versions of peridynamics, called "bond-based," the interactions are considered as elastic, with a strength cutoff. In the other version, called "state-based," an empirical tensorial constitutive law is embedded, based on the dyadic product of force interaction vectors. In peridynamics, the interatomic potential transmits forces not only to the first neighbor but also to the farther neighbors. However, at any material scale above the atomic scale, direct potential interactions of a particle with distant particles, skipping the intermediate particles, do not exist (Bažant *et al.*, 2016). Such interactions are mediated indirectly through the displacements, rotations, and contacts of the intermediate particles. See the critique in Bažant *et al.* (2016) which also documents other problems with peridynamics, such as excessive wave dispersion, and the inability to distinguish wave dispersions due the heterogeneity and due to inelastic damage, the inability to fit complex triaxial test data for damage behavior of specimen such as concrete having approximately the size of a FPZ, etc. Thus peridynamics, as it currently exists, is an unrealistic model for fracture (Bažant *et al.*, 2016).

8.7 Size Effect in Shear Failure of RC Beams

The first significant tests of size effect in shear failure of reinforced concrete (RC) beams were conducted in the 1960s by Leonhardt and Walther (Leonhardt and Walther, 1962), Kani (Kani, 1967) and Bhal (Bhal, 1968). These tests and the size-effect tests of full-scale beams up to 36 m span, conducted at Shimizu Co. in Tokyo in the 1980s, by Iguro, Shioya *et al.* (Iguro *et al.*, 1984; Shioya *et al.*, 1990; Shioya and Akiyama, 1994), convinced engineers that a large size effect on the nominal strength of structures indeed exists. But the question of its proper modeling remained a subject of polemics for decades.

Okamura and Higai (Okamura and Higai, 1980) pioneered the first design formula in 1980. It was incorporated in the mid-1980s into the design code of JSCE (Japan Society of Civil Engineers). It used the power-law $d^{-1/4}$, which was motivated by the Weibull size effect. Although this size effect does not apply to beam shear failure, no other size effect theory was known in 1980.

In the late 1980s, the importance of introducing a size effect factor was recognized in CEB (Comité européen de béton). An empirical size effect formula, similar to Carpinteri's "fractal size effect law" (Carpinteri, 1994; Carpinteri *et al.*, 1995 *a,b*), was introduced, despite serious criticism, into its design code.[1] Polemics about two other controversial ideas about beam shear continue despite having been rigorously dispelled.[2,3]

Afterthought

Many hot research subjects were closed in a few decades. But, like turbulence, fracture mechanics is different. This formidable subject has been researched for a century, and probably will be for another century.

---

[1]In the 1990s, an initially tantalizing idea on the size effect was to attribute it to fractality of the crack surface or to fractality of defect distribution in the material microstructure (Carpinteri, 1994; Carpinteri *et al.*, 1995*a,b*). However, while the fractality is certainly valid as one concept to characterize the crack surface roughness, the idea that the size effect somehow directly ensues from fractality was hazy, devoid of physical justification. It was shown to disagree with a multitude of experimental observations and be mathematically inconsistent (Bažant and Yavari, 2005, 2007).

[2]One such controversy concerns the so-called Modified Compression Field Theory (MCFT) of Collins, Bentz, *et al.*, which is embedded in the European Model Code 2000. This theory predicts the crack spacing, crack width and size effect from a certain strain calculate from the beam bending theory. In the log–log scale, their size effect has asymptotic form d^{-1}, which is thermodynamically inadmissible. Muttoni and et al., accepting Collins' premise of size effect prediction from strain and crack spacing, modified that approach in their Critical Shear Crack Theory (CSCT), which was critically compared in 2019 by Dönmez and Bažant Dönmez and Bažant (2019) to the energetic size effect theory. This criticism was disputed by in 2019 by Muttoni, Fernández Ruiz, Bentz, and Foster, whose arguments were refuted in full detail in (Dönmez *et al.*, 2020).

[3]Another protracted polemics has concerned the so-called boundary element model of Hu *et al.* A recent detailed critique by Carloni *et al.* (Carloni *et al.*, 2019; Dönmez *et al.*, 2020) showed this model, including its size effect equations, to be invalid.

Appendix A
Mathematical Proof of Path Independence of J-Integral

A particularly important property of the J-integral is its *path independence*, which was in Sec. 2.8 proven physically and will now be proven mathematically. Consider the closed contour $\Gamma = \Gamma_1 \cup \Gamma_{bc} \cup \Gamma_2 \cup \Gamma_{da}$ represented in Fig. A.1 by the path $abcd$. We first show that the J-integral calculated on this closed contour (e.g. $abcd$) is zero. To this end, let us rewrite the J-integral as follows:

$$J = \int_\Gamma \left(\bar{U} n_1 - n_j \sigma_{jk} u_{k,1} \right) \mathrm{d}s = \int_\Gamma \left(\bar{U} \delta_{1j} - \sigma_{jk} u_{k,1} \right) n_j \mathrm{d}s \tag{A.1}$$

Now, by applying Green's theorem, we get

$$J = \int_A \left(\bar{U} \delta_{1j} - \sigma_{jk} u_{k,1} \right)_{,j} \mathrm{d}A = \int_A \left(\bar{U}_{,1} - \sigma_{jk,j} u_{k,1} - \sigma_{jk} u_{k,1j} \right) \mathrm{d}A \tag{A.2}$$

In the absence of body forces, equilibrium requires that $\nabla \cdot \sigma = \sigma_{jk,j} = 0$. Furthermore, strains and displacements are related by $\varepsilon_{ij} = \frac{1}{2}(u_{i,j} + u_{j,i})$, and the strain energy density can be written as $\bar{U}_{,1} = (\partial \bar{U}/\partial \epsilon_{jk}) \epsilon_{jk,1} = \sigma_{jk} u_{k,1j}$. This leads to

$$J = \int_A \left(\sigma_{jk} u_{k,1j} - \sigma_{jk} u_{k,1j} \right) \mathrm{d}A = 0 \tag{A.3}$$

To demonstrate that J-integral is path-independent, let us consider the closed path $abcd$, indicated by Γ as represented in Fig. A.1. Now, the contour is the combination of the path Γ_1 (obtained by following the path ab counter-clockwise), Γ_2 (obtained by following the path cd clockwise) and the paths Γ_{bc} and Γ_{da} along the stress-free crack faces. So, as shown in the foregoing section, the J-integral on the closed path reads

$$J = \int_\Gamma \left(\bar{U} \, \mathrm{d}x_2 - t_i u_{i,1} \mathrm{d}s \right)$$
$$= J_{\Gamma_1} - J_{\Gamma_2} + \int_{\Gamma_{bc}} \left(\bar{U} \mathrm{d}x_2 - t_i u_{i,1} \mathrm{d}s \right) + \int_{\Gamma_{da}} \left(\bar{U} \mathrm{d}x_2 - t_i u_{i,1} \mathrm{d}s \right) = 0 \tag{A.4}$$

Along Γ_{bc} and Γ_{da}, we have $t_i = 0$ since the crack is stress free. Further, $\mathrm{d}x_2 = 0$, which makes the last two integrals in Eq. A.4 equal to zero. In view of Eq. A.4, this

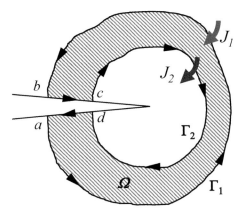

Fig. A.1 Path independence of J-integral.

provides a mathematical proof that J-integral is path-independent, that is, $J_{\Gamma_1} = J_{\Gamma_2}$. Q.E.D.

It should be mentioned that the mathematical background of Rice's J-integral is related to Noether's theorem on invariant variational principles (Noether, 1918). In fact, by using Noether's theorem and the principle of stationary potential energy, Knowles and Sternberg (Knowles and Sternberg, 1972) proposed two other types of path-independent integrals, the M- and L-integrals, in addition to the J-integral. Budiansky and Rice (Budiansky and Rice, 1973) showed that the L- and M-integrals are related to the energy release rates associated with cavity or crack rotation and expansion, respectively, while the J-integral describes the energy release rate for forward crack translation.

Appendix B
Derivation of Size Effect Equations by Dimensional Analysis and Asymptotic Matching

---

We will now show that the type 1 and 2 size effect equations (Eqs. 5.68 and 5.57) can also be derived by combining dimensional analysis and asymptotic matching of the first two terms of both small-size and large-size asymptotic expansions (Bažant, 2004). As discussed in Chapter 3, the simplest model of nonlinear fracture mechanics is the cohesive crack model. Consider the boundary value problem with a cohesive crack, where the cohesive law is expressed in a dimensionless form $\sigma/f_t = \varphi(w/w_f)$; here f_t = tensile strength, w_f = ultimate cohesive separation at which the cohesive stress drops to zero, and function φ describes the shape of the cohesive law. Let G_f denote the initial fracture energy.

It is clear that, for geometrically similar specimens, the nominal strength σ_N would depend on the specimen size D, the tensile strength f_t, and the fracture toughness $K_{1c} = \sqrt{EG_f}$, where E = Young's modulus. For characterizing the failure condition, we can replace the fracture toughness by Irwin's characteristic length $l_1 = K_{1c}^2/f_t^2$.

Therefore, we have four governing parameters σ_N, f_t, D, l_1, which have two independent metric dimensions (i.e. Nm^{-2} and m). The Vashy–Buckingham Π–theorem (Vashy, 1892; Buckingham, 1907, 1914; Barenblatt, 1996, 2003) implies that the failure condition can be expressed by two independent dimensionless parameters, Π_1 and Π_2, that is, $F(\Pi_1, \Pi_2) = 0$. We may assume that function $F(\Pi_1, \Pi_2)$ is sufficiently smooth.

To choose Π_1 and Π_2, we recall the small-size and large-size asymptotic behaviors, as discussed in Sec. 5.3:

1. When $D/l_1 \to 0$, the specimen is much smaller than a fully developed FPZ. So, G_f cannot matter. As indicated by Eq. 5.33, σ_N is a constant at this small-size limit.

2. When $D/l_1 \to \infty$, the FPZ becomes a point in dimensionless coordinates $\xi_i = x_i/D$ and the stress field approaches a singularity. So, f_t cannot matter. Therefore, the LEFM scaling, $\sigma_N \propto D^{-1/2}$, is the large-size asymptote of quasibrittle failure (Eq. 5.35).

The essential idea here is to choose Π_1 and Π_2 such that, in each asymptotic case, all Π_i vanish except one (Bažant, 2004). If consideration is limited to dimensionless monomials, this can be most generally achieved by the following:

$$\Pi_1 = \left(\frac{\sigma_N}{f_t}\right)^p \left(\frac{D}{l_1}\right)^u, \quad \Pi_2 = \left(\frac{\sigma_N}{f_t}\right)^q \left(\frac{D}{l_1}\right)^v \tag{B.1}$$

where exponents p, q, u, v are four unknown real constants. The next step is to determine these exponents based on the known first two terms of the small-size and large-size asymptotic expansions of size effect (Sec. 5.5.1, Eqs. 5.60–5.61).

B.1   Type 2 size effect

We first analyze the type 2 size effect. If we let $\Pi_1 = 0$ correspond to $D \to 0$, then $F(0, \Pi_2) = 0$. This implies that $\Pi_2 = \text{constant}$ or $\sigma_N^q D^v = \text{constant}$ for $D \to 0$, and so one has the case of no size effect; hence $v = 0$. If we let $\Pi_2 = 0$ correspond to $D \to \infty$, then $F(\Pi_1, 0) = 0$. This implies that $\Pi_1 = \text{constant}$ or $\sigma_N^p D^u = \text{constant}$, or $\sigma_N \propto D^{-u/p}$ for $D \to \infty$, and so one must have the linear elastic fracture mechanics (LEFM) scaling; hence $u/p = 1/2$ or $u = p/2$.

To determine p and q, we truncate the Taylor series expansion of F after the first-order terms, that is,

$$F(\Pi_1, \Pi_2) \approx F_0 + F_1 \Pi_1 + F_2 \Pi_2 = 0 \tag{B.2}$$

$$\text{or} \quad F_1 (\sigma_N \sqrt{D}/f_t \sqrt{l_1})^p + F_2 (\sigma_N/f_t)^q = -F_0 \tag{B.3}$$

where $F_i = \partial F/\partial \Pi_i$ ($i = 1, 2$) evaluated at $\Pi_1 = \Pi_2 = 0$, and $F_0 = F(0,0)$ ($F_0, F_1, F_2 \neq 0$).

As it appears, the interpretation of the last equation is easier if we consider its inverse, obtained by solving Eq. B.3 for D as a function of the given value of σ_N;

$$D = l_1 f_t^{\,2} (-F_0/F_1)^{2/p} \sigma_N^{-2} [1 + (F_2/F_0 f_t^{\,q}) \sigma_N^q]^{2/p} \tag{B.4}$$

This relation may be compared to the inverse expansion of the large-size asymptotic expansion of the cohesive crack model (for details, see Bažant (2005) pp. 189–191). It has the form

$$D = B_2 \sigma_N^{-2} (1 - C_2 \sigma_N^{\,2} + ...) \tag{B.5}$$

for $\sigma_N \to 0$ (where $B_2, C_2 = \text{positive constants}$). By matching of the first two terms of Eqs. B.4 and B.5, we obtain $p = q = 2$.

Finally, Eq. B.4 can be solved for σ_N, which yields

$$\sigma_N = \sigma_0 \left(1 + D/D_0\right)^{-1/2} \tag{B.6}$$

where $\sigma_0 = f_t (-F_0/F_2)^{-1/2}$ and $D_0 = l_1 F_2/F_1$. Eq. B.6 matches exactly the type 2 size effect equation (Eq. 5.57) derived via the equivalent LEFM.

B.2   Type 1 size effect

The foregoing analysis can also be applied, with little modification, to derive the type 1 size effect. Let us first consider the deterministic analysis, which predicts a vanishing size effect at the large-size limit (Eq. 5.67). In view of Eq. B.1, this large-size asymptote requires $F(0, \Pi_2) = 0$, which means $\Pi_2 = $ constant for $D \to \infty$. Therefore, we obtain $v = 0$. We may now solve D from Eq. B.2:

$$l_1/D = (-F_0/F_1)^{-1/u}[(f_t/\sigma_N)^p + (F_2/F_0)(f_t/\sigma_N)^{p-q}]^{-1/u} \tag{B.7}$$

Meanwhile, we note that the large-size expansion of the size effect equation (Eq. 5.67) may be generalized as

$$\sigma_N = (b_1 + rc_1 D^{-1} + ...)^{1/r} \tag{B.8}$$

where $r = $ arbitrary constant $\neq 0$. Like before, we write the inverse expansion of Eq. B.8, which is $1/D = (-b_1 + \sigma_N^r + ...)/rc_1$. Then, matching of Eq. B.7 obviously requires $u = -1$ and $p = q = -r$ in Eq. B.1. So we have $\Pi_1 = (f_t/\sigma_N)^r l_1/D$ and $\Pi_2 = (f_t/\sigma_N)^r$. Now Eq. B.2 can be solved for σ_N. This yields, and thus also verifies, Eq. 5.67, in which $D_b = F_1 l_1/rF_2$ and $f_{r\infty} = f_t(-F_2/F_0)^{1/r}$.

If we include the statistical component for the large-size asymptote, we again choose the large size asymptote to correspond to $F(0, \Pi_2) = 0$. However, the asymptotic scaling behavior now reads $\sigma_N = C_1 D^{-n/m}$ (Eq. 5.68), which indicates that $v/q = n/m$. Not to loose the deterministic limit, we keep $p = q = -r$ and $u = -1$. Eq. B.2 can now be solved for σ_N, which leads to the *mean* type 1 energetic-statistical size effect of the same form as Eq. 5.68:

$$\sigma_N = f_r^0 \left[(\lambda l_1/D)^{rn/m} + r\kappa(\lambda l_1/D) \right]^{1/r} \tag{B.9}$$

Here $f_r^0 = \lambda^{-n/m}(-F_2/F_0)^{1/r}$, $\kappa = \lambda^{rn/m-1} F_1/F_2 r$, and r is parameter of the order of 1, sensitive to structure geometry.

As mentioned earlier, Eq. B.9 predicts an infinite nominal strength as $D \to 0$. This is, of course, a mere mathematical abstraction. To impose a finite-value on the small-size asymptote, Eq. B.9 can be modified, but only without affecting the first two large-size asymptotic terms. To bridge the small-size and intermediate asymptotes, we could engage in similar arguments as we did for bridging the large-size and intermediate asymptotes. Suffice to say, the complete law for the mean size effect of type 1, matching the small-size asymptotics (Eq. 5.53) is obtained by replacing $\lambda l_1/D$ in Eq. B.9 with θ':

$$\sigma_N = f_r^0 \left(\theta'^{rn/m} + r\psi\kappa\theta' \right)^{1/r\psi}, \quad \theta' = (1 + D/\psi\eta l_0)^{-\psi} \lambda/\eta \tag{B.10}$$

Here $\eta = $ positive constant of the order of 1 and ψ may be taken as 1. The size effect predicted by Eq. B.10 consists of three asymptotes: 1) small-size asymptote ($\sigma_N \propto D^0$) when $D/l_1 \ll \eta$, 2) intermediate asymptote ($\sigma_N \propto D^{-1/r}$) when $\eta \ll D/l_1 \ll \lambda$, or $\lambda/\eta \to \infty$, and 3) large-size asymptote ($\sigma_N \propto D^{-n/m}$) when $D/l_1 \gg \lambda$.

As shown in Fig. 5.7b, the existing test data for concrete and composites show that the D values for which the difference between Eq. B.9 and Eq. B.10 is significant are smaller than the material inhomogeneities. This means that Eq. B.9 should mostly suffice in practice.

Appendix C
Universal Size Effect Law and Crack Length Effect

In Chapter 5, we discussed the size effect laws for two types of failure, type 1 and type 2. Formulation of a universal size effect law describing the transition between the type 1 and type 2 size effects is more complicated.

Of particular interest is the effect of crack length, a. Consider three-point-bend specimens with a centric crack of length a_0 (Fig. C.1). When a_0/D (D = beam depth) is sufficiently large, the size effect on the nominal strength will follow the type 2 size effect equation (Eq. 5.57), and when $a_0/D = 0$, the specimens will exhibit a type 1 size effect (Eq. 5.68). The question now is what is the scaling behavior for an arbitrary value of a_0/D. Clearly we need a universal size effect equation in which type 1 and type 2 size effects become the two limiting cases in terms of the crack length.

This question has recently been addressed experimentally and analytically in a series of studies (Hoover and Bažant, 2013; Hoover et al., 2013; Hoover and Bažant, 2014a,b). It was shown that, when the statistical type 1 component is negligible, the universal size effect law has the form:

$$\sigma_N = \left[\frac{E'G_f}{g_0 D + (1-\lambda)c_f g_0' + \lambda E' G_f/f_{r\infty}^2} \right]^{1/2} \left(1 + \frac{r\lambda D_b}{D + l_p} \right)^{1/r} \tag{C.1}$$

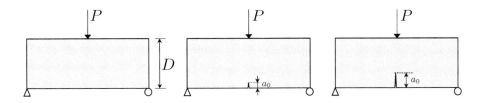

Fig. C.1 Three-point bend tests on beams with different notch lengths.

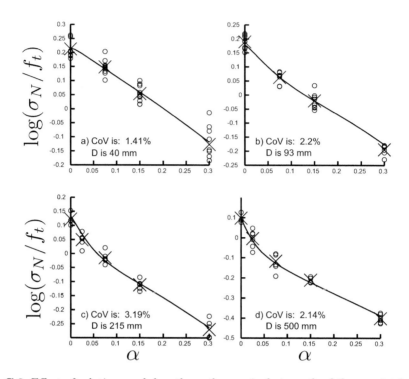

Fig. C.2 Effect of relative crack length on the nominal strength of three-point-bend beams optimally fitted by Eq. C.1.

More generally, when the Weibull statistical component is not negligible, it has the form:

$$\sigma_N = \left[\frac{E'G_f}{g_0 D + (1-\lambda)c_f g_0' + \lambda E' G_f / f_{r\infty}^2} \right]^{1/2} \left[\left(\frac{\lambda l_s}{D + l_s} \right)^{rn/m} + \frac{r\lambda D_b}{D + l_p} \right]^{1/r} \quad \text{(C.2)}$$

in which we introduce an empirical type 1 to 2 transition parameter defined as

$$\lambda = e^{-(\alpha_0^k / q)(D/d_a)^p} \quad \text{(C.3)}$$

where d_a is the maximum size of material inhomogeneity, α_0 is the initial relative crack depth, and $G_f, c_f, f_{r\infty}, D_b, r,\ k, p, q, l_p, l_s$ are parameters to be calibrated by data fitting. The only purpose of the transition parameter λ is to provide a smooth transition between the type 1 and 2 size effects. Thus, λ is justified as an asymptotic matching parameter, with asymptotic cases $\lambda = 1$ for no-notch specimens and $\lambda = 0$ for deep notch (or deep crack) specimens.

Eq. C.2 has been validated by a comprehensive series of size effect experiments on 142 beams (Hoover and Bažant, 2013; Hoover *et al.*, 2013; Hoover and Bažant,

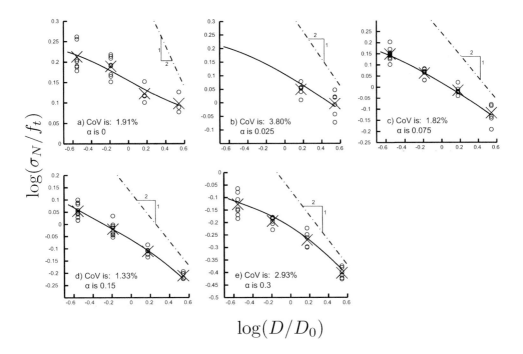

$$\log(D/D_0)$$

Fig. C.3 Effect of structure size on the nominal strength of three-point-bend beams optimally fitted by Eq. C.1.

2014a,b). In these tests, all the specimens were cast within a few hours from one and the same batch of modern ready-mix concrete into 142 forms, cured under the same environmental conditions, and tested at virtually the same age. In this way, a very low statistical scatter has been achieved (the coefficient of variation, CoV, of the deviations from the optimal fits, normalized by the data mean, was only 2.3%). All the specimens used in these tests were geometrically similar three-point bend beams of the same thickness (to eliminate any size effects of thickness). The beam depth varied over a rather broad range, $1:12.5$, and the notch depth varied from 0 to 30% of the cross-section depth. In total, 18 different combinations of notch depth and beam depth were used in these tests. The specimens were loaded under the crack mouth opening displacement (CMOD) control, using a stiff loading machine with fast hydraulics. For all the specimens, nearly complete postpeak softening curves could be recorded.

Figs. C.2 and C.3 show the measured effects of the specimen size and notch depth on the nominal strength of the beams, and the optimum fitting of the test data by Eq. C.1. The faint x-points represent the mean values of the measurements for the given specimen depth and notch depth. The CoV of errors for each curve is also indicated in these figures. The largest CoV occurs for the case of $\alpha = 0.025$, which is in the transition range. For the remaining cases, the CoV values are less than 3.2%. It should be pointed out that the current experimental verification of Eq. C.1 is limited to three-point bend beams. A complete check of Eq. C.1 would require testing also specimens of different geometries.

Bibliography

Achenbach, J. D. and Bažant, Z. P. (1975). Elastodynamic near-tip stress and displacement fields for rapidly propagating cracks in orthotropic materials. *J. Appl. Mech., ASME*, **42**, 183–9.

Achenbach, J. D., Bažant, Z. P., and Khetan, R. P. (1976a). Elastodynamic near-tip fields for a crack propagating along the interface of two orthotropic solids. *Int. J. Engrg. Sci.*, **14**, 811–8.

Achenbach, J. D., Bažant, Z. P., and Khetan, R. P. (1976b). Elastodynamic near-tip fields for a rapidly propagating interface crack. *Int. J. Engrg. Sci.*, **14**, 797–809.

ACI Committee 318 (2008). *Building code requirements for structural concrete ACI 318-08 and commentary 318R-08*. American Concrete Institute, Detroit, MI.

ACI Committee 349 (1989). *Code Requirements for Nuclear Safety*. American Concrete Institute, Detroit, MI.

Aeppli, A. (1924). *Zur Theorie Verketteter Wahrscheinlichkeiten*. Ph.D. thesis, ETH Zurich.

Aifantis, E. C. (1984). On the microstructural origin of certain inelastic models. *J. Eng. Mater. Tech. ASME*, **106**, 326–30.

Akisanya, A. R. and Fleck, N. A. (1997). Interfacial cracking from the free-edge of a long bi-material strip. *Int. J. Solids Struct.*, **34**(13), 1645–65.

Aliabadi, M. H. and Rooke, D. P. (1991). *Numerical Fracture Mechanics*. Volume 8, Solid Mechanics and Its Applications. Springer Netherlands, Southampton.

Allen, A. J., Thomas, J. J., and Jennings, H. M. (2007). Composition and density of nanoscale calcium-silicate-hydrate in cement. *Nat. Mater.*, **6**, 311–6.

Ang, A. H. S. and Tang, W. H. (1984). *Probability Concepts in Engineering Planning and Design. Volume II: Decision, Risk and Reliability*. John Wiley & Sons, New York, NY.

Argon, A. S. (1972). Fracture of composites. In *Treatise of Materials Science and Technology* (ed. H. Herman), Volume 1, pp. 79–114. Academic Press, New York, NY.

Ashrafi, B., Guan, J., Mirjalili, V., Zhang, Y., Chun, L., Hubert, P., Simard, B., Kingston, C. T., Bourne, O., and Johnston, A. (2011). Enhancement of mechanical performance of epoxy/carbon fiber laminate composites using single-walled carbon nanotubes. *Comp. Sci. Tech.*, **71**(13), 1569–78.

Askarinejad, S. and Rahbar, N. (2014). Toughening mechanisms in bioinspired multi-layered materials. *J. R. Soc. Interface*, **12**(102), 20140855.

Assur, A. (1963). Breakup of pack-ice floes. In *Ice and Snow: Properties, Processes and Applications* (ed. W. D. Kingery), Cambridge, MA. MIT Press.

ASTM, D5528-01 (2002). *Standard Test Method for Mode I Interlaminar Fracture Toughness of Unidirectional Fiber-Reinforced Polymer Matrix Composites.* ASTM International, West Conshohockenm, PA.

ASTM, E1820-17 (2017). *Standard Test Method for Measurement of Fracture Toughness.* ASTM International, West Conshohockenm, PA.

Aziz, M. J., Sabin, P. C., and Lu, G. Q. (1991). The activation strain tensor: Nonhydrostatic stress effects on crystal growth kinetics. *Phys. Rev. B*, **41**, 9812–16.

Ballarini, R. (1998). The Role of Mechanics in Microelectromechanical Systems (MEMS) Technology. Technical Report AFRL-ML-WP-TR-1998-4209, Air Force Research Laboratory, Dayton, OH.

Ballarini, R., Keer, L. M., and Shah, S. P. (1987). An analytical model for the pull-out of rigid anchors. *Int. J. Frac.*, **33**(2), 75–94.

Ballarini, R., Shah, S. P., and Keer, L. M. (1986). Failure characteristics of short anchor bolts embedded in a brittle material. *Proc. R. Soc. London A*, **404**, 35–54.

Ballarini, R. and Xie, Y. (2017). Fracture mechanics model of anchor group breakout. *J. Engrg. Mech. ASCE*, **143**(4), 04016125.

Ban, S., Hasegawa, J., and Anusavice, K. J. (1992). Effect of loading conditions on bi-axial flexure strength of dental cements. *Dent. Mater.*, **8**(2), 100–4.

Bao, G., Ho, S., Suo, Z., and Fan, B. (1992). The role of material orthotropy in fracture specimens for composites. *Int. J. Solids Struct.*, **29**(9), 1105–16.

Barenblatt, G. I. (1959). The formation of equilibrium cracks during brittle fracture, general ideas and hypothesis, axially symmetric cracks. *Prikl. Mat. Mech.*, **23**(3), 434–44.

Barenblatt, G. I. (1962). The mathematical theory of equilibrium cracks in brittle materials. *Advances. Appl. Mech.*, **7**, 55–129.

Barenblatt, G. I. (1964). On some general concepts of the mathematical theory of brittle fracture. *J. Appl. Math. Mech.*, **28**(4), 778–92.

Barenblatt, G. I. (1979). *Similarity, Self-Similarity and Intermediate Asymptotics.* Consultants Bureau, New York, NY.

Barenblatt, G. I. (1987). *Dimensional Analysis.* Gordon and Breach Science Publishers, New York, NY.

Barenblatt, G. I. (1996). *Scaling, Self-Similarity, and Intermediate Asymptotics.* Cambridge University Press, Cambridge.

Barenblatt, G. I. (2003). *Scaling.* Cambridge University Press, Cambridge.

Barenblatt, G. I., Entov, V. M., and Salganik, R. L. (1966). Kinetics of crack propagation–conditions of fracture and endurance limit(crack propagation in polymers and polymer type materials). *Inzhenernyi Zhurnal-Mekhanika Tverdogo Tela*, 76–80.

Barsoum, R. S. (1975). Further application of quadratic isoparameteric finite elements to linear fracture mechanics of plate bending and general shells. *Int. J. Frac.*, **11**, 167–9.

Barsoum, R. S. (1976). On the use of isoparameteric finite elements in linear fracture mechanics. *Int. J. Numer. Meth. Engrg.*, **10**, 25–37.

Barsoum, R. S. (2003). The best of both worlds: Hybrid ship hulls use composites & steel. *The AMPTIAC Quarterly*, **7**(3), 55–61.

Bartle, A. (1985). Four major dam failures re-examined. *Int. Water Power Dam Constr.*, **37**(11), 33–6, 41–6.

Bažant, Z. P. (1976). Instability, ductility, and size effect in strain-softening concrete. *J. Engrg. Mech. Div., ASCE*, **102, EM2**, 331–44.

Bažant, Z. P. (1982). Crack band model for fracture of geomaterials. In *Proc. 4th Int. Conf. on Num. Meth. in Geomechanics* (ed. Z. Eisenstein), Volume 3, Edmonton, Alberta, pp. 1137–52. A. A. Balkema.

Bažant, Z. P. (1984*a*). Imbricate continuum and progressive fracturing of concrete and geomaterials. *Meccanica*, **19**, 86–93.

Bažant, Z. P. (1984*b*). Size effect in blunt fracture: Concrete, rock, metal. *J. of Engrg. Mech. ASCE*, **110**(4), 518–35.

Bažant, Z. P. (1985*a*). *Fracture in concrete and reinforced concrete*, pp. 259–303. John Wiley & Sons, Chrichester and New York.

Bažant, Z. P. (1985*b*). Fracture mechanics and strain-softening in concrete. In *Preprints U.S.-Japan Seminar on Finite Element Analysis of Reinforced Concrete Structures* (ed. C. Meyer and H. Okamura), Volume 1, New York, NY, pp. 47–69. American Society of Civil Engineers.

Bažant, Z. P. (1985*c*). Mechanics of fracture and progressive cracking in concrete structures. In *Fracture Mechanics of Concrete: Structural Application and Numerical Calculation* (ed. G. C. Sih and A. DiTommaso), Dordrecht and Boston, pp. 1–94. Martinus Nijhoff.

Bažant, Z. P. (1986). Mechanics of distributed cracking. *Appl. Mech. Rev.*, **39**, 675–705.

Bažant, Z. P. (1990*a*). A critical appraisal of "no-tension" dam design: a fracture mechanics viewpoint. *Dam Engrg.*, **1**(4), 237–47.

Bažant, Z. P. (1990*b*). *Recent advances in failure localization and nonlocal models*, pp. 12–32. Elsevier, London.

Bažant, Z. P. (1992). Large-scale thermal bending fracture of sea ice plates. *J. Geophy. Res.*, **97**(C11), 17739–51.

Bažant, Z. P. (1993). Scaling laws in mechanics of fracture. *J. Engrg. Mech., ASCE*, **119**(9), 1828–44.

Bažant, Z. P. (1994). Nonlocal damage theory based on micromechanics of crack interactions. *J. Engrg. Mech., ASCE*, **120**(3), 593–617.

Bažant, Z. P. (1996). Is no-tension design of concrete or rock structures always safe? – Fracture analysis. *J. Struct. Engrg., ASCE*, **122**(1), 2–10.

Bažant, Z. P. (2001). Size effects in quasibrittle fracture: Aperçu of recent results. In *Fracture Mechanics of Concrete Structures (Proc. FraMCos-4 Int. Conf.)* (ed. R. de Borst, J. Mazars, G. Pijaudier-Cabot, and J. G. M. van Mier), Lisse, pp. 651–8. A. A. Balkema.

Bažant, Z. P. (2004). Scaling theory of quasibrittle structural failure. *Proc. Nat'l. Acad. Sci., USA*, **101**(37), 13400–7.

Bažant, Z. P. (2005). *Scaling of Structural Strength*. Elsevier, London.

Bažant, Z. P. (2019). Design of quasibrittle materials and structures to optimize strength and scaling at probability tail: An aperçu. *Proc. R. Soc. London A*, **475**, 20180617.

Bažant, Z. P. (2020). G.I. Barenblatt's contributions to fracture mechanics. Technical Report CEE Report No. 20-06/aro-g, Northwestern University, Evanston, IL.

Bažant, Z. P., Bai, S.-P., and Gettu, R. (1993*a*). Fracture of rock: Effect of loading rate. *Engrg. Frac. Mech.*, **45**(3), 393–8.

Bažant, Z. P. and Belytschko, T. (1985). Wave propagation in strain-softening bar: Exact solution. *J. Engrg. Mech., ASCE*, **111**(3), 381–9.

Bažant, Z. P., Belytschko, T., and Chang, T.-P. (1984). Continuum model for strain softening. *J. Engrg. Mech. ASCE*, **110**(12), 1666–92.

Bažant, Z. P., Caner, F. C., Carol, I., Adley, M. D., and Akers, S. A. (2000). Microplane model M4 for concrete. I: Formulation with work-conjugate deviatoric stress. *J. Engrg. Mech., ASCE*, **126**(9), 944–53.

Bažant, Z. P. and Cao, Z. (1986). Size effect of shear failure in prestressed concrete beams. *ACI J.*, **83**(3-4), 260–8.

Bažant, Z. P. and Cao, Z. (1987). Size effect in punching shear failure of slabs. *ACI Struct. J.*, **84**(1), 44–53.

Bažant, Z. P. and Cedolin, L. (1979). Blunt crack band propagation in finite element analysis. *J. Engrg. Mech. Div., ASCE*, **105**, 297–315.

Bažant, Z. P. and Cedolin, L. (1980). Fracture mechanics of reinforced concrete. *J. Engrg. Mech. Div., ASCE*, **106**, 1257–1306.

Bažant, Z. P. and Cedolin, L. (1991). *Stability of Structures: Elastic, Inelastic, Fracture and Damage Theories*. Oxford University Press, New York, NY.

Bažant, Z. P. and Chang, T.-P. (1984). Instability of nonlocal continuum and strain averaging. *J. Engrg. Mech., ASCE*, **110**(10), 1441–50.

Bažant, Z. P., Daniel, I. M., and Li, Z. (1996). Size effect and fracture characteristics of composite laminates. *J. Eng. Mater. Tech. ASME*, **118**(3), 317–24.

Bažant, Z. P. and Estenssoro, L. F. (1979). Surface singularity and crack propagation. *Int. J. Solids Struct.*, **15**, 405–26.

Bažant, Z. P. and Gettu, R. (1992). Rate effects and load relaxation: Static fracture of concrete. *ACI Mater. J.*, **89**(5), 456–68.

Bažant, Z. P., Gettu, R., and Kazemi, M. T. (1991). Identification of nonlinear fracture properties from size-effect tests and structural analysis based on geometry-dependent $R-$curves. *Int. J. Rock Mech. Min. Sci.*, **28**(1), 43–51.

Bažant, Z. P. and Grassl, P. (2007). Size effect of cohesive delamination fracture triggered by sandwich skin wrinkling. *J. Appl. Mech., ASME*, **74**(6), 1134–41.

Bažant, Z. P., Gu, W.-H., and Faber, K. T. (1995). Softening reversal and other effects of a change in loading rate on fracture of concrete. *ACI Mater. J.*, **92**(1), 3–9.

Bažant, Z. P. and Jirásek, M. (2002). Nonlocal integral formulations of plasticity and damage: Survey of progress. *J. Engrg. Mech., ASCE*, **128**(11), 1119–49.

Bažant, Z. P. and Kazemi, M. T. (1990). Determination of fracture energy, process zone length and brittleness number from size effect, with application to rock and concrete. *Int. J. Frac.*, **44**, 111–31.

Bažant, Z. P. and Kazemi, M. T. (1991). Size effect on diagonal shear failure of beams without stirrups. *ACI Struct. J.*, **88**, 268–76.

Bažant, Z. P., Kazemi, M. T., and Gettu, R. (1989). Recent studies of size effect in concrete structures. In *Proc. Tenth International Conference on Structural Mechanics in Reactor Technology* (ed. A. H. Hadjian), Volume H, Anaheim, CA, pp. 85–93. American Association for Structural Mechanics in Reactor Technology.

Bažant, Z. P. and Kim, J.-J. H. (1998*a*). Penetration fracture of sea ice plate with part-through cracks: I. Theory. *J. Engrg. Mech. ASCE.*, **124**(12), 1310–15.

Bažant, Z. P. and Kim, J.-J. H. (1998*b*). Penetration fracture of sea ice plate with part-through cracks: II. Results. *J. Eng. Mech., ASCE*, **124**(12), 1316–24.

Bažant, Z. P., Kim, J.-J. H., Daniel, I. M., Becq-Giraudon, E., and Zi, G. (1999). Size effect on compression strength of fiber composites failing by kink band propagation. *Int. J. Frac.*, **95**, 103–41.

Bažant, Z. P. and Kim, J.-K. (1984). Size effect in shear failure of longitudinally reinforced beams. *ACI J.*, **81**, 456–68.

Bažant, Z. P. and Le, J.-L. (2017). *Probabilistic Mechanics of Quasibrittle Structures: Strength, Lifetime, and Size Effect*. Cambridge University Press, Cambridge, U.K.

Bažant, Z. P., Le, J.-L., and Bažant, M. Z. (2009). Scaling of strength and lifetime distributions of quasibrittle structures based on atomistic fracture mechanics. *Proc. Nat'l. Acad. Sci., USA*, **106**, 11484–11489.

Bažant, Z. P., Le, J.-L., and Hoover, C. G. (2010). Nonlocal boundary layer model: Overcoming boundary condition problems in strength statistics and fracture analysis of quasibrittle materials. In *Fracture Mechanics of Concrete and Concrete Structures—Recent Advances in Fracture Mechanics of Concrete.*, Jeju, Korea, pp. 135–43.

Bažant, Z. P., Lee, S.-G., and Pfeiffer, P. A. (1987). Size effect tests and fracture characteristics of aluminum. *Engrg. Frac. Mech.*, **26**(1), 45–57.

Bažant, Z. P. and Li, Y.N. (1994). Cohesive crack model for geomaterials: stability analysis and rate effect. *Appl Mech Reviews*, **47**, 91–6.

Bažant, Z. P. and Li, Z. (1995). Modulus of rupture: size effect due to fracture initiation in boundary layer. *J. Struct. Engrg., ASCE*, **121**(4), 739–46.

Bažant, Z. P., Lin, F.-B., and Lippmann, H. (1993*b*). Fracture energy release and size effect in borehole breakout. *Int. J. Numer. Anal. Methods in Geomech.*, **17**, 1–14.

Bažant, Z. P., Luo, W., Chau, V. T., and Bessa, M. A. (2016). Wave dispersion and basic concepts of peridynamics compared to classical nonlocal models. *J. Appl. Mech. ASME*, **83**, 111004.

Bažant, Z. P. and Novák, D (2000*a*). Energetic-statistical size effect in quasibrittle failure at crack initiation. *ACI Mater. J.*, **97**(3), 381–92.

Bažant, Z. P. and Novák, D (2000*b*). Probabilistic nonlocal theory for quasibrittle fracture initiation and size effect. I. Theory. *J. Engrg. Mech. ASCE*, **126**(2), 166–74.

Bažant, Z. P. and Novák, D (2000*c*). Probabilistic nonlocal theory for quasibrittle fracture initiation and size effect. II. Application. *J. Engrg. Mech. ASCE*, **126**(2), 175–85.

Bažant, Z. P. and Oh, B.-H. (1983). Crack band theory for fracture of concrete. *Mater. Struc.*, **16**, 155–77.

Bažant, Z. P. and Ohtsubo, H. (1977). Stability conditions for propagation of a system of cracks in a brittle solid. *Mech. Res. Comm.*, **4**(5), 353–66.

Bažant, Z. P. and Ožbolt, J. (1990). Nonlocal microplane model for fracture, damage, and size effect in structures. *J. Engrg. Mech., ASCE*, **116**(11), 2485–505.

Bažant, Z. P. and Pang, S. D. (2006). Mechanics based statistics of failure risk of quasibrittle structures and size effect on safety factors. *Proc. Nat'l. Acad. Sci., USA*, **103**(25), 9434–39.

Bažant, Z. P. and Pang, S. D. (2007). Activation energy based extreme value statistics and size effect in brittle and quasibrittle fracture. *J. Mech. Phys. Solids.*, **55**(1), 91–134.

Bažant, Z. P. and Pijaudier-Cabot, G. (1988). Nonlocal continuum damage, localization instability and convergence. *J. Appl. Mech. ASME*, **55**, 287–93.

Bažant, Z. P. and Pijaudier-Cabot, G. (1989). Measurement of characteristic length of nonlocal continuum. *J. Engrg. Mech. ASCE*, **115**(4), 755–67.

Bažant, Z. P. and Planas, J. (1998). *Fracture and Size Effect in Concrete and Other Quasibrittle Materials.* CRC Press, Boca Raton, FL.

Bažant, Z. P. and Sun, H.-H. (1987). Size effect in diagonal shear failure: Influence of aggregate size and stirrups. *ACI Mater. J.*, **84**(4), 259–72.

Bažant, Z. P., Tabbara, M. R., Kazemi, M. T., and Pijaudier-Cabot, G (1990). Random particle model for fracture of aggregate or fiber composites. *J. Engrg. Mech., ASCE*, **116**(8), 1686–705.

Bažant, Z. P. and Vítek, J. L. (1999a). Compound size effect in composite beams with softening connectors. I: Energy approach. *J. Engrg. Mech., ASCE*, **125**(11), 1308–14.

Bažant, Z. P. and Vítek, J. L. (1999b). Compound size effect in composite beams with softening connectors. II: Differential equation and behavior. *J. Engrg. Mech., ASCE*, **125**(11), 1315–22.

Bažant, Z. P., Vořechovský, M., and Novák, D. (2007a). Asymptotic prediction of energetic-statistical size effect from deterministic finite element solutions. *J. Engrg. Mech., ASCE*, **128**(2), 153–62.

Bažant, Z. P. and Xi, Y. (1991). Statistical size effect in quasi-brittle structures: II. Nonlocal theory. *J. Engrg. Mech., ASCE*, **117**(7), 2623–40.

Bažant, Z. P. and Xiang, Y. (1997). Size effect in compression fracture: Splitting crack band propagation. *J. Engrg. Mech., ASCE*, **123**(2), 162–72.

Bažant, Z. P. and Yavari, A. (2005). Is the cause of size effect on structural strength fractal or energetic-statistical? *Engrg. Fract. Mech.*, **72**, 1–31.

Bažant, Z. P. and Yavari, A. (2007). Response to A. Carpinteri, B. Chiaia, P. Cornetti and S. Puzzi's comments on "is the cause of size effect on structural strength fractal or energetic-statistical? *Engrg. Fract. Mech.*, **74**, 2897–910.

Bažant, Z. P. and Yu, Q. (2005a). Designing against size effect on shear strength of reinforced concrete beams without stirrups: I. Formulation. *J. Struct. Engrg., ASCE*, **131**(12), 1877–85.

Bažant, Z. P. and Yu, Q. (2005*b*). Desigining against size effect on shear strength of reinforced concrete beams without stirrups: II. Verification and calibration. *J. Struct. Engrg., ASCE*, **131**(12), 1886–97.

Bažant, Z. P. and Yu, Q. (2008). Minimizing statistical bias to identify size effect from beam shear database. *ACI Struct. J.*, **105**(6), 685–91.

Bažant, Z. P. and Yu, Q. (2009). Universal size effect law and effect of crack depth on quasi-brittle structure strength. *J. Engrg. Mech. ASCE*, **135**(2), 78–84.

Bažant, Z. P. and Yu, Q. (2011). Size effect testing of cohesive fracture parameters and non-uniqueness of work-of-fracture method. *J. Engrg. Mech. ASCE*, **137**(8), 580–8.

Bažant, Z. P., Yu, Q., Gerstle, W., Hanson, J., and Ju, J. W. (2007*b*). Justification of ACI-446 proposal for updating ACI code provisions for shear design of reinforced concrete beams. *ACI Struct. J.*, **104**(5), 601–10.

Bažant, Z. P., Yu, Q., and Zi, G. (2002). Choice of standard fracture test for concrete and its statistical evaluation. *Int. J. Frac.*, **118**(4), 303–37.

Bažant, Z. P., Zhou, Y., Daniel, I. M., Caner, F. C., and Yu, Q. (2006). Size effect on strength of laminate-foam sandwich plates. *J. Eng. Mater. Tech. ASME*, **28**(3), 366–74.

Belytschko, T., Bažant, Z. P., Hyun, Y. W., and Chang, T.-P. (1986). Strain-softening materials and finite elemnt solutions. *Comput. Struct.*, **23**(2), 163–80.

Benthem, J. P. (1977). State of stress at the vertex of a quarter-infinite crack in a half-space. *Int. J. Solids Struct.*, **13**(5), 479–92.

Beremin, F. M. (1983). A local criterion for cleavage fracture of a nuclear pressure vessel steel. *Metall. Trans.*, **114A**, 2277–87.

Bhal, N. S. (1968). *Über den Einfluss der Balkenhöhe aud Schubtragfähighkeit von einfeldrigen Stalbetonbalken mit und ohne Schubbewehrung.* Ph.D. thesis, Stuttgart Universität, Stuttgart, Germany.

Bhat, H. S., Rosakis, A. J., and Sammis, C. G. (2012). A micromechanics based constitutive model for brittle failure at high strain rates. *J. Appl. Mech. ASME*, **79**(5), 031016.

Biggs, W. D. (1960). *The Brittle Fracture of Steel.* MacDonald and Evans, London.

Bilby, R. A., Cottrell, A. H., and Swinden, K.H. (1963). The spread of plastic zone from a notch. *Proc. R. Soc. London A*, **272**, 304–14.

Bogy, D. B. (1971). Two edge-bonded elastic wedges of different materials and wedge angles under surface tractions. *J. Appl. Mech. ASME*, **38**, 377–85.

Bolander, J. E., Hong, G. S., and Yoshitake, K. (2000). Structural concrete analysis using rigid-body-spring networks. *J. Comput. Aided Civil Infra. Eng.*, **15**, 120–33.

Bolander, J. E. and Saito, S. (1998). Fracture analysis using spring network with random geometry. *Engrg. Fract. Mech.*, **61**(5–6), 569–91.

Borden, M. J., Hughes, T. J. R., Landis, C. M., and Verhoosel, C. V. (2014). A higher-order phase-field model for brittle fracture: Formulation and analysis within the isogeometric analysis framework. *Comp. Methods in Appl. Mech. Engrg.*, **273**, 100–18.

Borden, M. J., Verhoosel, C. V., Scott, M. A., Hughes, T. J. R., and Landis, C. M. (2012). A phase-field description of dynamic brittle fracture. *Comp. Methods in Appl. Mech. Engrg.*, **217-220**, 77–95.

Borino, G., Failla, B., and Parrinello, F. (2003). A symmetric nonlocal damage theory. *Int. J. Solids Struct.*, **40**, 3621–45.

Bouchaud, J.-P. and Potters, M. (2000). *Theory of Financial Risks: From Statistical Physics to Risk Management.* Cambridge University Press, Cambridge.

Bresler, B. and Wollack, E. (1952). Shear strength of concrete. Technical report, Dept. of Civil Eng., University of California, Berkeley, Berkeley, CA.

British Standards Institute (1979). *BS 5782 Crack Opening Displacement (COD) Testing.* BSI, London.

Broek, D. (1986). *Elementary Engineering Fracture Mechanics* (4th edn). Martinus Nijhoff, Dordrecht.

Buckingham, E. (1907). Studies on the movement of soil moisture. Bulletin 38, US Dept. of Agriculture, Bureau of Soils, Washington, DC.

Buckingham, E. (1914). On physically linear systems; illustration of the use of dimensional equations. *Phys. Rev. Ser. 2*, **IV**(4), 345–76.

Buckingham, E. (1915). Model experiments and the form of empirical equations. *Trans. ASME*, **37**, 263–96.

Budiansky, B. (1983). Micromechanics. *Comp. Struct.*, **16**(1-4), 3–12.

Budiansky, B. and Fleck, N. A. (1993). Compressive failure of fiber composites. *J. Mech. Phys. Solids*, **41**(1), 183–211.

Budiansky, B. and Rice, J. R. (1973). Conservation laws and energy-release rates. *J. Appl. Mech., ASME*, **40**, 201–3.

Buehler, M. J. and Gao, H. (2006). Dynamical fracture instabilities due to local hyperelasticity at crack tips. *Nature*, **439**(19), 307–10.

Buehler, M. J. and Keten, S. (2010). Colloquium: Failure of molecules, bones, and the Earth itself. *Rev. Mod. Phys.*, **82**(2), 1459–87.

Byskov, E. (1970). The calculation of stress intensity factors using the finite element method with cracked elements. *Int. J. Frac.*, **6**, 159–67.

Caner, F. C. and Bažant, Z. P. (2013). Microplane model M7 for plain concrete. I: Formulation. *J. Engrg. Mech. ASCE*, **139**(12), 1724–35.

Carloni, C., Cusatis, G., Salviato, M., Le, J.-L., Hoover, C. G., and Bažant, Z. P. (2019). Critical comparison of the boundary effect model with cohesive crack model and size effect law. *Engrg. Frac. Mech.*, **215**, 193–210.

Carolan, D., Ivankovic, A., Kinloch, A. J., Sprenger, S., and Taylor, A. C. (2016). Toughening of epoxy-based hybrid nanocomposites. *Polymer*, **97**, 179–90.

Carpinteri, A. (1994). Fractal nature of materials microstructure and size effects on apparent material properties. *Mech. Mater.*, **18**, 89–101.

Carpinteri, A., Chiaia, B., and Ferro, G. (1995a). Multifractal scaling law: An extensive application to nominal strength size effect of concrete structures. Technical Report 50, Atti del Dipartimento di Ingegneria Strutturale, Politecnico de Torino.

Carpinteri, A., Chiaia, B., and Ferro, G. (1995*b*). Size effect on nominal tensile strength of concrete structures: multifractality of material ligament and dimensional transition from order to disorder. *Mater. Struc.*, **28**, 311–17.

Carter, B. C. (1992). Size and stress gradient effects on fracture around cavities. *Rock Mech. Rock Eng.*, **25**, 221–36.

Carter, B. C., Lajtai, E. Z., and Yuan, Y. (1992). Tensile fracture from circular cavities loaded in compression. *Int. J. Frac.*, **57**, 221–36.

Cedolin, L. and Bažant, Z. P. (1980). Effect of finite element choice in blunt crack band analysis. *Comp. Meth. Appl. Mech. Eng.*, **24**, 305–16.

Červenka, J. (1998). Applied brittle analysis of concrete structures. In *Proc. 3rd International Conference on Fracture Mechanics of Concrete Structures* (ed. H. Mihashi and K. Rokugo), Freiburg, pp. 1–15. Aedificatio Publishers.

Červenka, J., Bažant, Z. P., and Wierer, M. (2005). Equivalent localization element for crack band approach to mesh-sensitivity in microplane model. *Int. J. Numer. Methods in Engrg.*, **62**, 700–26.

Chandrasekaran, S., Sato, N., Tölle, F., Mülhaupt, R., Fiedler, B., and Schulte, K. (2014). Fracture toughness and failure mechanism of graphene based epoxy composites. *Comp. Sci. Tech.*, **97**, 90–9.

Chen, L., Ballarini, R., and Heuer, A. H. (2007). A bioinspired micro-composite structure. *J. Mater. Res.*, **22**(1), 124–31.

Cherepanov, G. P. (1962). The stress state in a heterogenous plate with slits. *Izvestia ANN SSSR, OTN, Mekhan. i Mashin.*, **1**, 131–7.

Chhetri, S., Adak, N. C., Samanta, P., Murmu, N. C., and Kuila, T. (2017). Functionalized reduced graphene oxide/epoxy composites with enhanced mechanical properties and thermal stability. *Polymer Testing*, **63**, 1–11.

Clough, R. W. (1962, August). The stress distribution of Norfork Dam. Technical Report 100(19), University of California, Berkeley, Berkeley, CA.

Coleman, B. D. (1958*a*). On the strength of classical fibers and fiber bundles. *J. Mech. Phys. Solids*, **7**, 60–70.

Coleman, B. D. (1958*b*). The statistics and time dependent of mechanical breakdown in fibers. *J. Appl. Phys.*, **29**(6), 968–83.

Comité Euro-International du Beton (1997). *Design of Fastenings in Concrete, Design Guide*. Thomas Telford, London.

Cornell, C. A. (1969). A probability-based structural code. *J. Amer. Concrete Inst.*, **66**(12), 974–85.

Cotterell, B. (2002). The past, present, and future of fracture mechanics. *Engrg. Frac. Mech.*, **69**(5), 533–53.

Cottrell, A. H. (1961). Structural Processes in Creep, ISI Special Report. Technical Report 70, Iron and Steel Institute, London.

Cundall, P. A. (1971). A computer model for simulating progressive large scale movements in blocky rock systems. In *Proc. Int. Symp. Rock Fracture*, Volume 1, Nancy.

Cundall, P. A. and Strack, O. (1979). A discrete numerical model for granular assemblies. *Geotechnique*, **29**(1), 47–65.

Cusatis, G., Bažant, Z. P., and Cedolin, L. (2003a). Confinement-shear lattice model for concrete damage in tension and compression: I. Theory. *J. Engrg. Mech., ASCE*, **129**(12), 1439–48.

Cusatis, G., Bažant, Z. P., and Cedolin, L. (2003b). Confinement-shear lattice model for concrete damage in tension and compression: II. Computation and validation. *J. Engrg. Mech., ASCE*, **129**(12), 1449–58.

Cusatis, G., Pelessone, D., and Mencarelli, A. (2011). Lattice discrete particle model (LDPM) for failure behavior of concrete. I. Theory. *Cem. Concr. Comp.*, **33**(9), 881–90.

Cusatis, G. and Schauffert, E. A. (2009). Cohesive crack analysis of size effect. *Engrg. Fract. Mech.*, **76**, 2163–73.

Daniels, H. E. (1945). The statistical theory of the strength of bundles and threads. *Proc. R. Soc. London A.*, **183**, 405–35.

Dempsey, J. P., Adamson, R. M., and Mulmule, S. V. (1999). Scale effects on the in-situ tensile strength and fracture of ice. Part II: First-year sea ice at Resolute, N.W.T. *Int. J. Frac.*, **95**, 347.

Desmorat, R. and Leckie, F. A. (1998). Singularities in bimaterials: Parametric study of an isotropic/anisotropic joint. *Eur. J. Mcch., A/Solids*, **17**, 33–52.

Dittanet, P. and Pearson, R. A. (2012). Effect of silica nanoparticle size on toughening mechanisms of filled epoxy. *Polymer*, **53**(9), 1890–905.

Dolbow, J. and Belytschko, T. (1999). A finite element method for crack growth without remeshing. *Int J Numer Meth Eng*, **46**, 131–50.

Dönmez, A. and Bažant, Z. P. (2017). Size effect on punching strength of reinforced concrete slabs with and without shear reinforcement. *ACI Struct. J.*, **114**(4), 875–86.

Dönmez, A. and Bažant, Z. P. (2019). Critique of critical shear crack theory for fib model code articles on shear strength and size effect of reinforced concrete beams. *Struct. Concr. (fib)*, **20**(4), 1451–63.

Dönmez, A. and Bažant, Z. P. (2020). Size effect on branched sideways cracks in orthotropic fiber composites. *Int. J. Frac.*, **222**, 155–69.

Dönmez, A., Carloni, C., Cusatis, G., and Bažant, Z. P. (2020). Size effect on shear strength of reinforced concrete: Is Muttoni et al.'s CSCT or Collins et al.'s MCFT a viable alternative to energy-based design code? *J. Engrg. Mech., ASCE*, **146**(10), 04020110.

dos Santos, C., Strecker, K., Piorino Neto, F., de Macedo Silva, O. M., Baldacum, S. A., and da Silva., C. R. M. (2003). Evaluation of the reliability of Si_3N_4-Al_2O_3-CTR_2O_3 ceramics through weibull analysis. *Materials Research*, **6**(4), 463–7.

Duarte, C. A., Hamzeh, O. N., Liszka, T. J., and Tworzydlo, W. W. (2001). A generalized finite element method for the simulation of three-dimensional dynamic crack propagation. *Comput Meth Appl Mech & Eng*, **190**, 2227–62.

Duffaut, P. (2013). The traps behind the failure of Malpsset arch dam, France, in 1959. *J. Rock Mech. Geotech. Engrg.*, **5**, 335–341.

Dugdale, D. S. (1960). Yielding of steel sheets containing slits. *J. Mech. Phys. Solids*, **9**, 100–4.

Dundurs, J. (1969). Edge-bonded dissimilar orthogonal elastic wedges. *J. Appl. Mech., ASME*, **36**, 650–2.

Dutta, A. and Tekalur, S. A. (2014). Crack tortuousity in the nacreous layer–Topological dependence and biomimetic design guideline. *Int. J. Solids Struct.*, **51**(2), 325–35.

Dutta, A., Tekalur, S. A., and Miklavcic, M. (2013). Optimal overlap length in staggered architecture composites under dynamic loading conditions. *J. Mech. Phys. Solids*, **61**(1), 145–60.

Eliáš, J., Vořechovský, M., Skoček, J., and Bažant, Z. P. (2015). Stochastic discrete meso-scale simulations of concrete fracture: Comparison to experimental data. *Engrg. Fract. Mech.*, **135**(1), 1–16.

Eligehausen, R. and Sawade, G. (1989). Analysis of anchorage behaviour (literature review). In *Fracture mechanics of concrete structures: From theory to Applications* (ed. L. Elfgren), London, pp. 263–80. Chapman & Hall.

England, A. H. (1965). A crack between dissimilar media. *J. Appl. Mech., ASME*, **32**, 400–2.

Erdogan, F. (1965). Stress distribution in bonded dissimilar materials with cracks. *J. Appl. Mech., ASME*, **32**, 403–10.

Eringen, A. C. (1966). A unified theory of thermomechanical materials. *Int. J. Engrg. Sci.*, **4**, 179–202.

Eringen, A. C. (1972). Linear theory of nonlocal elasticity and dispersion of plane waves. *Int. J. Engrg. Sci.*, **10**, 425–35.

Eringen, A. C. and Edelen, D. G. B. (1972). On nonlocal elasticity. *Int. J. Engrg. Sci.*, **10**, 233–48.

Eshelby, J. D. (1956). The continuum theory of lattice defects. *Solid State Physics*, **3**, 79–144.

Espinosa, H. D., Peng, B., Moldovan, N., Friedmann, T. A., Xiao, X., Mancini, D. C., Auciello, O., Carlisle, J., and Zorman, C. A. (2005). A comparison of mechanical properties of three MEMS materials-silicon carbide, ultrananocrystalline diamond, and hydrogen-free tetrahedral amorphous carbon (ta-c). In *Proc. 11th Int. Conf. on Fracture*, Volume 5, Red Hook, NY, pp. 3806–11. Curran Associates.

Evans, R. H. and Marathe, M. S. (1968). Microcracking and stress-strain curves for concrete in tension. *Mater. Struc.*, **1**(1), 61–4.

Eyring, H. (1936). Viscosity, plasticity, and diffusion as examples of absolute reaction rates. *J. Chem. Phys.*, **4**, 283–91.

Feddersen, C. E. (1966). *Discussion*, pp. 77–79. American Society for Testing and Materials, Philadelphia, PA.

Fisher, R. A. and Tippet, L. H. C. (1928). Limiting form of the frequency distribution the largest and smallest number of a sample. *Proc. Cambridge. Philos. Soc.*, **24**(2), 180–90.

Fitzgerald, A. M., Pierce, D. M., Huigens, B. M., and White, C. D. (2009). A general methodology to predict the reliability of single-crystal silicon MEMS devices. *J. Microelectromech. Syst.*, **18**(4), 962–70.

Fleck, N. A. (1997). Compressive failure of fiber composites. *Adv. Appl. Mech.*, **33**, 43–117.

Fleck, N. A. and Shu, J. Y. (1995). Microbuckle initiation in fibre composites: a finite element study. *J. Mech. Phys. Solids*, **43**, 1887–918.

Fréchet, M. (1927). Sur la loi de probailité de l'écart maximum. *Ann. Soc. Polon. Math. (Cracow)*, **6**, 93.

Freudenthal, A. M. (1956). Safety and probability of structural failure. *ASCE Trans.*, **121**, 1337–1397.

Freudenthal, A. M. (1968). Statistical approach to brittle fracture. In *Fracture: An Advanced Treatise, Volume 2 –Mathematical Fundamentals* (ed. H. Liebowitz), New York, NY, pp. 591–619. Academic Press.

Fuchs, W., Eligehausen, R., and Breen, J. E. (1995). Concrete capacity design (ccd) approach for fastening to concrete. *ACI Struct. J.*, **92**(1), 73–94.

Fuller, R. B. (1961). Octet truss. US Patent Serial No. 2,986,241.

Galilei, G. (1638). *Discorsi e Dimostrazioni Matematiche Intorno a Due Nuove Scienze*. Elsevirii, Leiden.

Gao, H., Ji, B., Jäger, I. L., Arzt, E., and Fratz, P. (2003). Materials become insensitive to flaws at nanoscale: Lessons from nature. *Proc. Nat'l. Acad. Sci., USA*, **100**(10), 5597–600.

Glasstone, S., Laidler, K. J., and Eyring, H. (1941). *The Theory of Rate Processes*. McGraw-Hill, New York, NY.

Gnedenko, B. V. (1943). Sur la distribution limite du terme maximum d'une serie aleatoire. *Ann. Math.*, **44**, 423–53.

Gorgogianni, A., Eliáš, J., and Le, J.-L. (2020). Mechanism-based energy regularization in computation modeling of quasibrittle fracture. *J. Appl. Mech. ASME*, **87**(9), 091003.

Grassl, P. and Bažant, Z. P. (2009). Random lattice-particle simulation of statistical size effect in quasi-brittle structures failing at crack initiation. *J. Engrg. Mech., ASCE*, **135(2)**, 85–92.

Griffith, A. A. (1921). The phenomenon of rupture in solids. *Phil. Trans.*, **221A**, 582–93.

Gross, B. (1996). Least squares best fit method for the three parameter weibull distribution: analysis of tensile and bend specimens with volume or surface flaw failure. *NASA Technical Report*, **TM-4721**, 1–21.

Gumbel, E. J. (1958). *Statistics of Extremes*. Columbia University Press, New York.

Gustafsson, P.-J. (1985). Fracture mechanics studies of non-yielding materials like concrete: modelling of tensile fracture and applied strength analyses. *Report TVBM 1007*.

Hadamard, J. (1903). *Leçons sur la propagation des ondes*. Hermann, Paris.

Haimson, S. C. and Herrick, C. G. (1989). In-situ stress calculation from borehole breakout experimental studies. In *Proc. 26th U.S. Symp. Rock Mech.* (ed. E. Ashworth), Boston, pp. 1207–1218. Balkema.

Haldar, A. and Mahadevan, S. (2000). *Probability, Reliability, and Statistical Methods in Engineering Design*. Wiley, New York, NY.

Hall, E. O. (1951). The deformation and ageing of mild steel: III Discussion of results. *Proc. Phys. Soc. Sec. B*, **62**(9), 747–53.

Harlow, D. G. and Phoenix, S. L. (1978*a*). The chain-of-bundles probability model for the strength of fibrous materials I: Analysis and conjectures. *J. Comp. Mater.*, **12**, 195–214.

Harlow, D. G. and Phoenix, S. L. (1978*b*). The chain-of-bundles probability model for the strength of fibrous materials II: A numerical study of convergence. *J. Comp. Mater.*, **12**, 314–34.

Harlow, D. G., Smith, R. L., and Taylor, H. M. (1983). Lower tail analysis of the distribution of the strength of load-sharing systems. *J. Appl. Prob.*, **20**, 358–67.

Havlásek, P., Grassl, P., and Jirásek, M. (2016). Analysis of size effect on strength of quasi-brittle materials using integral-type nonlocal models. *Eng. Frac. Mech.*, **157**, 72–85.

Hayashi, K. and Nemat-Nasser, S. (1981). Energy-release rate and crack kinking under combined loading. *J. Appl. Mech., ASME*, **48**, 520–4.

Hazra, S. S., Baker, M. S., Beuth, J. L., and de Boer, M. P. (2009). Demonstration of an in-situ on-chip tester. *J. Micromech. Microeng.*, **19**, 082001 (5 pp.).

He, M.-Y., Bartlett, A., Evans, A. G., and Hutchinson, J. W. (1991). Kinking of crack out of an interface: role of in-plane stress. *J. Am. Ceram. Soc.*, **74**, 767–771.

He, M.-Y. and Hutchinson, J. W. (1989). Kinking of crack out of an interface. *J. Appl. Mech., ASME*, **56**, 270–8.

Heilmann, H. G., Finsterwalder, K., and Hilsdorf, H. K. (1969*a*). *Festigkeit und verformung von beton unter zugspannungen*. Ernst & Sohn.

Heilmann, H. G., Hilsdorf, H. K., and Finsterwalder, K. (1969*b*). Festigkeit und verformung von beton unter zugspannungen. Heft 203, Deutscher Ausschu s für Stahlabeton.

Henshell, R. D. and Shaw, K. G. (1975). Crack tip finite elements are unnecessary. *Int. J. Numer. Meth. Engrg.*, **9**, 495–507.

Higgs, W. A. J., Lucksanasombool, P., Higgs, R. J. E. D., and Swain, M. V. (2001). Evaluating acrylic and glass-ionomer cement strength using the biaxial flexure test. *Biomater.*, **22**(12), 1583–90.

Hill, R. (1963). Elastic properties of reinforced solids: some theoretical principles. *J. Mech. Phys. Solids*, **11**, 357–62.

Hillerborg, A. (1985*a*). *Numerical methods to simulate softening and fracture of concrete*, pp. 141–70. Springer Netherlands, Dordrecht.

Hillerborg, A. (1985*b*). The theoretical basis of a method to determine the fracture energy G_F of concrete. *Mater. Struc.*, **18**(4), 291–6.

Hillerborg, A., Modéer, M., and Petersson, P. E. (1976). Analysis of crack formation and crack growth in concrete by means of fracture mechanics and finite elements. *Cem. Concr. Res.*, **6**(6), 773–82.

Hoover, C. G. and Bažant, Z. P. (2013). Comprehensive concrete fracture tests: Size effects of types 1 & 2, crack length effect and postpeak. *Engrg. Frac. Mech.*, **110**, 281–9.

Hoover, C. G. and Bažant, Z. P. (2014*a*). Cohesive crack, size effect, crack band and work-of-fracture models compared to comprehensive concrete fracture tests. *Int. J. Frac.*, **187**(1), 133–43.

Hoover, C. G. and Bažant, Z. P. (2014*b*). Universal size-shape effect law based on comprehensive concrete fracture tests. *J. Engrg. Mech., ASCE*, **140**(3), 473–9.

Hoover, C. G., Bažant, Z. P., Vorel, J., Wendner, R., and Hubler, M. H. (2013). Comprehensive concrete fracture tests: Description and results. *Engrg. Frac. Mech.*, **114**, 92–103.

Hrennikoff, A. (1941). Solution of problems of elasticity by the framework method. *J. Appl. Mech.*, **12**, 169–75.

Hudson, J. A., Brown, E. T., and Fairhurst, C. (1971). Optimizing the control of rock failure in servo-controlled laboratory tests. *Rock Mech.*, **3**, 217–24.

Hughes, B. P. and Chapman, G. P. (1966). The complete stress-strain curve for concrete in direct tension. *Matls & Structures, Res & Testing* (30), 95–7.

Hutchinson, J. W. (1968). Singular behavior at the end of a tensile crack tip in a hardening material. *J. Mech. Phys. Solids*, **16**, 13–31.

Hutchinson, J. W. and Suo, Z. (1992). Mixed-mode cracking in layered materials. *Adv. Appl. Mech.*, **29**, 63–191.

Iguro, M., Shioya, T., Nojiri, Y., and Akiyama, H. (1984). Experimental studies on shear strength of large reinforced concrete beams under uniformly distributed load. *Proc. JSCE*, **345**(V1), 137–46.

Inglis, C. E. (1913). Stresses in plates due to the presence of cracks and sharp corners. *Trans. Inst. Naval Arch.*, **55**, 219–41.

Ironside, J. G. and Swain, M. V. (1998). Ceramics in dental restorations– a review and critical issues. *J. Australasian Ceram. Soc.*, **34**(2), 78–91.

Irwin, G. R. (1948). Fracture dynamics. In *Fracturing of Metals*, Cleveland, OH. American Society for Metals.

Irwin, G. R. (1957). Analysis of stresses and strains near the end of a crack transversing a plate. *J. Appl. Mech.-T. ASME*, **24**, 361–4.

Irwin, G. R. (1958). *Fracture*, Volume 6, pp. 551–90. Spinger-Verlag, Berlin.

Irwin, G. R. (1960). Fracture mechanics. In *Structural Mechanics* (ed. J. N. Goodier and N. J. Hoff), New York, NY. Pergamon Press.

Japanese Society of Civil Engineers (1991). *Standard Specication for Design and Construction of Concrete Structures, Part I: Design*. Japanese Society of Civil Engineers, Tokyo.

Jiang, L. and Spearing, S. M. (2012). A reassessment of materials issues in microelectromechanical systems (MEMS). *J. Indian Inst. Sci.*, **87**(3), 363.

Jiang, T., Kuila, T., Kim, N. H., Ku, B.-C., and Lee, J. H. (2013). Enhanced mechanical properties of silanized silica nanoparticle attached graphene oxide/epoxy composites. *Comp. Sci. Tech.*, **79**, 115–25.

Jirásek, M. and Bažant, Z. P. (1995*a*). Macroscopic fracture characteristics of random particle systems. *Int. J. Frac.*, **69**, 201–28.

Jirásek, M. and Bažant, Z. P. (1995*b*). Particle model for quasibrittle fracture and application to sea ice. *J. Engrg. Mech. ASCE*, **121**, 1016–25.

Kani, G. N. J. (1967). Basic facts concerning shear failure. *ACI J., Proc.*, **64**, 128–41.

Kaplan, M. F. (1961). Crack propagation and the fracture concrete. *ACI J.*, **58**(11), 591–610.

Karp, S. N. and Karal, F. C. (1962). The elastic field behavior in the neighbourhood of a crack of arbitrary angle. *Commun. Pur. Appl. Math.*, **15**, 413–21.

Kawai, T. (1978). New discrete element models and their application to seismic response analysis of structures. *Nucl. Eng. Des.*, **48**, 207–29.

Kaxiras, E. (2003). *Atomic and Electronic Structure of Solids*. Cambridge University Press, Cambridge.

Kemeny, J. M. and Cook, N. G. W. (1987). Crack models for the failure of rock under compression. In *Proc. 2nd Int. Conf. on Constitutive Laws for Engineering Materials* (ed. C. S. Desai, E. Krempl, P. D. Kiousis, and T. Kundu), Volume 2, New York, NY, pp. 879–887. Elsevier.

Kemeny, J. M. and Cook, N. G. W. (1991). Micromechanics of deformation in rock. In *Toughening Mechanisms in Quasibrittle Materials* (ed. S. P. Shah), Netherlands, pp. 155–88. Kluwer.

Kesler, C. E., Naus, D. J., and Lott, J. L. (1972). Fracture mechanics – its applicability to concrete. In *Proc. Int. Conf. on the Mechanical Behavior of Materials*, Kyoto, Japan, pp. 113–24. The Society of Material Science.

Kim, B. C., Park, S. W., and Lee, D. G. (2008). Fracture toughness of the nano-particle reinforced epoxy composite. *Comp. Struct.*, **86**(1-3), 69–77.

Kim, K. T., Bažant, Z. P., and Yu, Q. (2013). Non-uniqueness of cohesive-crack stress-separation law of human and bovine bones and remedy by size effect tests. *Int. J. Frac.*, **181**, 67–81.

Kirane, K., Bažant, Z. P., and Zi, G. (2014). Fracture and size effect on strength of plain concrete disks under biaxial flexure analyzed by microplane model M7. *J. Eng. Mech., ASCE*, **140**(3), 604–13.

Kirane, K., Salviato, M., and Bažant, Z. P. (2016a). Microplane-triad model for elastic and fracturing behavior of woven composites. *J. Appl. Mech. ASME*, **83**(4), 041006.

Kirane, K., Salviato, M., and Bažant, Z. P. (2016b). Microplane triad model for simple and accurate prediction of orthotropic elastic constants of woven fabric composites. *J. Comp. Mater.*, **50**(9), 1247–60.

Kirane, K., Singh, K. D., and Bažant, Z. P. (2016c). Size effect in torsional strength of plain and reinforced concrete. *ACI Struct. J.*, **113**(6), 1253–62.

Kirsch, E. G. (1898). Die theorie der elastizität und die bedürfnisse der festigkeitslehre. *Zeitschrift des Vereines deutscher Ingenieure*, **42**, 797–807.

Klinger, R. E. and Mendonca, J. A. (1982). Tensile capacity of short anchor bolts and welded studs: A literature review. *ACI J.*, **79**(27), 270–9.

Knein, M. (1927). Zue theorie des druckversuchs. *Abhandlungen aus dem Aerodynamischen Insitut an der Technische Hochschule Aachen*, **7**, 43–62.

Knowles, J. K. and Sternberg, E. (1972). On a class of conservation laws in linearized and finite elastostatics. *Arch. Ration. Mech. Anal.*, **44**, 187–211.

Ko, S., Davey, J., Douglass, S., Yang, J., Tuttle, M. E., and Salviato, M. (2019*a*). Effect of the thickness on the fracturing behavior of discontinuous fiber composite structures. *Comp. Part A: Appl. Sci. & Manu.*, **125**, 105520.

Ko, S., Yang, J., Tuttle, M. E., and Salviato, M. (2019*b*). Effect of the platelet size on the fracturing behavior and size effect of discontinuous fiber composite structures. *Comp. Struct.*, **227**, 111245.

Konnola, R., Reghunadhan Nair, C. P., and Joseph, K. (2016). High strength toughened epoxy nanocomposite based on poly (ether sulfone)-grafted multi-walled carbon nanotube. *Polymers for Advanced Technologies*, **27**(1), 82–9.

Kotz, S. and Nadarajah, S. (2000). *Extreme Value Distributions: Theory and Applications*. Imperial College Press, London.

Krafft, J. M., Sullivan, A. M., and Boyle, R. W. (1961). Effect of dimensions on fast fracture instability of notched sheets. In *Proc. Crack Propagation Symposium*, Volume 1, Cranfield, pp. 8–29. College of Aeronautics and the Royal Aeronautical Society.

Kramers, H. A. (1940). Brownian motion in a field of force and the diffusion model of chemical reaction. *Physica*, **7**, 284–304.

Krausz, A. S. and Krausz, K. (1988). *Fracture Kinetics of Crack Growth*. Kluwer Academic Publisher, Netherlands.

Krayani, A., Pijaudier-Cabot, G., and Dufour, F. (2009). Boundary effect on weight function in nonlocal damage model. *Engrg. Fract. Mech.*, **76**(14), 2217–31.

Kröner, E. (1966). Continuum mechanics and range of atomic cohesion forces. In *Proc. 1st Int. Conf. on Fracture* (ed. T. Yokobori, T. Kawasaki, and J. Swedlow), Sendai, Japan, p. 27. Japanese Society for Strength and Fracture of Materials.

Kröner, E. (1967). Elasticity theory of materials with long range cohesive forces. *Int. J. Solids Struct.*, **3**, 731–42.

Kumar, A., Li, S., Roy, S., King, J. A., and Odegard, G. M. (2015). Fracture properties of nanographene reinforced epon 862 thermoset polymer system. *Comp. Sci. Tech.*, **114**, 87–93.

Kumar, A. and Roy, S. (2018). Characterization of mixed mode fracture properties of nanographene reinforced epoxy and mode i delamination of its carbon fiber composite. *Comp. Part B: Engrg.*, **134**, 98–105.

Labossiere, P. E. W., Dunn, M. L., and Cunningham, S. J. (2002). Application of bimaterial interface corner failure mechanics to silicon/glass anodic bonds. *J. Mech. Phys. Solids*, **50**, 405–33.

Lazzarin, P., Zappalorto, M., and Yates, J. R. (2007). Analytical study of stress distributions due to semi-elliptic notches in shafts under torsion loading. *Int. J. Engrg. Sci.*, **45**(2-8), 308–28.

Le, J.-L. (2011). General size effect on strength of bi-material quasibrittle structures. *Int. J. Frac.*, **172**, 151–60.

Le, J.-L. (2015). Size effect on reliability indices and safety factors of quasibrittle structures. *Struct. Saf.*, **52**, 20–8.

Le, J.-L. (2020). Level excursion analysis of probabilistic quasibrittle fracture. *Sci. China: Tech. Sci.*, **63**, 1141–53.

Le, J.-L., Ballarini, R., and Zhu, Z. (2015). Modeling of probabilistic failure of poly-crystalline silicon MEMS structures. *J. Amer. Cer. Soc.*, **98**(6), 1685–97.

Le, J.-L. and Bažant, Z. P. (2012). Scaling of static fracture of quasi-brittle structures: Strength, lifetime, and fracture kinetics. *J. Appl. Mech. ASME*, **79**(5), 031006.

Le, J.-L. and Bažant, Z. P. (2020). Failure probability of concrete specimens of uncertain mean strength in large database. *J. Eng. Mech., ASCE*, **146**(6), 04020039.

Le, J.-L., Bažant, Z. P., and Bažant, M. Z. (2011). Unified nano-mechanics based probabilistic theory of quasibrittle and brittle structures: I. Strength, crack growth, lifetime and scaling. *J. Mech. Phys. Solids.*, **59**, 1291–321.

Le, J.-L., Bažant, Z. P., and Yu, Q. (2010). Scaling of strength of metal-composite joints: II. Interface fracture analysis. *J. Appl. Mech. ASME*, **77**, 011012.

Le, J.-L., Cannone Falchetto, A., and Marasteanu, M. O. (2013). Determination of strength distribution of quasibrittle structures from mean size effect analysis. *Mech. Mater.*, **66**, 79–87.

Le, J.-L. and Eliáš, J. (2016). A probabilistic crack band model for quasibrittle fracture. *J. Appl. Mech. ASME*, **83**(5), 051005.

Le, J.-L., Eliáš, J., and Bažant, Z. P. (2012). Computation of probability distribution of strength of quasibrittle structures failing at macro-crack initiation. *J. Engrg. Mech. ASCE*, **138(7)**, 888–899.

Le, J.-L., Eliáš, J., Gorgogianni, A., Vievering, J., and Květoň, J. (2018*a*). Rate-dependent scaling of dynamic tensile strength of quasibrittle structures. *J. Appl. Mech. ASME*, **2**, 021003.

Le, J.-L., Xu, Z., and Eliáš, J. (2018*b*). Intrinsic length scale of weakest-link statistical model of quasibrittle fracture. *J. Eng. Mech., ASCE*, **144**(4), 04018017.

Le, J.-L. and Xue, B. (2013). Energetic-statistical size effect in fracture of bimaterial hybrid structures. *Engrg. Frac. Mech.*, **111**, 106–115.

Leadbetter, M. R., Lindgren, G., and Rootzaen, H. (2012). *Extremes and Related Properties of Random Sequences and Processes*. Springer Science & Business Media, New York, NY.

Leicester, R. H. (1969). The size effect of notches. In *Proc. 2nd Australasian Conf. on Mech. of Struct. Mater.*, Melbourne, pp. 4.1–4.20.

Lekhnitskii, S. G. (1963). *Theory of Elasticity of An Anisotropic Body*. Holden-Day, San Francisco, CA.

Lennard-Jones, J. E. (1924). On the determination of molecular fields II. From the equation of state of a gas. *Proc. R. Soc. London A*, **106**(738), 463–77.

Leonhardt, F. and Walther, R. (1962). Beiträge zur behandlung der schubprobleme in stahlbetonbau. *Beton- und Stahlbetonbau (Berlin)*, **57**(3), 54–64.

Levy, M. and Salvadori, M. (1992). *Why buildings fall down?* Norton, New York, NY.

Li, F. Z., Shih, C. F., and Needleman, A. (1985). A comparison of methods for calculating energy release rates. *Eng. Frac. Mech.*, **21**(2), 405–421.

Li, K. K. L. (2000). Influence of size on punching shear strength of concrete slabs. Master's thesis, Department of Civil Engineering and Applied Mechanics, McGill University, Montréal.

Li, W. (2018). *Computational and experimental characterization of the behaviors of anisotropic quasi-brittle materials: Shale and textile composites.* Ph.D. thesis, Northwestern University, Evanston, IL.

Li, W., Qiao, Y., Fenner, J., Warren, K., Salviato, M., Bažant, Z. P., and Cusatis, G. (2021). Elastic and fracture behavior of three-dimensional ply-to-ply angle interlock woven composites: Through-thickness, size effect, and multiaxial tests. *Compos Part C: Open Access*, **4**, 100098.

Li, W., Salviato, M., and Cusatis, G. (2017). Spectral stiffness microplane modeling of fracture and damage of 3D woven composites. In *Proc. American Society for Composites 32nd Technical Conference* (ed. W. Yu, R. B. Pipes, and J. Goodsell). DEStech Publications.

Li, Y.N. and Bažant, Z. P. (1994). Eigenvalue analysis of size effect for cohesive crack model. *Int. J. Fracture*, **66**(3), 213–26.

Liu, D. and Fleck, N. A. (1999). Scale effect in the initiation of cracking of a scarf joint. *Int. J. Frac.*, **95**, 66–88.

Liu, H.-Y., Wang, G.-T., Mai, Y.-W., and Zeng, Y. (2011). On fracture toughness of nano-particle modified epoxy. *Comp. Part B: Engrg.*, **42**(8), 2170–5.

Lo, Y.-S., Borden, M. J., Ravi-Chandar, K., and Landis, C. M. (2019). A phase-field model for fatigue crack growth. *J. Mech. Phys. Solids*, **132**, 103684.

Lohbauer, U., Petchelt, A., and Greil, P. (2002). Lifetime prediction of CAD/CAM dental ceramics. *J. Biomedical Mater. Res.*, **63**(6), 780–5.

Luo, W. and Bažant, Z. P. (2017a). Fishnet model for failure probability tail of nacre-like imbricated lamellar materials. *Proc. Nat'l. Acad. Sci., USA*, **114**(49), 12900–5.

Luo, W. and Bažant, Z. P. (2017b). Fishnet statistics for probabilistic strength and scaling of nacreous imbricated lamellar materials. *J. Mech. Phys. Solids*, **109**, 264–87.

Luo, W. and Bažant, Z. P. (2018). Fishnet model with order statistics for tail probability of failure of nacreous biomimetic materials with softening interlaminar links. *J. Mech. Phys. Solids*, **121**, 281–95.

Luo, W. and Bažant, Z. P. (2019). Fishnet statistical size effect on strength of materials with nacreous microstructure. *J. Appl. Mech. ASME*, **86**(8), 081006.

Luo, W. and Bažant, Z. P. (2020). General fishnet statistics of strength: Nacreous, biomimetic, concrete, octet-truss, and other architected or quasibrittle materials. *J. Appl. Mech. ASME*, **87**(3), 031015.

Luo, W., Le, J.-L., Rasoolinejad, M., and Bažant, Z. P. (2021). Coefficient of variation of shear strength of rc beams and size effect. *J. Engrg. Mech. ASCE*, **147**(2), 04020144.

Mahesh, S. and Phoenix, S. L. (2004). Lifetime distributions for unidirectional fibrous composites under creep-rupture loading. *Int. J. Frac.*, **127**, 303–60.

Mariotte, E. (1686). *Traité du mouvement des eaux, posthumously edited by M. de la Hire; Eng. Transl. by J. T. Desvaguliers, London (1718), p. 249; also Marriotte's collected works, 2nd ed., The Hague (1740).*

Mefford, C. H., Qiao, Y., and Salviato, M. (2017). Failure behavior and scaling of graphene nanocomposites. *Comp. Struct.*, **176**, 961–72.

Mindlin, R. D. (1964). Microstructure in linear elasticity. *Arch. Ration. Mech. Anal.*, **16**, 51–78.

Mindlin, R. D. (1965). Second gradient of strain and surface tension in linear elasticity. *Int. J. Solids Struct.*, **1**, 417–38.

Mindlin, R. D. and Tiersten, H. F. (1962). Effects of couple stresses in linear elasticity. *Arch. Ration. Mech. Anal.*, **11**, 415–48.

Mirjalili, V., Ramachandramoorthy, R., and Hubert, P. (2014). Enhancement of fracture toughness of carbon fiber laminated composites using multi wall carbon nanotubes. *Carbon*, **79**, 413–23.

Modéer, M. (1979). A fracture mechanics approach to failure analyses of concrete materials. Technical report, Division of Building Materials, University of Lund Sweden.

Mörsch, E. (1922). *De Eisenbetonbau, Seine Theorie und Anwendung* (5th edn). Volume 1. Wittwer, Stuttgart.

Morse, P. M. (1929). Diatomic molecules according to the wave mechanics. II. Vibrational levels. *Phys. Rev.*, **34**, 57–64.

Murakami, Y. (1986). *Stress Intensity Factors Handbook* (1st edn). Pergamon Press, Oxford.

Nairn, J. A. (1988). Fracture mechanics of unidirectional composites using the shear-lag model II: experiment. *J. Comp. Mater.*, **22**(6), 589–600.

Nakayama, J. (1965). Direct measurement of fracture energies of brittle heterogeneous materials. *J. Am. Cer. Soc.*, **48**(11), 583–7.

Nalla, R. K., Kruzic, J. J., Kinney, J. H., and Ritchie, R. O. (2005). Mechanistic aspects of fracture and r-curve behavior in human cortical bone. *Biomater.*, **26**(2), 217–31.

Nalla, R. K., Kruzic, J. J., and Ritchie, R. O. (2004). On the origin of the toughness of mineralized tissue: micro cracking or crack bridging? *Bone*, **34**(5), 790–8.

Nesetova, V. and Lajtai, E. Z. (1973). Fracture from compressive stress concentration around elastic flaws. *Int. J. Rock Mech. Min. Sci.*, **10**, 265–84.

Neuber, H. (1985). *Kerbspannungslehre: Theorie der Spannungskonzentration Genaue Berechnung der Festigkeit*. Springer-Verlag, Berlin and Heidelberg.

Newman Jr., J. C. (1971). An improved method of collocation for the stress analysis of cracked plates with various shaped boundaries. Technical Note TN D-6376, NASA.

Nguyen, H., Pathirage, M., Cusatis, G., and Bažant, Z. P. (2020a). Gap test of crack-parallel stress effect on quasibrittle fracture and its consequences. *J. Appl. Mech. ASME*, **87**(7), 071012–1–11.

Nguyen, H., Pathirage, M., Rezaei, M., Issa, M., Cusatis, G., and Bažant, Z. P. (2020b). New perspective of fracture mechanics inspired by novel tests with crack-parallel compression. *Proc. Nat'l. Acad. Sci., USA*, **117**(25), 14015–20.

Nguyen, H. T., Dönmez, A., and Bažant, Z. P. (2021). Structural strength scaling law for fracture of plastic-hardening metals and testing of fracture properties. *Extreme Mech. Lett*, **43**, 101141.

Nitka, M. and Tejchman, J. (2015). Modelling of concrete behaviour in uniaxial compression and tension with DEM. *Granular Matt.*, **17**(1), 145–164.

Noether, E. (1918). Invariante variationsprobleme. *Nachr. D. König. Gesellsch. D. Wiss. Zu Göttingen, Math-phys. Klasse*, 235–57.

Norman, T. L., Vashishth, D., and Burr, D. B. (1995). Fracture toughness of human bone under tension. *J. Biomech.*, **28**(3), 309–20.

Oglesby, J. J. and Lamackey, O. (1972). An evaluation of finite element methods for the computation of elastic stress intensity factors. Technical Report No. 3751, NSRDC.

Okabe, N. and Hirata, H. (1995). *High temperature fatigue properties for some types of SiC and Si$_3$N$_4$ and the unified strength estimation method*, pp. 245–76. Elsevier Science B. V. and The Society of Materials Science, Japan.

Okamura, H. and Higai, T. (1980). Proposed design equation for shear strength of reinforced concrete beams without web reinforcement. *Proc., Japanese Soc. of Civil Engrs.*, **300**, 131–41.

Orowan, E. (1945). Notch brittleness and strength of solids. *Trans. Inst. Eng. Shipbuilders Scotland*, **89**, 165–215.

Orowan, E. (1949). Fracture and strength of solids. *Reports on Progress in Physics*, **12**(1), 185–232.

Otsuka, K. and Date, H. (2000). Fracture process zone in concrete tension specimen. *Eng. Fract. Mech.*, **65**(2-3), 111–31.

Ožbolt, J. and Bažant, Z. P. (1996). Numerical smeared fracture analysis: Nonlocal microcrack interaction approach. *Int. J. Numer. Methods in Engrg.*, **39**, 635–661.

Ožbolt, J. and Eligehausen, R. (1992). Fastening elements in concrete structures–Numerical solutions. In *Proc. 2nd Int. Conf. Fracture of Concrete and Rock* (ed. H. P. Rossmanith), London, pp. 527–547. E & FN Spon.

Ožbolt, J., Eligehausen, R., and Reinhardt, H. W. (1999). Size effect on the concrete cone pull-out load. *Int. J. Frac.*, **95**(1-4), 391–404.

Palmer, A.C. and Rice, J. R. (1973). The growth of slip surfaces in the progressive failure of overconsolidated clay. *Proc. R. Soc. London A*, **332**, 527–48.

Pastor, J. Y., Guinea, G., Planas, J., and Elices, M. (1995). Nueva expresión del factor de intensidad de tensiones para la probeta de flexón en tres puntos. *Anales de Mecánica de la Fractura*, **12**, 85–90.

Peerlings, R. H. J., de Borst, R., Brekelmans, W. A. M., and de Vree, J. H. P. (1996). Gradient enhanced damage for quasi-brittle materials. *Int. J. Numer. Methods in Engrg.*, **39**(19), 3391–403.

Peerlings, R. H. J., Geers, M. G. D., de Borst, R., and Brekelmans, W. A. M. (2001). A critical comparison of nonlocal and gradient-enhanced softening continua. *Int. J. Solids Struct.*, **38**(44-45), 7723–46.

Petch, N. J. (1954). The cleavage strength of polycrystals. *J. Iron and Steel Inst.*, **174**, 25–8.

Petersson, P. E. (1981). Crack growth and development of fracture zones in plain concrete and similar materials. Report TVBM-1006, Div. of Building Materials, Lun Inst. of Tech., Lund, Sweden.

Phoenix, S. L. (1978a). The asymptotic time to failure of a mechanical system of parallel members. *SIAM J. Appl. Maths.*, **34**(**2**), 227–46.

Phoenix, S. L. (1978*b*). Stochastic strength and fatigue of fiber bundles. *Int. J. Frac.*, **14(3)**, 327–44.

Phoenix, S. L., Ibnabdeljalil, M., and Hui, C.-Y. (1997). Size effects in the distribution for strength of brittle matrix fibrous composites. *Int. J. Solids Struct.*, **34(5)**, 545–68.

Phoenix, S. L. and Tierney, L.-J. (1983). A statistical model for the time dependent failure of unidirectional composite materials under local elastic load-sharing among fibers. *Engrg. Fract. Mech.*, **18(1)**, 193–215.

Piccinin, R., Ballarini, R., and Cattaneo, S. (2010). Linear elastic fracture mechanics pullout analyses of headed anchors in stressed concrete. *J. Engrg. Mech. ASCE*, **136**(6), 761–8.

Piccinin, R., Ballarini, R., and Cattaneo, S. (2012). Pullout capacity of headed anchors in prestressed concrete. *J. Engrg. Mech. ASCE*, **138**(7), 877–87.

Pietruszczak, St and Mróz, Z. (1981). Finite element analysis of deformation of strain-softening materials. *Int. J. Numer. Methods in Engrg.*, **17**, 327–34.

Pijaudier-Cabot, G. and Bažant, Z. P. (1987). Nonlocal damage theory. *J. Engrg. Mech. ASCE*, **113**(10), 1512–33.

Planas, J. and Elices, M. (1992). Asymptotic analysis of a cohesive crack: 1. Theoretical background. *Int. J. Frac.*, **55**(2), 153–77.

Planas, J. and Elices, M. (1993). Asymptotic analysis of a cohesive crack: 2. Influence of the softening curve. *Int. J. Frac.*, **64**(3), 221–37.

Podgorniak-Stanik, B. A. (1998). *The Influence of Concrete Strength, Distribution of Longitudinal Reinforcement, Amount of Transverse Reinforcement and Member Size on Shear Strength of Reinforced Concrete Members*. Ph.D. thesis, University of Toronto.

Qian, Z. and Akisanya, A. R. (1998). An experimental investigation of failure initiation in bonded joints. *Acta Mater.*, **46**, 4895–904.

Qiao, Y. and Salviato, M. (2019*a*). Strength and cohesive behavior of thermoset polymers at the microscale: A size-effect study. *Eng. Fract. Mech.*, **213**, 100–17.

Qiao, Y. and Salviato, M. (2019*b*). Study of the fracturing behavior of thermoset polymer nanocomposites via cohesive zone modeling. *Compos. Struct.*, **220**, 127–47.

Quaresimin, M., Salviato, M., and Zappalorto, M. (2012). Fracture and interlaminar properties of clay-modified epoxies and their glass reinforced laminates. *Eng. Frac. Mech.*, **81**, 80–93.

Rabczuk, T., Bordas, S., and Zi, G. (2007). Three-dimensional meshfree method for continuous crack initiation, nucleation and propagation in statics and dynamics. *Comput Mech*, **40**, 473–95.

Rahimi-Aghdam, S., Chau, V. T., Lee, H., Nguyen, H., Li, W., Karra, S., Rougier, E., Viswanathan, H., Srinivasan, G., and Bažant, Z. P. (2019). Branching of hydraulic cracks enabling permeability of gas or oil shale with closed natural fractures. *Proc. Nat'l. Acad. Sci., USA*, **116**(5), 1532–37.

Rashid, J. (1968). Ultimate strength analysis of prestressed concrete pressure vessels. *Nucl. Eng. Des.*, **7**(4), 334–44.

Rasoolinejad, M. and Bažant, Z. P. (2019). Size effect of squat shear walls extrapolated by microplane model M7. *ACI Struct. J.*, **116**(3), 75–84.

Read, H. E. and Hegemier, G. A. (1984). Strain softening of rock, soil, and concrete - A review article. *Mech. Mater.*, **3**, 271–94.

Redner, S. (2001). *A Guide to First-Passage Processes*. Cambridge University Press, Cambridge.

Reedy Jr., E. D. (2000). Comparison between interface corner and interfacial fracture analysis of an adhesively-bonded butt joint. *Int. J. Solids Struct.*, **37**, 2429–42.

Reedy, Jr., E. D., Boyce, B. L., Foulk, III, J. W., Field Jr., R. V., de Boer, M. P., and Hazra, S. S. (2011). Predicting fracture in micrometer-scale polycrystalline silicon MEMS structures. *J. Microelectromech. Syst.*, **20**(4), 922–32.

Rezakhani, R. and Cusatis, G. (2016). Asymptotic expansion homogenization of discrete fine-scale models with rotational degrees of freedom for the simulation of quasi-brittle materials. *J. Mech. Phys. Solids*, **88**, 320–45.

Rice, J. R. (1966). An examination of the fracture mechanics energy balance from the point of view of continuum mechanics. In *Proc. 1st International Conference on Fracture* (ed. T. Yokobori, T. Kawasaki, and J. L. Swedlow), Volume 1, pp. 309–40. Japanese Society for Strength and Fracture of Materials.

Rice, J. R. (1968*a*). *Mathematical analysis in the mechanics of fracture*, Volume 2, pp. 191–311. Academic Press, New York, NY.

Rice, J. R. (1968*b*). Path independent integral and approximate analysis of strain concentrations by notches and cracks. *J. Appl. Mech. ASME*, **35**, 379–86.

Rice, J. R. (1988). Elastic fracture mechanics concepts for interface cracks. *J. Appl. Mech. ASME*, **55**, 98–103.

Rice, J. R. and Levy, N. (1972). The part-through surface crack in an elastic plate. *J. Appl. Mech. ASME*, **39**, 185–94.

Rice, J. R. and Rosengren, G. F. (1968). Plane strain deformation near a crack tip in a power law hardening material. *J. Mech. Phys. Solids*, **16**, 1–12.

Rice, J. R. and Sih, G. C. (1965). Plane problems of cracks in dissimilar media. *J. Appl. Mech., ASME*, **32**, 418–23.

Risken, H. (1989). *The Fokker–Planck Equation*. Springer-Verlag, Berlin.

Ritter, W. (1899). Die bauweise hennebique. *Schweiz. Bauzeitung Zürich*, **33**(7), 59–61.

Rosen, B. W. (1965). *Mechanics of composite strengthening*, Chapter 3, pp. 37–75. American Society for Metals, Metals Park, OH.

Rudnicki, J. W. and Rice, J. R. (1975). Conditions for the localization of deformation in pressure-sensitive dilatant materials. *J. Mech. Phys. Solids*, **23**, 371–94.

Rüsch, H. and Hilsdorf, H. K. (1963). Deformation characteristics of concrete under axial tension. Voruntersuchungen bericht (preliminary report), Munich.

Saleh, M. E., Beuth, J. L., and de Boer, M. P. (2014). Validated prediction of the strength size effect in polycrystalline silicon using the three-parameter Weibull function. *J. Amer. Cer. Soc.*, **97**(12), 3982–90.

Salem, J. A., Nemeth, N. N., Powers, L. P., and Choi, S. R. (1996). Reliability analysis of uniaxially ground brittle materials. *J. Engrg. Gas Turbines & Power*, **118**, 863–71.

Sallam, S. and Simitses, G. J. (1985). Delamination buckling and growth of flat, cross-ply laminates. *Compos. Struct.*, **4**, 361–81.

Sallam, S. and Simitses, G. J. (1987). Delamination buckling of cylinderical shells under axial compression. *Compos. Struct.*, **8**, 83–101.

Salviato, M., Ashari, S. E., and Cusatis, G. (2016*a*). Spectral stiffness microplane model for damage and fracture of textile composites. *Comp. Struct.*, **137**, 170–84.

Salviato, M. and Bažant, Z. P. (2014). The asymptotic stochastic strength of bundles of elements exhibiting general stress-strain laws. *Prob. Engrg. Mech.*, **36**, 1–7.

Salviato, M., Chau, V. T., Li, W., Bažant, Z. P., and Cusatis, G. (2016*b*). Direct testing of gradual postpeak softening of fracture specimens of fiber composites stabilized by enhanced grip stiffness and mass. *J. Appl. Mech. ASME*, **83**(11), 111003.

Salviato, M., Kirane, K., Ashari, S. E., Bažant, Z. P., and Cusatis, G. (2016*c*). Experimental and numerical investigation of intra-laminar energy dissipation and size effect in two-dimensional textile composites. *Compos. Sci. Tech.*, **135**, 67–75.

Salviato, M., Kirane, K., Bažant, Z. P., and Cusatis, G. (2019). Mode i and ii interlaminar fracture in laminated composites: a size effect study. *J. Appl. Mech.*, **86**(9), 091008.

Salviato, M. and Zappalorto, M. (2016). A unified solution approach for a large variety of antiplane shear and torsion notch problems: Theory and examples. *Int. J. Solids Struct.*, **102**, 10–20.

Salviato, M., Zappalorto, M., and Maragoni, L. (2018). Exact solution for the mode iii stress fields ahead of cracks initiated at sharp notch tips. *Eur. J. Mech., A/Solids*, **72**, 88–96.

Sandler, I. S. (1984). Strain-softening for static and dynamic problems. In *Proc., Symp. on Constitutive Equations: Macro and Computational Aspects* (ed. K. J. Willam), New York, NY, pp. 217–231. ASME.

Schauffert, E. A. and Cusatis, G. (2011). Lattice discrete particle model for fiber-reinforced concrete. I: Theory. *J. Engrg. Mech. ASCE*, **138**(7), 826–33.

Schauffert, E. A., Cusatis, G., Pelessone, D., O'Daniel, J. L., and Baylot, J. T. (2011). Lattice discrete particle model for fiber-reinforced concrete. II: Tensile fracture and multiaxial loading behavior. *J. Engrg. Mech. ASCE*, **138**(7), 834–41.

Schlangen, E. and van Mier, J. G. M. (1992). Experimental and numerical analysis of micromechanisms of fracture of cement-based composites. *Cem. Concr. Res.*, **14**, 105–18.

Schultheisz, C. R. and Waas, A. M. (1996). Compressive failure of composites, part I: testing and micromechanical theories. *Prog. Aero. Sci.*, **32**, 1–42.

Shao, Y., Zhao, H. P., Feng, X. Q., and Gao, H. (2012). Discontinuous crack-bridging model for fracture toughness analysis of nacre. *J. Mech. Phys. Solids*, **60**(8), 1400–19.

Shih, C. F., Moran, B., and Nakamura, T. (1986). Energy release rate along a three-dimensional crack front in a thermally stressed body. *Int. J. Frac.*, **30**, 79–102.

Shioya, T. and Akiyama, H. (1994). *Application to design of size effect in reinforced concrete structures*, pp. 409–16. E & FN Spon, London.

Shioya, T., Iguro, M., Nojiri, Y., Akiyama, H., and Okada, T. (1990). Shear strength of large reinforced concrete beams. *ACI Special Publication*, **118**(1), 259–80.

Sih, G. C. (1966). On the westergaard method of crack analysis. *Int. J. Frac. Mech.*, **2**(4), 628–31.

Silling, S. A. (2000). Reformulation of elasticity theory for discontinuities and long-range forces. *J. Mech. Phys. Solids*, **48**(1), 175–209.

Silling, S. A. and Askari, E. (2005). A meshfree method based on the peridynamic model of solid mechanics. *Comput. Struct.*, **83**(17), 1526–35.

Silling, S. A., Epton, M., Weckner, O., Xu, J., and Askari, E. (2007). Peridynamic states and constitutive modeling. *J. Elasticity*, **88**(2), 151–84.

Silling, S. A. and Lehoucq, R. B. (2010). Peridynamic theory of solid mechanics. *Adv. Appl. Mech.*, **44**, 73–168.

Slepyan, L. I. (1990). Modeling of fracture of sheet ice. *Izvestia ANN SSSR, Mekh. Tverd. Tela.*, **25**, 151–7.

Smith, J., Cusatis, G., Pelessone, D., Landis, E., O'Daniel, J. L., and Baylot, J. T. (2014). Discrete modeling of ultra-high-performance concrete with application to projectile penetration. *Int. J. Impact Engrg.*, **65**, 13–32.

Smith, R. L. (1982). The asymptotic distribution of the strength of a series-parallel system with equal load sharing. *Ann Probab.*, **10**(1), 137–71.

Sniegowski, J. J. and de Boer, M. P. (2000, August). IC-compatible polysilicon surface micromachining. *Annu. Rev. Mater. Sci.*, **30**, 299–333.

Soutis, C., Curtis, P. T., and Fleck, N. A. (1993). Compressive failure of notched fibre composites. *Proc. R. Soc. London A*, **440**, 241–56.

Srawley, J. E. (1976). Wide range stress intensity factor expressions for astm e-399 standard fracture toughness specimens. *Int. J. Frac.*, **12**, 475–6.

Stillinger, F. H. and Weber, T. A. (1985). Computer simulation of local order in condensed phases of silicon. *Phys. Rev. B*, **31**, 5256–71.

Stroh, A. N. (1958). Dislocations and cracks in anisotropic elasticity. *Phil. Mag.*, **3**, 625–46.

Sutcliffe, M. P. F. and Fleck, N. A. (1994). Microbuckle propagation in carbon fibre-epoxy composites. *Acta Metall.*, **42**(7), 2219–31.

Syroka-Korol, E. and Tejchman, J. (2014). Experimental investigation of size effect in reinforced concrete beams failing by shear. *Engrg. Struct.*, **58**, 63–78.

Tada, H., Paris, P. C., and Irwin, G. R. (1973). *The Stress Analysis of Cracks Handbook* (1 edn). Del Research Corporation, Hellertown, PA.

Tada, H., Paris, P. C., and Irwin, G. R. (2000). *The Stress Analysis of Cracks Handbook*. ASME Press, New York, NY.

Tankasala, H. C., Deshpande, V. S., and Fleck, N. A. (2018). Notch sensitivity of orthotropic solids: interaction of tensile and shear damage zones. *Int. J. Frac.*, **212**(2), 123–42.

Tersoff, J. (1988). New empirical approach for the structure and energy of covalent systems. *Phys. Rev. B*, **37**, 6991.

Thomas, T. Y. (1961). *Plastic flow and fracture of solids.* Academic, New York, NY.

Tinschert, J., Zwez, D., Marx, R., and Anusavice, K. J. (2000). Structural reliability of alumina-, feldspar-, leucite-, mica- and zirconia-based ceramics. *J. Dent.*, **28**, 529–35.

Tobolsky, A. and Erying, H. (1943). Mechanical properties of polymeric materials. *J. Chem. Phys.*, **11**, 125–34.

Tracey, D. M. (1971). Finite elements for determination of crack tip elastic stress intensity factors. *Eng. Frac. Mech.*, **7**, 255–66.

van Mier, J. G. M. and Schlangen, E. (1993). An experimental and numerical study mode I (tensile) and mode II (shear) fracture in concrete. *J. Mech. Behav. Mater.*, **4**, 179–90.

Vanmarcke, E. (2010). *Random Fields Analysis and Synthesis.* World Scientific Publishers, Singapore.

Vashy, A. (1892). Sur les lois de similitude en physique. *Annales télégraphiques*, **19**, 25–8.

Vaziri, H. S., Abadyan, M., Nouri, M., Omaraei, I. A., Sadredini, Z., and Ebrahimnia, M. (2011). Investigation of the fracture mechanism and mechanical properties of polystyrene/silica nanocomposite in various silica contents. *J. Mater. Res.*, **46**(17), 5628–38.

Voyiadjis, G. and Yaghoobi, M. (2019). *Size Effect in Plasticity: From Macro to Nano.* Academic Press, London.

Walsh, P. F. (1972). Fracture of plain concrete. *Indian Concrete Journal*, **46**(11), 469–470, 476.

Walsh, P. F. (1976). Crack initiation in plain concrete. *Mag. Concr. Res.*, **28**, 37–41.

Wang, R. Z., Suo, Z., Evans, A. G., Yao, N., and Aksay, I. A. (2001). Deformation mechanisms in nacre. *J. Mater. Res.*, **16**(9), 2485–93.

Waversik, W.R. and Fairhurst, C. (1970). A study of brittle rock fracture in laboratory compression experiments. *Int. J. Rock Mech. Min. Sci.*, **7**(5), 561–75.

Wei, X., Filleter, T., and Espinosa, H. D. (2015). Statistical shear lag model: unraveling the size effect in hierarchical composites. *Acta Biomater.*, **18**, 206–12.

Weibull, W. (1939). The phenomenon of rupture in solids. *Proc. Royal Sweden Inst. Engrg. Res.*, **153**, 1–55.

Weibull, W. (1951). A statistical distribution function of wide applicability. *J. Appl. Mech. ASME*, **153**(18), 293–7.

Wells, A. A. (1961). Unstable crack propagation in metals: Cleavage and fast fracture. In *Proc. the Crack Propagation Symposium*, Volume 1, Cranfield, U.K., p. Paper 84.

Wells, A. A. (1963). Application of fracture mechanics at and beyond general yielding. *Brit. Weld. J.*, **10**, 563–70.

Westergaard, H. M. (1939). Bearing pressures and cracks. *J. Appl. Mech., ASME*, **6**, A49–53.

Williams, E. (1951). Some observations of Leonardo, Galileo, Mariotte and others relative to size effect. *Ann. Sci.*, **13**, 23–9.

Williams, M. L. (1952). Stress singularities resulting from various boundary conditions in angular corners of plates in extension. *J. Appl. Mech.*, **74**, 526–8.

Wilson, W. K. (1971). Crack tip finite elements for plane elasticity. Technical Report 71-1E7-FM-PWR-P2, Westinghouse.

Wu, J.-Y. and Nguyen, V. P. (2018). A length scale insensitive phase-field damage model for brittle fracture. *J. Mech. Phys. Solids*, **119**, 20–42.

Xu, Z., Ballarini, R., and Le, J.-L. (2019). A renewal weakest-link model of strength distribution of polycrystalline silicon MEMS structures. *J. Appl. Mech. ASME*, **86**(8), 081005.

Xu, Z. and Le, J.-L. (2017). A first passage model for probabilistic failure of polycrystalline silicon MEMS structures. *J. Mech. Phys. Solids*, **99**, 225–41.

Xu, Z. and Le, J.-L. (2018). On power-law tail distribution of strength statistics of brittle and quasibrittle structures. *Engrg. Frac. Mech.*, **197**, 80–91.

Xu, Z. and Le, J.-L. (2019). A simplified probabilistic model for nanocrack propagation and its implications for tail distribution of structural strength. *Fizicheskaya Mezomekhanika*, **21**(6), 83–92.

Yin, W.-L., Sallam, S., and Simitses, G. J. (1986). Ultimate axial capacity of a delaminated beam-plate. *AIAA J.*, **24**(1), 123–8.

Young, T. (1807). *A course of lectures on natural philosophy and the mechanical arts.* Volume I. Joseph Johnson, London.

Yu, Q. and Bažant, Z. P. (2011). Can stirrups suppress size effect on shear strength of RC beams? *J. Struct. Engrg., ASCE*, **137**(5), 607–17.

Yu, Q., Bažant, Z. P., Bayldon, J. M., Le, J.-L., Caner, F. C., Ng, W. H., and Waas, A. M. (2010*a*). Scaling of strength of metal-composite joints: I. Experimental investigation. *J. Appl. Mech. ASME*, **77**, 011011.

Yu, Q., Bažant, Z. P., and Le, J.-L. (2013). Scaling of strength of metal-composite joints: III. numerical simulation. *J. Appl. Mech. ASME*, **80**, 054593.

Yu, Q., Le, J.-L., Hoover, C. G., and Bažant, Z. P. (2010*b*). Problems with hu-duan boundary effect model and its comparison to size-shape effect law for quasibrittle fracture. *J. Engrg. Mech., ASCE*, **136(1)**, 40–50.

Yu, Q., Le, J.-L., Hubler, M. H., Wendner, R., Cusatis, G., and Bažant, Z. P. (2016). Comparison of main models for size effect on shear strength of reinforced and prestressed concrete beams. *Structural Concrete (fib)*, **17**(5), 778–89.

Zamanian, M., Mortezaei, M., Salehnia, B., and Jam, J. E. (2013). Fracture toughness of epoxy polymer modified with nanosilica particles: Particle size effect. *Eng. Frac. Mech.*, **97**, 193–206.

Zappalorto, M., Salviato, M., and Maragoni, L. (2019). Analytical study on the mode iii stress fields due to blunt notches with cracks. *Fatigue of Engrg. Mater. & Structures*, **42**(3), 612–26.

Zappalorto, M., Salviato, M., Pontefisso, A., and Quaresimin, M. (2013*a*). Notch effect in clay-modified epoxy: a new perspective on nanocomposite properties. *Composite Interfaces*, **20**(6), 405–19.

Zappalorto, M., Salviato, M., and Quaresimin, M. (2013*b*). Mixed mode (i+ ii) fracture toughness of polymer nanoclay nanocomposites. *Eng. Frac. Mech.*, **111**, 50–64.

Zhang, H., Tang, L.-C., Zhang, Z., Friedrich, K., and Sprenger, S. (2008). Fracture behaviours of in situ silica nanoparticle-filled epoxy at different temperatures. *Polymer*, **49**(17), 3816–25.

Zheltov, Y. P. and Kristianovich, S. A. (1955). The mechanism of hydraulic fracture of oil-bearing strata. *Izvestia Adademii Nauk SSSR, Otd Tekh Nauk* (11).

Zhurkov, S. N. (1965). Kinetic concept of the strength of solids. *Int. J. Frac. Mech.*, **1**(4), 311–23.

Zhurkov, S. N. and Korsukov, V. E. (1974). Atomic mechanism of fracture of solid polymer. *J. Polym. Sci.*, **12**(2), 385–98.

Zi, G., Kim, J., and Bažant, Z. P. (2014). Size effect on biaxial flexural strength of concrete. *ACI Mater. J.*, **111**(1), 1–8.

Zi, G., Oh, H., and Park, S. K. (2008). A novel indirect tensile test method to measure the biaxial tensile strength of concretes and other quasibrittle material. *Cem. Concr. Res.*, **6**, 751–6.

Zienkiewicz, O. C., Owen, D. R. J., Phillips, D. V., and Nayak, G. C. (1972). Finite element methods in the analysis of reactor vessels. *Nucl. Eng. Des.*, **20**(2), 507–41.

Zubelewicz, A. and Bažant, Z. P. (1987). Interface modeling of fracture in aggregate composites. *J. Engrg. Mech., ASCE*, **113**(11), 1619–30.

Author index

Subject index